STUDENT SOLUTIONS MANUAL

TO ACCOMPANY

Fifth Edition

BY CHARLES P. MCKEAGUE

Ross Rueger
College of the Sequoias

Australia • Canada • Mexico • Singapore • Spain
United Kingdom • United States

Printed in the United States of America

1 2 3 4 5 6 7 09 08 07

ISBN: 0-03-022678-3

For more information about our products, contact us at:
Thomson Learning Academic Resource Center
1-800-423-0563

For permission to use material from this text, contact us by:
Phone: 1-800-730-2214
Fax: 1-800-731-2215
Web: www.thomsonrights.com

Asia
Thomson Learning
60 Albert Complex, #15-01
Alpert Complex
Singapore 189969

Australia
Nelson Thomson Learning
102 Dodds Street
South Street
South Melbourne, Victoria 3205
Australia

Canada
Nelson Thomson Learning
1120 Birchmount Road
Toronto, Ontario M1K 5G4
Canada

Europe/Middle East/South Africa
Thomson Learning
Berkshire House
168-173 High Holborn
London WC1 V7AA
United Kingdom

Latin America
Thomson Learning
Seneca, 53
Colonia Polanco
11560 Mexico D.F.
Mexico

Spain
Paraninfo Thomson Learning
Calle/Magallanes, 25
28015 Madrid, Spain

Preface

This Student's Solutions Manual contains complete solutions to every odd-numbered problem in the exercise sets and review exercises for Beginning Algebra, A Text/Workbook, Fifth Edition by Charles P. McKeague. It also contains solutions to every problem in the chapter tests. I have attempted to format solutions for readability and accuracy, and apologize to you for any errors that you may encounter. If you have any comments, suggestions, error corrections, or alternative solutions please feel free to drop me a note. If you prefer, you can send me e-mail with corrections or comments (address at the bottom of this page).

Please use this manual with some degree of caution. It is intended as a reference for solutions, and should only be referred to after you have attempted the problems on your own. Remember that you can only learn algebra by doing it, and practicing it, and that mistakes and incorrect solutions are part of the learning of algebra. When you do look up solutions, try to find the error in your own work, so that future problems of the same type do not result in the same error.

I would like to thank Carol Loyd of Saunders College Publishing for her valuable assistance, and Charles McKeague for writing this textbook and allowing me to work on this supplement. Most especially, I would like to thank Charles Heuer for his meticulous error-checking of my solutions.

This manual was prepared using a Power Macintosh 8500, using Microsoft Word, Design Science MathType, Adobe Illustrator, and Wolfram Mathematica.

In memory of Tom Riddle and John Schaeffer.

Ross Rueger
College of the Sequoias
915 South Mooney Boulevard
Visalia, CA 93277
email: matmanross@aol.com

May, 1999

Contents

Chapter 1
The Basics

1.1 Notation and Symbols

1. The equivalent expression is $x+5=14$.
3. The equivalent expression is $5y<30$.
5. The equivalent expression is $3y \le y+6$.
7. The equivalent expression is $\frac{x}{3}=x+2$.
9. Expanding the expression: $3^2=3 \cdot 3=9$
11. Expanding the expression: $7^2=7 \cdot 7=49$
13. Expanding the expression: $2^3=2 \cdot 2 \cdot 2=8$
15. Expanding the expression: $4^3=4 \cdot 4 \cdot 4=64$
17. Expanding the expression: $2^4=2 \cdot 2 \cdot 2 \cdot 2=16$
19. Expanding the expression: $10^2=10 \cdot 10=100$
21. Expanding the expression: $11^2=11 \cdot 11=121$
23. Using the order of operations: $2 \cdot 3+5=6+5=11$
25. Using the order of operations: $2(3+5)=2(8)=16$
27. Using the order of operations: $5+2 \cdot 6=5+12=17$
29. Using the order of operations: $(5+2) \cdot 6=7 \cdot 6=42$
31. Using the order of operations: $5 \cdot 4+5 \cdot 2=20+10=30$
33. Using the order of operations: $5(4+2)=5(6)=30$
35. Using the order of operations: $8+2(5+3)=8+2(8)=8+16=24$
37. Using the order of operations: $(8+2)(5+3)=(10)(8)=80$
39. Using the order of operations: $20+2(8-5)+1=20+2(3)+1=20+6+1=27$
41. Using the order of operations: $5+2(3 \cdot 4-1)+8=5+2(12-1)+8=5+2(11)+8=5+22+8=35$
43. Using the order of operations: $8+10 \div 2=8+5=13$
45. Using the order of operations: $4+8 \div 4-2=4+2-2=4$
47. Using the order of operations: $3+12 \div 3+6 \cdot 5=3+4+30=37$
49. Using the order of operations: $3 \cdot 8+10 \div 2+4 \cdot 2=24+5+8=37$
51. Using the order of operations: $(5+3)(5-3)=(8)(2)=16$
53. Using the order of operations: $5^2-3^2=5 \cdot 5-3 \cdot 3=25-9=16$
55. Using the order of operations: $(4+5)^2=9^2=9 \cdot 9=81$
57. Using the order of operations: $4^2+5^2=4 \cdot 4+5 \cdot 5=16+25=41$
59. Using the order of operations: $3 \cdot 10^2+4 \cdot 10+5=300+40+5=345$
61. Using the order of operations: $2 \cdot 10^3+3 \cdot 10^2+4 \cdot 10+5=2000+300+40+5=2345$
63. Using the order of operations: $10-2(4 \cdot 5-16)=10-2(20-16)=10-2(4)=10-8=2$
65. Using the order of operations: $4[7+3(2 \cdot 9-8)]=4[7+3(18-8)]=4[7+3(10)]=4(7+30)=4(37)=148$

67. Using the order of operations: $5(7-3)+8(6-4)=5(4)+8(2)=20+16=36$
69. Using the order of operations: $3(4\bullet5-12)+6(7\bullet6-40)=3(20-12)+6(42-40)=3(8)+6(2)=24+12=36$
71. Using the order of operations: $3^4+4^2\div2^3-5^2=81+16\div8-25=81+2-25=58$
73. Using the order of operations: $5^2+3^4\div9^2+6^2=25+81\div81+36=25+1+36=62$
75. The number of shares can be represented by $2(10+4)=2(14)=28$ shares.
77. The amount of money can be represented by $3\bullet50-14=150-14=136$ dollars.
79. There are $5\bullet2=10$ cookies in the package.
81. The total number of calories is $210\bullet2=420$ calories.
83. There are $7\bullet32=224$ chips in the bag.
85. For a person eating 3,000 calories per day, the recommended amount of fat would be $80+15=95$ grams.
87. The next number is 5.
89. The next number is 10.
91. The next number is $5^2=25$.
93. Since $2+2=4$ and $2+4=6$, the next number is $4+6=10$.

1.2 Real Numbers

1. Labeling the point:

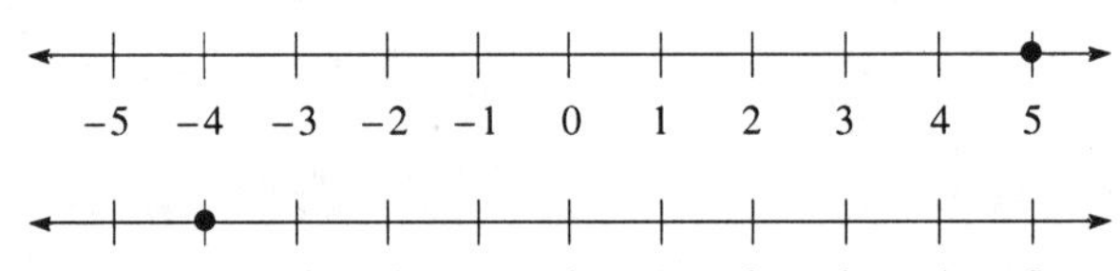

3. Labeling the point:

5. Labeling the point:

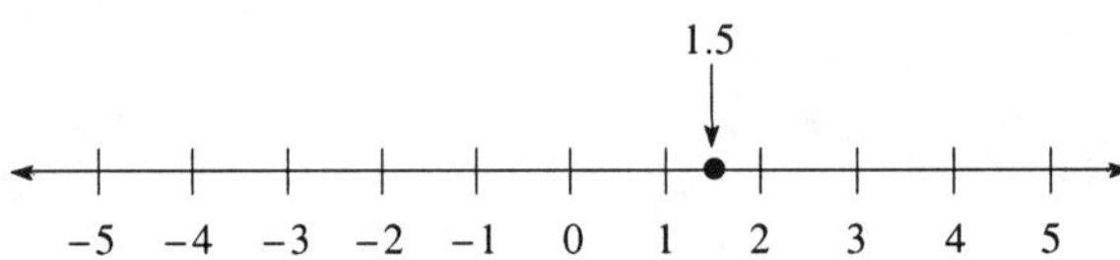

7. Labeling the point:

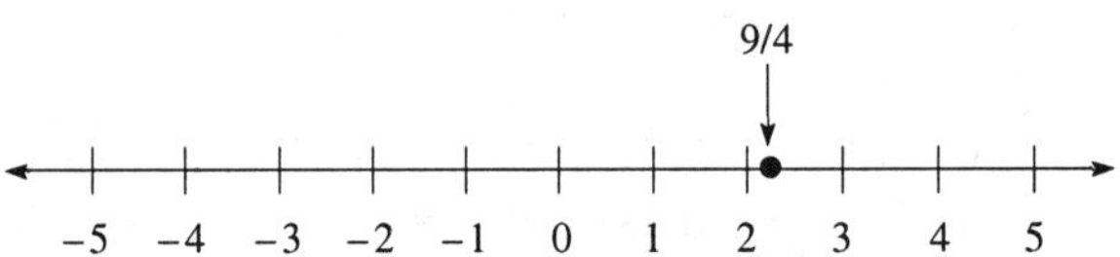

9. Building the fraction: $\frac{3}{4}=\frac{3}{4}\bullet\frac{6}{6}=\frac{18}{24}$
11. Building the fraction: $\frac{1}{2}=\frac{1}{2}\bullet\frac{12}{12}=\frac{12}{24}$
13. Building the fraction: $\frac{5}{8}=\frac{5}{8}\bullet\frac{3}{3}=\frac{15}{24}$
15. Building the fraction: $\frac{3}{5}=\frac{3}{5}\bullet\frac{12}{12}=\frac{36}{60}$
17. Building the fraction: $\frac{11}{30}=\frac{11}{30}\bullet\frac{2}{2}=\frac{22}{60}$
19. The opposite of 10 is –10, the reciprocal is $\frac{1}{10}$, and the absolute value is $|10|=10$.
21. The opposite of $\frac{3}{4}$ is $-\frac{3}{4}$, the reciprocal is $\frac{4}{3}$, and the absolute value is $\left|\frac{3}{4}\right|=\frac{3}{4}$.
23. The opposite of $\frac{11}{2}$ is $-\frac{11}{2}$, the reciprocal is $\frac{2}{11}$, and the absolute value is $\left|\frac{11}{2}\right|=\frac{11}{2}$.
25. The opposite of –3 is 3, the reciprocal is $-\frac{1}{3}$, and the absolute value is $|-3|=3$.
27. The opposite of $-\frac{2}{5}$ is $\frac{2}{5}$, the reciprocal is $-\frac{5}{2}$, and the absolute value is $\left|-\frac{2}{5}\right|=\frac{2}{5}$.

29. The opposite of x is $-x$, the reciprocal is $\frac{1}{x}$, and the absolute value is $|x|$.
31. The correct symbol is <: $-5<-3$
33. The correct symbol is >: $-3>-7$
35. Since $|-4|=4$ and $-|-4|=-4$, the correct symbol is >: $|-4|>-|-4|$
37. Since $-|-7|=-7$, the correct symbol is >: $7>-|-7|$
39. The correct symbol is <: $-\frac{3}{4}<-\frac{1}{4}$
41. The correct symbol is <: $-\frac{3}{2}<-\frac{3}{4}$
43. Simplifying the expression: $|8-2|=|6|=6$
45. Simplifying the expression: $\left|5 \cdot 2^3-2 \cdot 3^2\right|=|5 \cdot 8-2 \cdot 9|=|40-18|=|22|=22$
47. Simplifying the expression: $|7-2|-|4-2|=|5|-|2|=5-2=3$
49. Simplifying the expression: $10-|7-2(5-3)|=10-|7-2(2)|=10-|7-4|=10-|3|=10-3=7$
51. Simplifying the expression:

$$\begin{aligned}15-|8-2(3 \cdot 4-9)|-10&=15-|8-2(12-9)|-10\\&=15-|8-2(3)|-10\\&=15-|8-6|-10\\&=15-|2|-10\\&=15-2-10\\&=3\end{aligned}$$

53. Multiplying the fractions: $\frac{2}{3} \cdot \frac{4}{5}=\frac{8}{15}$
55. Multiplying the fractions: $\frac{1}{2}(3)=\frac{1}{2} \cdot \frac{3}{1}=\frac{3}{2}$
57. Multiplying the fractions: $\frac{1}{4}(5)=\frac{1}{4} \cdot \frac{5}{1}=\frac{5}{4}$
59. Multiplying the fractions: $\frac{4}{3} \cdot \frac{3}{4}=\frac{12}{12}=1$
61. Multiplying the fractions: $6\left(\frac{1}{6}\right)=\frac{6}{1} \cdot \frac{1}{6}=\frac{6}{6}=1$
63. Multiplying the fractions: $3 \cdot \frac{1}{3}=\frac{3}{1} \cdot \frac{1}{3}=\frac{3}{3}=1$
65. Expanding the exponent: $\left(\frac{3}{4}\right)^2=\frac{3}{4} \cdot \frac{3}{4}=\frac{9}{16}$
67. Expanding the exponent: $\left(\frac{2}{3}\right)^3=\frac{2}{3} \cdot \frac{2}{3} \cdot \frac{2}{3}=\frac{8}{27}$
69. Expanding the exponent: $\left(\frac{1}{10}\right)^4=\frac{1}{10} \cdot \frac{1}{10} \cdot \frac{1}{10} \cdot \frac{1}{10}=\frac{1}{10,000}$
71. The next number is $\frac{1}{9}$.
73. The next number is $\frac{1}{5^2}=\frac{1}{25}$.
75. The perimeter is $4(1 \text{ in.})=4 \text{ in.}$, and the area is $(1 \text{ in.})^2=1 \text{ in.}^2$.
77. The perimeter is $2(1.5 \text{ in.})+2(0.75 \text{ in.})=3.0 \text{ in.}+1.5 \text{ in.}=4.5 \text{ in.}$, and the area is $(1.5 \text{ in.})(0.75 \text{ in.})=1.125 \text{ in.}^2$.

79. The perimeter is $2.75\text{ cm}+4\text{ cm}+3.5\text{ cm}=10.25\text{ cm}$, and the area is $\frac{1}{2}(4\text{ cm})(2.5\text{ cm})=5.0\text{ cm}^2$.

81. A loss of 8 yards corresponds to –8 on a number line. The total yards gained corresponds to –2 yards.

83. The temperature can be represented as –64°. The new (warmer) temperature corresponds to –54°.

85. His position corresponds to –100 feet. His new (deeper) position corresponds to –105 feet.

87. The area is given by: $8\frac{1}{2}\bullet 11=\frac{17}{2}\bullet\frac{11}{1}=\frac{187}{2}=93.5\text{ in.}^2$

The perimeter is given by: $2\left(8\frac{1}{2}\right)+2(11)=17+22=39\text{ in.}$

89. The calories consumed would be: $2(544)+299=1,387$ calories

91. The calories consumed by the 180 lb person would be $3(653)=1,959$ calories, while the calories consumed by the 120 lb person would be $3(435)=1,305$ calories. Thus the 180 lb person consumed $1959-1306=654$ more calories.

1.3 Addition of Real Numbers

1. Adding all positive and negative combinations of 3 and 5:

$3+5=8$

$3+(-5)=-2$

$-3+5=2$

$(-3)+(-5)=-8$

3. Adding all positive and negative combinations of 15 and 20:

$15+20=35$

$15+(-20)=-5$

$-15+20=5$

$(-15)+(-20)=-35$

5. Adding the numbers: $6+(-3)=3$

7. Adding the numbers: $13+(-20)=-7$

9. Adding the numbers: $18+(-32)=-14$

11. Adding the numbers: $-6+3=-3$

13. Adding the numbers: $-30+5=-25$

15. Adding the numbers: $-6+(-6)=-12$

17. Adding the numbers: $-9+(-10)=-19$

19. Adding the numbers: $-10+(-15)=-25$

21. Performing the additions: $5+(-6)+(-7)=5+(-13)=-8$

23. Performing the additions: $-7+8+(-5)=-12+8=-4$

25. Performing the additions: $5+[6+(-2)]+(-3)=5+4+(-3)=9+(-3)=6$

27. Performing the additions: $[6+(-2)]+[3+(-1)]=4+2=6$

29. Performing the additions: $20+(-6)+[3+(-9)]=20+(-6)+(-6)=20+(-12)=8$

31. Performing the additions: $-3+(-2)+[5+(-4)]=-3+(-2)+1=-5+1=-4$

33. Performing the additions: $(-9+2)+[5+(-8)]+(-4)=-7+(-3)+(-4)=-14$

35. Performing the additions: $[-6+(-4)]+[7+(-5)]+(-9)=-10+2+(-9)=-19+2=-17$

37. Performing the additions: $(-6+9)+(-5)+(-4+3)+7=3+(-5)+(-1)+7=10+(-6)=4$

39. Using order of operations: $-5+2(-3+7)=-5+2(4)=-5+8=3$

41. Using order of operations: $9+3(-8+10)=9+3(2)=9+6=15$

43. Using order of operations: $-10+2(-6+8)+(-2)=-10+2(2)+(-2)=-10+4+(-2)=-12+4=-8$

45. Using order of operations: $2(-4+7)+3(-6+8)=2(3)+3(2)=6+6=12$

47. The pattern is to add 5, so the next two terms are $18+5=23$ and $23+5=28$.

49. The pattern is to add 5, so the next two terms are $25+5=30$ and $30+5=35$.

51. The pattern is to add –5, so the next two terms are $5+(-5)=0$ and $0+(-5)=-5$.

53. The pattern is to add –6, so the next two terms are $-6+(-6)=-12$ and $-12+(-6)=-18$.

55. The pattern is to add –4, so the next two terms are $0+(-4)=-4$ and $-4+(-4)=-8$.

57. Yes, since each successive odd number is 2 added to the previous one.

59. The expression is: $5+9=14$

61. The expression is: $[-7+(-5)]+4=-12+4=-8$
63. The expression is: $[-2+(-3)]+10=-5+10=5$
65. The number is 3, since $-8+3=-5$.
67. The number is –3, since $-6+(-3)=-9$.
69. The expression is $-12°+4°=-8°$.
71. The expression is $\$10+(-\$6)+(-\$8)=\$10+(-\$14)=-\$4$.
73. The new balance is $-\$30+\$40=\$10$.
75. The sequence of her wages is: \$6.50, \$6.75, \$7.00, \$7.25, \$7.50, \$7.75, \$8.00.
This is an arithmetic sequence, since \$0.25 is added to each wage to result in the new wage.

1.4 Subtraction of Real Numbers

1. Subtracting the numbers: $5-8=5+(-8)=-3$
3. Subtracting the numbers: $3-9=3+(-9)=-6$
5. Subtracting the numbers: $5-5=5+(-5)=0$
7. Subtracting the numbers: $-8-2=-8+(-2)=-10$
9. Subtracting the numbers: $-4-12=-4+(-12)=-16$
11. Subtracting the numbers: $-6-6=-6+(-6)=-12$
13. Subtracting the numbers: $-8-(-1)=-8+1=-7$
15. Subtracting the numbers: $15-(-20)=15+20=35$
17. Subtracting the numbers: $-4-(-4)=-4+4=0$
19. Using order of operations: $3-2-5=3+(-2)+(-5)=3+(-7)=-4$
21. Using order of operations: $9-2-3=9+(-2)+(-3)=9+(-5)=4$
23. Using order of operations: $-6-8-10=-6+(-8)+(-10)=-24$
25. Using order of operations: $-22+4-10=-22+4+(-10)=-32+4=-28$
27. Using order of operations: $10-(-20)-5=10+20+(-5)=30+(-5)=25$
29. Using order of operations: $8-(2-3)-5=8-(-1)-5=8+1+(-5)=9+(-5)=4$
31. Using order of operations: $7-(3-9)-6=7-(-6)-6=7+6+(-6)=13+(-6)=7$
33. Using order of operations: $5-(-8-6)-2=5-(-14)-2=5+14+(-2)=19+(-2)=17$
35. Using order of operations: $-(5-7)-(2-8)=-(-2)-(-6)=2+6=8$
37. Using order of operations: $-(3-10)-(6-3)=-(-7)-3=7+(-3)=4$
39. Using order of operations: $16-[(4-5)-1]=16-(-1-1)=16-(-2)=16+2=18$
41. Using order of operations: $5-[(2-3)-4]=5-(-1-4)=5-(-5)=5+5=10$
43. Using order of operations:
$$21-[-(3-4)-2]-5=21-[-(-1)-2]-5=21-(1-2)-5=21-(-1)-5=21+1+(-5)=22+(-5)=17$$
45. Using order of operations: $2\cdot8-3\cdot5=16-15=16+(-15)=1$
47. Using order of operations: $3\cdot5-2\cdot7=15-14=15+(-14)=1$
49. Using order of operations: $5\cdot9-2\cdot3-6\cdot2=45-6-12=45+(-6)+(-12)=45+(-18)=27$
51. Using order of operations: $3\cdot8-2\cdot4-6\cdot7=24-8-42=24+(-8)+(-42)=24+(-50)=-26$
53. Using order of operations: $2\cdot3^2-5\cdot2^2=2\cdot9-5\cdot4=18-20=18+(-20)=-2$
55. Using order of operations: $4\cdot3^3-5\cdot2^3=4\cdot27-5\cdot8=108-40=108+(-40)=68$
57. Writing the expression: $-7-4=-7+(-4)=-11$
59. Writing the expression: $12-(-8)=12+8=20$
61. Writing the expression: $-5-(-7)=-5+7=2$
63. Writing the expression: $[4+(-5)]-17=-1-17=-1+(-17)=-18$
65. Writing the expression: $8-5=8+(-5)=3$
67. Writing the expression: $-8-5=-8+(-5)=-13$
69. Writing the expression: $8-(-5)=8+5=13$
71. The number is 10, since $8-10=8+(-10)=-2$.
73. The number is –2, since $8-(-2)=8+2=10$.
75. The expression is $\$1,500-\$730=\$770$.
77. The expression is $-\$35+\$15-\$20=-\$35+(-\$20)+\$15=-\$55+\$15=-\$40$.

79. The expression is $\$98-\$65-\$53=\$98+(-\$65)+(-\$53)=\$98+(-\$118)=-\$20$.
81. The sequence of values is \$4500, \$3950, \$3400, \$2850, and \$2300. This is an arithmetic sequence, since –\$550 is added to each value to obtain the new value.
83. The difference is $1000 \text{ feet} - 231 \text{ feet} = 769 \text{ feet}$.
85. He is 439 feet from the starting line.
87. 2 seconds have gone by.
89. The difference is $149 \text{ lb} - 68 \text{ lb} = 81 \text{ lb}$.
91. The lamb gains the most weight in the first month.
93. The angles add to 90°, so $x=90°-55°=35°$.
95. The angles add to 180°, so $x=180°-120°=60°$.

1.5 Properties of Real Numbers

1. commutative property (of addition)
3. multiplicative inverse property
5. commutative property (of addition)
7. distributive property
9. commutative and associative properties (of addition)
11. commutative and associative properties (of addition)
13. commutative property (of addition)
15. commutative and associative properties (of multiplication)
17. commutative property (of multiplication)
19. additive inverse property
21. The expression should read $3(x+2)=3x+6$.
23. The expression should read $9(a+b)=9a+9b$.
25. The expression should read $3(0)=0$.
27. The expression should read $3+(-3)=0$.
29. The expression should read $10(1)=10$.
31. Simplifying the expression: $4+(2+x)=(4+2)+x=6+x$
33. Simplifying the expression: $(x+2)+7=x+(2+7)=x+9$
35. Simplifying the expression: $3(5x)=(3 \bullet 5)x=15x$
37. Simplifying the expression: $9(6y)=(9 \bullet 6)y=54y$
39. Simplifying the expression: $\frac{1}{2}(3a)=\left(\frac{1}{2} \bullet 3\right)a=\frac{3}{2}a$
41. Simplifying the expression: $\frac{1}{3}(3x)=\left(\frac{1}{3} \bullet 3\right)x=1x=x$
43. Simplifying the expression: $\frac{1}{2}(2y)=\left(\frac{1}{2} \bullet 2\right)y=1y=y$
45. Simplifying the expression: $\frac{3}{4}\left(\frac{4}{3}x\right)=\left(\frac{3}{4} \bullet \frac{4}{3}\right)x=1x=x$
47. Simplifying the expression: $\frac{6}{5}\left(\frac{5}{6}a\right)=\left(\frac{6}{5} \bullet \frac{5}{6}\right)a=1a=a$
49. Applying the distributive property: $8(x+2)=8 \bullet x+8 \bullet 2=8x+16$
51. Applying the distributive property: $8(x-2)=8 \bullet x-8 \bullet 2=8x-16$
53. Applying the distributive property: $4(y+1)=4 \bullet y+4 \bullet 1=4y+4$
55. Applying the distributive property: $3(6x+5)=3 \bullet 6x+3 \bullet 5=18x+15$
57. Applying the distributive property: $2(3a+7)=2 \bullet 3a+2 \bullet 7=6a+14$
59. Applying the distributive property: $9(6y-8)=9 \bullet 6y-9 \bullet 8=54y-72$
61. Applying the distributive property: $\frac{1}{2}(3x-6)=\frac{1}{2} \bullet 3x-\frac{1}{2} \bullet 6=\frac{3}{2}x-3$
63. Applying the distributive property: $\frac{1}{3}(3x+6)=\frac{1}{3} \bullet 3x+\frac{1}{3} \bullet 6=x+2$
65. Applying the distributive property: $3(x+y)=3x+3y$

67. Applying the distributive property: $8(a-b)=8a-8b$
69. Applying the distributive property: $6(2x+3y)=6 \cdot 2x+6 \cdot 3y=12x+18y$
71. Applying the distributive property: $4(3a-2b)=4 \cdot 3a-4 \cdot 2b=12a-8b$
73. Applying the distributive property: $\frac{1}{2}(6x+4y)=\frac{1}{2} \cdot 6x+\frac{1}{2} \cdot 4y=3x+2y$
75. Applying the distributive property: $4(a+4)+9=4a+16+9=4a+25$
77. Applying the distributive property: $2(3x+5)+2=6x+10+2=6x+12$
79. Applying the distributive property: $7(2x+4)+10=14x+28+10=14x+38$
81. No. The man cannot reverse the order of putting on his socks and putting on his shoes.
83. No. The skydiver must jump out of the plane before pulling the rip cord.
85. Division is not a commutative operation. For example, $8 \div 4=2$ while $4 \div 8=\frac{1}{2}$.
87. Computing the yearly take-home pay:

$$12(2400-480)=12(1920)=\$23,040$$

$$12 \cdot 2400-12 \cdot 480=28,800-5,760=\$23,040$$

89. Rewriting the formula: $P=2w+2l=2(w+l)$

1.6 Multiplication of Real Numbers

1. Finding the product: $7(-6)=-42$
3. Finding the product: $-7(3)=-21$
5. Finding the product: $-8(2)=-16$
7. Finding the product: $-3(-1)=3$
9. Finding the product: $-11(-11)=121$
11. Using order of operations: $-3(2)(-1)=6$
13. Using order of operations: $-3(-4)(-5)=-60$
15. Using order of operations: $-2(-4)(-3)(-1)=24$
17. Using order of operations: $(-7)^2=(-7)(-7)=49$
19. Using order of operations: $(-3)^3=(-3)(-3)(-3)=-27$
21. Using order of operations: $-2(2-5)=-2(-3)=6$
23. Using order of operations: $-5(8-10)=-5(-2)=10$
25. Using order of operations: $(4-7)(6-9)=(-3)(-3)=9$
27. Using order of operations: $(-3-2)(-5-4)=(-5)(-9)=45$
29. Using order of operations: $-3(-6)+4(-1)=18+(-4)=14$
31. Using order of operations: $2(3)-3(-4)+4(-5)=6+12+(-20)=18+(-20)=-2$
33. Using order of operations: $4(-3)^2+5(-6)^2=4(9)+5(36)=36+180=216$
35. Using order of operations: $7(-2)^3-2(-3)^3=7(-8)-2(-27)=-56+54=-2$
37. Using order of operations: $6-4(8-2)=6-4(6)=6-24=6+(-24)=-18$
39. Using order of operations: $9-4(3-8)=9-4(-5)=9+20=29$
41. Using order of operations: $-4(3-8)-6(2-5)=-4(-5)-6(-3)=20+18=38$
43. Using order of operations: $7-2[-6-4(-3)]=7-2(-6+12)=7-2(6)=7-12=7+(-12)=-5$
45. Using order of operations:

$$7-3[2(-4-4)-3(-1-1)]=7-3[2(-8)-3(-2)]=7-3(-16+6)=7-3(-10)=7+30=37$$

47. Using order of operations:

$$8-6[-2(-3-1)+4(-2-3)]=8-6[-2(-4)+4(-5)]=8-6[8+(-20)]=8-6(-12)=8+72=80$$

49. Multiplying the fractions: $-\frac{2}{3} \cdot \frac{5}{7}=-\frac{2 \cdot 5}{3 \cdot 7}=-\frac{10}{21}$
51. Multiplying the fractions: $-8\left(\frac{1}{2}\right)=-\frac{8}{1} \cdot \frac{1}{2}=-\frac{8}{2}=-4$
53. Multiplying the fractions: $-\frac{3}{4}\left(-\frac{4}{3}\right)=-\frac{3}{4} \cdot\left(-\frac{4}{3}\right)=\frac{12}{12}=1$

55. Multiplying the fractions: $\left(-\frac{3}{4}\right)^2 = \left(-\frac{3}{4}\right)\left(-\frac{3}{4}\right) = \frac{9}{16}$

57. Multiplying the fractions: $\left(-\frac{2}{3}\right)^3 = \left(-\frac{2}{3}\right)\left(-\frac{2}{3}\right)\left(-\frac{2}{3}\right) = -\frac{8}{27}$

59. Multiplying the expressions: $-2(4x) = (-2 \bullet 4)x = -8x$

61. Multiplying the expressions: $-7(-6x) = [-7 \bullet (-6)]x = 42x$

63. Multiplying the expressions: $-\frac{1}{3}(-3x) = \left[-\frac{1}{3} \bullet (-3)\right]x = 1x = x$

65. Multiplying the expressions: $-4\left(-\frac{1}{4}x\right) = \left[-4 \bullet \left(-\frac{1}{4}\right)\right]x = 1x = x$

67. Simplifying the expression: $-4(a+2) = -4a + (-4)(2) = -4a - 8$

69. Simplifying the expression: $-\frac{1}{2}(3x-6) = -\frac{3}{2}x - \frac{1}{2}(-6) = -\frac{3}{2}x + 3$

71. Simplifying the expression: $-3(2x-5) - 7 = -6x + 15 - 7 = -6x + 8$

73. Simplifying the expression: $-5(3x+4) - 10 = -15x - 20 - 10 = -15x - 30$

75. Writing the expression: $3(-10) + 5 = -30 + 5 = -25$

77. Writing the expression: $2(-4x) = -8x$

79. Writing the expression: $-9 \bullet 2 - 8 = -18 + (-8) = -26$

81. The pattern is to multiply by 2, so the next number is $4 \bullet 2 = 8$.

83. The pattern is to multiply by –2, so the next number is $40 \bullet (-2) = -80$.

85. The pattern is to multiply by $\frac{1}{2}$, so the next number is $\frac{1}{4} \bullet \frac{1}{2} = \frac{1}{8}$.

87. The pattern is to multiply by $\frac{1}{2}$, so the next number is $2 \bullet \frac{1}{2} = 1$.

89. The pattern is to multiply by –2, so the next number is $12 \bullet (-2) = -24$.

91. The amount lost is: $20(\$3) = \$60$

93. The temperature is: $25° - 4(6°) = 25° - 24° = 1°$

95. The sequence of values is \$500, \$1000, \$2000, \$4000, \$8000, and \$16000. Yes, this is a geometric sequence, since each value is 2 times the preceeding value.

1.7 Division of Real Numbers

1. Finding the quotient: $\frac{8}{-4} = -2$

3. Finding the quotient: $\frac{-48}{16} = -3$

5. Finding the quotient: $\frac{-7}{21} = -\frac{1}{3}$

7. Finding the quotient: $\frac{-39}{-13} = 3$

9. Finding the quotient: $\frac{-6}{-42} = \frac{1}{7}$

11. Finding the quotient: $\frac{0}{-32} = 0$

13. Performing the operations: $-3 + 12 = 9$

15. Performing the operations: $-3 - 12 = -3 + (-12) = -15$

17. Performing the operations: $-3(12) = -36$

19. Performing the operations: $-3 \div 12 = \frac{-3}{12} = -\frac{1}{4}$

21. Dividing and reducing: $\frac{4}{5} \div \frac{3}{4} = \frac{4}{5} \bullet \frac{4}{3} = \frac{16}{15}$

23. Dividing and reducing: $-\frac{5}{6}\div\left(-\frac{5}{8}\right)=-\frac{5}{6}\bullet\left(-\frac{8}{5}\right)=\frac{40}{30}=\frac{4}{3}$

25. Dividing and reducing: $\frac{10}{13}\div\left(-\frac{5}{4}\right)=\frac{10}{13}\bullet\left(-\frac{4}{5}\right)=-\frac{40}{65}=-\frac{8}{13}$

27. Dividing and reducing: $-\frac{5}{6}\div\frac{5}{6}=-\frac{5}{6}\bullet\frac{6}{5}=-\frac{30}{30}=-1$

29. Dividing and reducing: $-\frac{3}{4}\div\left(-\frac{3}{4}\right)=-\frac{3}{4}\bullet\left(-\frac{4}{3}\right)=\frac{12}{12}=1$

31. Using order of operations: $\frac{3(-2)}{-10}=\frac{-6}{-10}=\frac{3}{5}$

33. Using order of operations: $\frac{-5(-5)}{-15}=\frac{25}{-15}=-\frac{5}{3}$

35. Using order of operations: $\frac{-8(-7)}{-28}=\frac{56}{-28}=-2$

37. Using order of operations: $\frac{27}{4-13}=\frac{27}{-9}=-3$

39. Using order of operations: $\frac{20-6}{5-5}=\frac{14}{0}=$ undefined

41. Using order of operations: $\frac{-3+9}{2\bullet 5-10}=\frac{6}{10-10}=\frac{6}{0}=$ undefined

43. Using order of operations: $\frac{15(-5)-25}{2(-10)}=\frac{-75-25}{-20}=\frac{-100}{-20}=5$

45. Using order of operations: $\frac{27-2(-4)}{-3(5)}=\frac{27+8}{-15}=\frac{35}{-15}=-\frac{7}{3}$

47. Using order of operations: $\frac{12-6(-2)}{12(-2)}=\frac{12+12}{-24}=\frac{24}{-24}=-1$

49. Using order of operations: $\frac{5^2-2^2}{-5+2}=\frac{25-4}{-3}=\frac{21}{-3}=-7$

51. Using order of operations: $\frac{8^2-2^2}{8^2+2^2}=\frac{64-4}{64+4}=\frac{60}{68}=\frac{15}{17}$

53. Using order of operations: $\frac{(5+3)^2}{-5^2-3^2}=\frac{8^2}{-25-9}=\frac{64}{-34}=-\frac{32}{17}$

55. Using order of operations: $\frac{(8-4)^2}{8^2-4^2}=\frac{4^2}{64-16}=\frac{16}{48}=\frac{1}{3}$

57. Using order of operations: $\frac{-4\bullet 3^2-5\bullet 2^2}{-8(7)}=\frac{-4\bullet 9-5\bullet 4}{-56}=\frac{-36-20}{-56}=\frac{-56}{-56}=1$

59. Using order of operations: $\frac{3\bullet 10^2+4\bullet 10+5}{345}=\frac{300+40+5}{345}=\frac{345}{345}=1$

61. Using order of operations: $\frac{7-[(2-3)-4]}{-1-2-3}=\frac{7-(-1-4)}{-6}=\frac{7-(-5)}{-6}=\frac{7+5}{-6}=\frac{12}{-6}=-2$

63. Using order of operations: $\frac{6(-4)-2(5-8)}{-6-3-5}=\frac{-24-2(-3)}{-14}=\frac{-24+6}{-14}=\frac{-18}{-14}=\frac{9}{7}$

65. Using order of operations: $\frac{3(-5-3)+4(7-9)}{5(-2)+3(-4)}=\frac{3(-8)+4(-2)}{-10+(-12)}=\frac{-24+(-8)}{-22}=\frac{-32}{-22}=\frac{16}{11}$

67. Using order of operations: $\frac{|3-9|}{3-9}=\frac{|-6|}{-6}=\frac{6}{-6}=-1$

69. The quotient is $\frac{-12}{-4}=3$.

71. The number is –10, since $\frac{-10}{-5}=2$.

73. The number is –3, since $\frac{27}{-3}=-9$.

75. The expression is: $\frac{-20}{4}-3=-5-3=-8$

77. Each person would lose: $\frac{13600-15000}{4}=\frac{-1400}{4}=-350=\$350$ loss

79. The change per hour is: $\frac{61^\circ-75^\circ}{4}=\frac{-14^\circ}{4}=-3.5^\circ$ per hour

1.8 Subsets of the Real Numbers

1. The whole numbers are: 0, 1
3. The rational numbers are: –3, –2.5, 0, 1, $\frac{3}{2}$
5. The real numbers are: –3, –2.5, 0, 1, $\frac{3}{2}$, $\sqrt{15}$
7. The integers are: –10, –8, –2, 9
9. The irrational numbers are: π
11. true
13. false
15. false
17. true
19. This number is composite: $48=6\cdot 8=(2\cdot 3)\cdot(2\cdot 2\cdot 2)=2^4\cdot 3$
21. This number is prime.
23. This number is composite: $1023=3\cdot 341=3\cdot 11\cdot 31$
25. Factoring the number: $144=12\cdot 12=(3\cdot 4)\cdot(3\cdot 4)=(3\cdot 2\cdot 2)\cdot(3\cdot 2\cdot 2)=2^4\cdot 3^2$
27. Factoring the number: $38=2\cdot 19$
29. Factoring the number: $105=5\cdot 21=5\cdot(3\cdot 7)=3\cdot 5\cdot 7$
31. Factoring the number: $180=10\cdot 18=(2\cdot 5)\cdot(3\cdot 6)=(2\cdot 5)\cdot(3\cdot 2\cdot 3)=2^2\cdot 3^2\cdot 5$
33. Factoring the number: $385=5\cdot 77=5\cdot(7\cdot 11)=5\cdot 7\cdot 11$
35. Factoring the number: $121=11\cdot 11=11^2$
37. Factoring the number: $420=10\cdot 42=(2\cdot 5)\cdot(7\cdot 6)=(2\cdot 5)\cdot(7\cdot 2\cdot 3)=2^2\cdot 3\cdot 5\cdot 7$
39. Factoring the number: $620=10\cdot 62=(2\cdot 5)\cdot(2\cdot 31)=2^2\cdot 5\cdot 31$
41. Reducing the fraction: $\frac{105}{165}=\frac{3\cdot 5\cdot 7}{3\cdot 5\cdot 11}=\frac{7}{11}$
43. Reducing the fraction: $\frac{525}{735}=\frac{3\cdot 5\cdot 5\cdot 7}{3\cdot 5\cdot 7\cdot 7}=\frac{5}{7}$
45. Reducing the fraction: $\frac{385}{455}=\frac{5\cdot 7\cdot 11}{5\cdot 7\cdot 13}=\frac{11}{13}$
47. Reducing the fraction: $\frac{322}{345}=\frac{2\cdot 7\cdot 23}{3\cdot 5\cdot 23}=\frac{2\cdot 7}{3\cdot 5}=\frac{14}{15}$
49. Reducing the fraction: $\frac{205}{369}=\frac{5\cdot 41}{3\cdot 3\cdot 41}=\frac{5}{3\cdot 3}=\frac{5}{9}$
51. Reducing the fraction: $\frac{215}{344}=\frac{5\cdot 43}{2\cdot 2\cdot 2\cdot 43}=\frac{5}{2\cdot 2\cdot 2}=\frac{5}{8}$
53. Factoring into prime numbers: $6^3=(2\cdot 3)^3=2^3\cdot 3^3$
55. Factoring into prime numbers: $9^4\cdot 16^2=(3\cdot 3)^4\cdot(2\cdot 2\cdot 2\cdot 2)^2=3^4\cdot 3^4\cdot 2^2\cdot 2^2\cdot 2^2\cdot 2^2=2^8\cdot 3^8$
57. Simplifying and factoring: $3\cdot 8+3\cdot 7+3\cdot 5=24+21+15=60=6\cdot 10=(2\cdot 3)\cdot(2\cdot 5)=2^2\cdot 3\cdot 5$

59. They are not a subset of the irrational numbers.
61. 8, 21, and 34 are Fibonacci numbers that are composite numbers.

1.9 Addition and Subtraction with Fractions

1. Combining the fractions: $\frac{3}{6}+\frac{1}{6}=\frac{4}{6}=\frac{2}{3}$

3. Combining the fractions: $\frac{3}{8}-\frac{5}{8}=-\frac{2}{8}=-\frac{1}{4}$

5. Combining the fractions: $-\frac{1}{4}+\frac{3}{4}=\frac{2}{4}=\frac{1}{2}$

7. Combining the fractions: $\frac{x}{3}-\frac{1}{3}=\frac{x-1}{3}$

9. Combining the fractions: $\frac{1}{4}+\frac{2}{4}+\frac{3}{4}=\frac{6}{4}=\frac{3}{2}$

11. Combining the fractions: $\frac{x+7}{2}-\frac{1}{2}=\frac{x+7-1}{2}=\frac{x+6}{2}$

13. Combining the fractions: $\frac{1}{10}-\frac{3}{10}-\frac{4}{10}=-\frac{6}{10}=-\frac{3}{5}$

15. Combining the fractions: $\frac{1}{a}+\frac{4}{a}+\frac{5}{a}=\frac{10}{a}$

17. Combining the fractions: $\frac{1}{8}+\frac{3}{4}=\frac{1}{8}+\frac{3\cdot 2}{4\cdot 2}=\frac{1}{8}+\frac{6}{8}=\frac{7}{8}$

19. Combining the fractions: $\frac{3}{10}-\frac{1}{5}=\frac{3}{10}-\frac{1\cdot 2}{5\cdot 2}=\frac{3}{10}-\frac{2}{10}=\frac{1}{10}$

21. Combining the fractions: $\frac{4}{9}+\frac{1}{3}=\frac{4}{9}+\frac{1\cdot 3}{3\cdot 3}=\frac{4}{9}+\frac{3}{9}=\frac{7}{9}$

23. Combining the fractions: $2+\frac{1}{3}=\frac{2\cdot 3}{1\cdot 3}+\frac{1}{3}=\frac{6}{3}+\frac{1}{3}=\frac{7}{3}$

25. Combining the fractions: $-\frac{3}{4}+1=-\frac{3}{4}+\frac{1\cdot 4}{1\cdot 4}=-\frac{3}{4}+\frac{4}{4}=\frac{1}{4}$

27. Combining the fractions: $\frac{1}{2}+\frac{2}{3}=\frac{1\cdot 3}{2\cdot 3}+\frac{2\cdot 2}{3\cdot 2}=\frac{3}{6}+\frac{4}{6}=\frac{7}{6}$

29. Combining the fractions: $\frac{5}{12}-\left(-\frac{3}{8}\right)=\frac{5}{12}+\frac{3}{8}=\frac{5\cdot 2}{12\cdot 2}+\frac{3\cdot 3}{8\cdot 3}=\frac{10}{24}+\frac{9}{24}=\frac{19}{24}$

31. Combining the fractions: $-\frac{1}{20}+\frac{8}{30}=-\frac{1\cdot 3}{20\cdot 3}+\frac{8\cdot 2}{30\cdot 2}=-\frac{3}{60}+\frac{16}{60}=\frac{13}{60}$

33. First factor the denominators to find the LCM:

$30=2\cdot 3\cdot 5$

$42=2\cdot 3\cdot 7$

$\text{LCM}=2\cdot 3\cdot 5\cdot 7=210$

Combining the fractions: $\frac{17}{30}+\frac{11}{42}=\frac{17\cdot 7}{30\cdot 7}+\frac{11\cdot 5}{42\cdot 5}=\frac{119}{210}+\frac{55}{210}=\frac{174}{210}=\frac{2\cdot 3\cdot 29}{2\cdot 3\cdot 5\cdot 7}=\frac{29}{5\cdot 7}=\frac{29}{35}$

35. First factor the denominators to find the LCM:

$84=2\cdot 2\cdot 3\cdot 7$

$90=2\cdot 3\cdot 3\cdot 5$

$\text{LCM}=2\cdot 2\cdot 3\cdot 3\cdot 5\cdot 7=1260$

Combining the fractions: $\frac{25}{84}+\frac{41}{90}=\frac{25\cdot 15}{84\cdot 15}+\frac{41\cdot 14}{90\cdot 14}=\frac{375}{1260}+\frac{574}{1260}=\frac{949}{1260}$

37. First factor the denominators to find the LCM:

$$126 = 2 \cdot 3 \cdot 3 \cdot 7$$
$$180 = 2 \cdot 2 \cdot 3 \cdot 3 \cdot 5$$
$$\text{LCM} = 2 \cdot 2 \cdot 3 \cdot 3 \cdot 5 \cdot 7 = 1260$$

Combining the fractions:

$$\frac{13}{126} - \frac{13}{180} = \frac{13 \cdot 10}{126 \cdot 10} - \frac{13 \cdot 7}{180 \cdot 7} = \frac{130}{1260} - \frac{91}{1260} = \frac{39}{1260} = \frac{3 \cdot 13}{2 \cdot 2 \cdot 3 \cdot 3 \cdot 5 \cdot 7} = \frac{13}{2 \cdot 2 \cdot 3 \cdot 5 \cdot 7} = \frac{13}{420}$$

39. Combining the fractions: $\frac{3}{4} + \frac{1}{8} + \frac{5}{6} = \frac{3 \cdot 6}{4 \cdot 6} + \frac{1 \cdot 3}{8 \cdot 3} + \frac{5 \cdot 4}{6 \cdot 4} = \frac{18}{24} + \frac{3}{24} + \frac{20}{24} = \frac{41}{24}$

41. Combining the fractions: $\frac{1}{2} + \frac{1}{3} + \frac{1}{4} + \frac{1}{6} = \frac{1 \cdot 6}{2 \cdot 6} + \frac{1 \cdot 4}{3 \cdot 4} + \frac{1 \cdot 3}{4 \cdot 3} + \frac{1 \cdot 2}{6 \cdot 2} = \frac{6}{12} + \frac{4}{12} + \frac{3}{12} + \frac{2}{12} = \frac{15}{12} = \frac{5}{4}$

43. The sum is given by: $\frac{3}{7} + 2 + \frac{1}{9} = \frac{3 \cdot 9}{7 \cdot 9} + \frac{2 \cdot 63}{1 \cdot 63} + \frac{1 \cdot 7}{9 \cdot 7} = \frac{27}{63} + \frac{126}{63} + \frac{7}{63} = \frac{160}{63}$

45. The difference is given by: $\frac{7}{8} - \frac{1}{4} = \frac{7}{8} - \frac{1 \cdot 2}{4 \cdot 2} = \frac{7}{8} - \frac{2}{8} = \frac{5}{8}$

47. The pattern is to add $-\frac{1}{3}$, so the fourth term is: $-\frac{1}{3} + \left(-\frac{1}{3}\right) = -\frac{2}{3}$

49. The pattern is to add $\frac{2}{3}$, so the fourth term is: $\frac{5}{3} + \frac{2}{3} = \frac{7}{3}$

51. The pattern is to multiply by $\frac{1}{5}$, so the fourth term is: $\frac{1}{25} \cdot \frac{1}{5} = \frac{1}{125}$

Chapter 1 Review

1. The expression is: $-7 + (-10) = -17$

3. The expression is: $(-3 + 12) + 5 = 9 + 5 = 14$

5. The expression is: $9 - (-3) = 9 + 3 = 12$

7. The expression is: $(-3)(-7) - 6 = 21 - 6 = 15$

9. The expression is: $2[-8(3x)] = 2(-24x) = -48x$

11. The expression is: $\frac{-40}{8} - 7 = -5 - 7 = -5 + (-7) = -12$

13. Labeling the point:

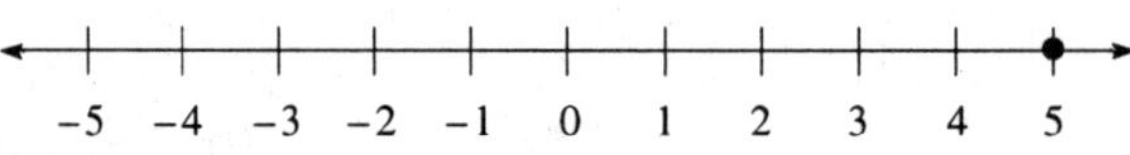

15. Labeling the point (note that $\frac{24}{8} = 3$):

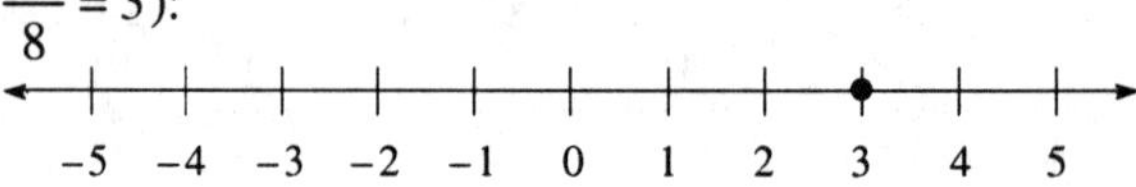

17. Labeling the point:

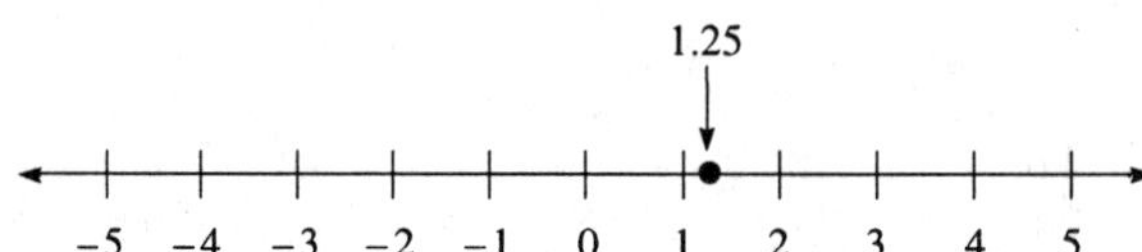

19. The absolute value is: $|12| = 12$

21. The absolute value is: $\left|-\frac{4}{5}\right| = \frac{4}{5}$

23. Simplifying: $|-1.8| = 1.8$

25. The opposite is –6, and the reciprocal is $\frac{1}{6}$.

27. The opposite is 9, and the reciprocal is $-\frac{1}{9}$.

29. Multiplying the fractions: $\left(\frac{2}{5}\right)\left(\frac{3}{7}\right)=\frac{6}{35}$

31. Multiplying the fractions: $\left(-\frac{4}{5}\right)\left(\frac{25}{16}\right)=-\frac{100}{80}=-\frac{5}{4}$

33. Adding: $-18+(-20)=-38$

35. Adding: $(-5)+(-10)+(-7)=-22$

37. Adding: $(-21)+40+(-23)+5=-44+45=1$

39. Subtracting: $14-(-8)=14+8=22$

41. Subtracting: $4-9-15=4+(-9)+(-15)=4+(-24)=-20$

43. Simplifying: $5-(-10-2)-3=5-(-12)-3=5+12+(-3)=17+(-3)=14$

45. Simplifying: $20-[-(10-3)-8]=20-(-7-8)-7=20-(-15)-7=20+15+(-7)=35+(-7)=28$

47. Multiplying: $4(-3)=-12$

49. Multiplying: $(-1)(-3)(-1)(-4)=12$

51. Finding the quotient: $\frac{-9}{36}=-\frac{1}{4}$

53. Simplifying using order of operations: $4\cdot 5+3=20+3=23$

55. Simplifying using order of operations: $2^3-4\cdot 3^2+5^2=8-4\cdot 9+25=8-36+25=33-36=-3$

57. Simplifying using order of operations: $20+8\div 4+2\cdot 5=20+2+10=32$

59. Simplifying using order of operations: $-4(-5)+10=20+10=30$

61. Simplifying using order of operations:

$$3(4-7)^2-5(3-8)^2=3(-3)^2-5(-5)^2=3\cdot 9-5\cdot 25=27-125=27+(-125)=-98$$

63. Simplifying using order of operations: $\frac{4(-3)}{-6}=\frac{-12}{-6}=2$

65. Simplifying using order of operations: $\frac{15-10}{6-6}=\frac{5}{0}=$ undefined

67. Simplifying using order of operations: $\frac{2(-7)+(-11)(-4)}{7-(-3)}=\frac{-14+44}{7+3}=\frac{30}{10}=3$

69. multiplicative identity property

71. additive inverse property

73. additive identity property

75. distributive property

77. Simplifying the expression: $4(7a)=(4\cdot 7)a=28a$

79. Simplifying the expression: $\frac{4}{5}\left(\frac{5}{4}y\right)-\left(\frac{4}{5}\cdot\frac{5}{4}\right)y-1y-y$

81. Applying the distributive property: $3(2a-4)=3\cdot 2a-3\cdot 4=6a-12$

83. Applying the distributive property: $-\frac{1}{2}(3x-6)=-\frac{1}{2}\cdot 3x-\left(-\frac{1}{2}\right)\cdot 6=-\frac{3}{2}x+3$

85. The whole numbers are: 0, 5

87. The integers are: 0, 5, –3

89. Factoring the number: $840=84\cdot 10=(21\cdot 4)\cdot(2\cdot 5)=(3\cdot 7\cdot 2\cdot 2)\cdot(2\cdot 5)=2^3\cdot 3\cdot 5\cdot 7$

91. First factor the denominators to find the LCM:

$70=2\cdot 5\cdot 7$

$84=2\cdot 2\cdot 3\cdot 7$

$\text{LCM}=2\cdot 2\cdot 3\cdot 5\cdot 7=420$

Combining the fractions: $\frac{9}{70}+\frac{11}{84}=\frac{9\cdot 6}{70\cdot 6}+\frac{11\cdot 5}{84\cdot 5}=\frac{54}{420}+\frac{55}{420}=\frac{109}{420}$

93. The pattern is to multiply by –3, so the next number is: $-270\cdot(-3)=810$

95. The pattern is to add 2, so the next number is: $10+2=12$

97. The pattern is to multiply by $-\frac{1}{2}$, so the next number is: $-\frac{1}{8}\cdot\left(-\frac{1}{2}\right)=\frac{1}{16}$

Chapter 1 Test

1. Translating into symbols: $x+3=8$
2. Translating into symbols: $5y=15$
3. Simplifying using order of operations: $5^2+3(9-7)+3^2=5^2+3(2)+3^2=25+6+9=40$
4. Simplifying using order of operations: $10-6\div3+2^3=10-6\div3+8=10-2+8=18-2=16$
5. The opposite of –4 is 4, the reciprocal is $-\frac{1}{4}$, and the absolute value is $|-4|=4$.
6. The opposite of $\frac{3}{4}$ is $-\frac{3}{4}$, the reciprocal is $\frac{4}{3}$, and the absolute value is $\left|\frac{3}{4}\right|=\frac{3}{4}$.
7. Adding: $3+(-7)=-4$
8. Adding: $|-9+(-6)|+|-3+5|=|-15|+|2|=15+2=17$
9. Subtracting: $-4-8=-4+(-8)=-12$
10. Subtracting: $9-(7-2)-4=9-5-4=9+(-5)+(-4)=9+(-9)=0$
11. c (associative property of addition)
12. e (distributive property)
13. d (associative property of multiplication)
14. a (commutative property of addition)
15. Multiplying: $-3(7)=-21$
16. Multiplying: $-4(8)(-2)=64$
17. Multiplying: $8\left(-\frac{1}{4}\right)=\frac{8}{1}\cdot\left(-\frac{1}{4}\right)=-\frac{8}{4}=-2$
18. Multiplying: $\left(-\frac{2}{3}\right)^3=\left(-\frac{2}{3}\right)\cdot\left(-\frac{2}{3}\right)\cdot\left(-\frac{2}{3}\right)=-\frac{8}{27}$
19. Simplifying using order of operations: $-3(-4)-8=12-8=4$
20. Simplifying using order of operations: $5(-6)^2-3(-2)^3=5(36)-3(-8)=180+24=204$
21. Simplifying using order of operations: $7-3(2-8)=7-3(-6)=7+18=25$
22. Simplifying using order of operations:
$$4-2[-3(-1+5)+4(-3)]=4-2[-3(4)+4(-3)]=4-2(-12-12)=4-2(-24)=4+48=52$$
23. Simplifying using order of operations: $\frac{4(-5)-2(7)}{-10-7}=\frac{-20-14}{-17}=\frac{-34}{-17}=2$
24. Simplifying using order of operations: $\frac{2(-3-1)+4(-5+2)}{-3(2)-4}=\frac{2(-4)+4(-3)}{-6-4}=\frac{-8-12}{-10}=\frac{-20}{-10}=2$
25. Simplifying: $3+(5+2x)=(3+5)+2x=8+2x$
26. Simplifying: $-2(-5x)=[-2(-5)]x=10x$
27. Multiplying: $2(3x+5)=2\cdot3x+2\cdot5=6x+10$
28. Multiplying: $-\frac{1}{2}(4x-2)=-\frac{1}{2}(4x)-\left(-\frac{1}{2}\right)(2)=-2x+1$
29. The integers are 1 and –8.
30. The rational numbers are 1, 1.5, $\frac{3}{4}$, and –8.
31. The irrational numbers are $\sqrt{2}$.
32. The real numbers are 1, 1.5, $\sqrt{2}$, $\frac{3}{4}$, and –8.
33. Factoring the number: $592=4\cdot148=(2\cdot2)\cdot(4\cdot37)=(2\cdot2)\cdot(2\cdot2\cdot37)=2^4\cdot37$
34. Factoring the number: $1340=10\cdot134=(2\cdot5)\cdot(2\cdot67)=2^2\cdot5\cdot67$

35. First factor the denominators to find the LCM:

$$15 = 3 \cdot 5$$
$$42 = 2 \cdot 3 \cdot 7$$
$$\text{LCM} = 2 \cdot 3 \cdot 5 \cdot 7 = 210$$

Combining the fractions: $\frac{5}{15} + \frac{11}{42} = \frac{5 \cdot 14}{15 \cdot 14} + \frac{11 \cdot 5}{42 \cdot 5} = \frac{70}{210} + \frac{55}{210} = \frac{125}{210} = \frac{5 \cdot 25}{5 \cdot 42} = \frac{25}{42}$

36. Combining the fractions: $\frac{5}{x} + \frac{3}{2} = \frac{5 \cdot 2}{x \cdot 2} + \frac{3 \cdot x}{2 \cdot x} = \frac{10}{2x} + \frac{3x}{2x} = \frac{10 + 3x}{2x}$

37. The expression is: $8 + (-3) = 5$

38. The expression is: $-24 - 2 = -24 + (-2) = -26$

39. The expression is: $-5(-4) = 20$

40. The expression is: $\frac{-24}{-2} = 12$

41. The pattern is to add 5, so the next term is $7 + 5 = 12$.

42. The pattern is to multiply by $-\frac{1}{2}$, so the next term is $-1\left(-\frac{1}{2}\right) = \frac{1}{2}$.

Chapter 2
Linear Equations and Inequalities

2.1 Simplifying Expressions

1. Simplifying the expression: $3x-6x=(3-6)x=-3x$
3. Simplifying the expression: $-2a+a=(-2+1)a=-a$
5. Simplifying the expression: $7x+3x+2x=(7+3+2)x=12x$
7. Simplifying the expression: $3a-2a+5a=(3-2+5)a=6a$
9. Simplifying the expression: $4x-3+2x=4x+2x-3=6x-3$
11. Simplifying the expression: $3a+4a+5=7a+5$
13. Simplifying the expression: $2x-3+3x-2=2x+3x-3-2=5x-5$
15. Simplifying the expression: $3a-1+a+3=3a+a-1+3=4a+2$
17. Simplifying the expression: $-4x+8-5x-10=-4x-5x+8-10=-9x-2$
19. Simplifying the expression: $7a+3+2a+3a=7a+2a+3a+3=12a+3$
21. Simplifying the expression: $5(2x-1)+4=10x-5+4=10x-1$
23. Simplifying the expression: $7(3y+2)-8=21y+14-8=21y+6$
25. Simplifying the expression: $-3(2x-1)+5=-6x+3+5=-6x+8$
27. Simplifying the expression: $5-2(a+1)=5-2a-2=-2a-2+5=-2a+3$
29. Simplifying the expression: $6-4(x-5)=6-4x+20=-4x+20+6=-4x+26$
31. Simplifying the expression: $-9-4(2-y)+1=-9-8+4y+1=4y+1-9-8=4y-16$
33. Simplifying the expression: $-6+2(2-3x)+1=-6+4-6x+1=-6x-6+4+1=-6x-1$
35. Simplifying the expression: $(4x-7)-(2x+5)=4x-7-2x-5=4x-2x-7-5=2x-12$
37. Simplifying the expression: $8(2a+4)-(6a-1)=16a+32-6a+1=16a-6a+32+1=10a+33$
39. Simplifying the expression: $3(x-2)+(x-3)=3x-6+x-3=3x+x-6-3=4x-9$
41. Simplifying the expression: $4(2y-8)-(y+7)=8y-32-y-7=8y-y-32-7=7y-39$
43. Simplifying the expression: $-9(2x+1)-(x+5)=-18x-9-x-5=-18x-x-9-5=-19x-14$
45. Evaluating when $x=2$: $3x-1=3(2)-1=6-1=5$
47. Evaluating when $x=2$: $-2x-5=-2(2)-5=-4-5=-9$
49. Evaluating when $x=2$: $x^2-8x+16=(2)^2-8(2)+16=4-16+16=4$
51. Evaluating when $x=2$: $(x-4)^2=(2-4)^2=(-2)^2=4$
53. Evaluating when $x=-5$: $7x-4-x-3=7(-5)-4-(-5)-3=-35-4+5-3=-42+5=-37$
    Now simplifying the expression: $7x-4-x-3=7x-x-4-3=6x-7$
    Evaluating when $x=-5$: $6x-7=6(-5)-7=-30-7=-37$
    Note that the two values are the same.
55. Evaluating when $x=-5$: $5(2x+1)+4=5[2(-5)+1]+4=5(-10+1)+4=5(-9)+4=-45+4=-41$
    Now simplifying the expression: $5(2x+1)+4=10x+5+4=10x+9$
    Evaluating when $x=-5$: $10x+9=10(-5)+9=-50+9=-41$
    Note that the two values are the same.
57. Evaluating when $x=-3$ and $y=5$: $x^2-2xy+y^2=(-3)^2-2(-3)(5)+(5)^2=9+30+25=64$
59. Evaluating when $x=-3$ and $y=5$: $(x-y)^2=(-3-5)^2=(-8)^2=64$

61. Evaluating when $x=-3$ and $y=5$: $x^2+6xy+9y^2=(-3)^2+6(-3)(5)+9(5)^2=9-90+225=144$

63. Evaluating when $x=-3$ and $y=5$: $(x+3y)^2=[-3+3(5)]^2=(-3+15)^2=(12)^2=144$

65. Evaluating when $x=\frac{1}{2}$: $12x-3=12\left(\frac{1}{2}\right)-3=6-3=3$

67. Evaluating when $x=\frac{1}{4}$: $12x-3=12\left(\frac{1}{4}\right)-3=3-3=0$

69. Evaluating when $x=\frac{3}{2}$: $12x-3=12\left(\frac{3}{2}\right)-3=18-3=15$

71. Evaluating when $x=\frac{3}{4}$: $12x-3=12\left(\frac{3}{4}\right)-3=9-3=6$

73. Substituting the values for n:

$n=1$: $2(1)+3=2+3=5$
$n=2$: $2(2)+3=4+3=7$
$n=3$: $2(3)+3=6+3=9$
$n=4$: $2(4)+3=8+3=11$

75. Substituting the values for n:

$n=1$: $1^3=1$
$n=2$: $2^3=8$
$n=3$: $3^3=27$
$n=4$: $4^3=64$

77. Substituting the values for n:

$n=1$: $(1)^2+1=1+1=2$
$n=2$: $(2)^2+1=4+1=5$
$n=3$: $(3)^2+1=9+1=10$
$n=4$: $(4)^2+1=16+1=17$

79. Substituting $n=1, 2, 3, 4$:

$n=1$: $3(1)-2=3-2=1$
$n=2$: $3(2)-2=6-2=4$
$n=3$: $3(3)-2=9-2=7$
$n=4$: $3(4)-2=12-2=10$

The sequence is 1, 4, 7, 10, ..., which is an arithmetic sequence.

81. Substituting $n=1, 2, 3, 4$:

$n=1$: $(1)^2-2(1)+1=1-2+1=0$
$n=2$: $(2)^2-2(2)+1=4-4+1=1$
$n=3$: $(3)^2-2(3)+1=9-6+1=4$
$n=4$: $(4)^2-2(4)+1=16-8+1=9$

The sequence is 0, 1, 4, 9, ..., which is a sequence of squares.

83. Translating into an algebraic expression: $x+5$
Evaluating when $x=-2$: $x+5=(-2)+5=3$

85. Translating into an algebraic expression: $x-5$
Evaluating when $x=-2$: $x-5=(-2)-5=-7$

87. Translating into an algebraic expression: $2(x+10)$
Evaluating when $x=-2$: $2(x+10)=2(-2+10)=2(8)=16$

89. Translating into an algebraic expression: $\frac{10}{x}$

Evaluating when $x=-2$: $\frac{10}{x}=\frac{10}{-2}=-5$

91. Translating into an algebraic expression, and simplifying: $[3x+(-2)]-5=3x-2-5=3x-7$
Evaluating when $x=-2$: $3x-7=3(-2)-7=-6-7=-13$

93. Subtracting: $-3-\frac{1}{2}=\frac{-3}{1}-\frac{1}{2}=\frac{-3\cdot 2}{1\cdot 2}-\frac{1}{2}=\frac{-6}{2}-\frac{1}{2}=-\frac{7}{2}$

95. Adding: $\frac{4}{5}+\frac{1}{10}+\frac{3}{8}=\frac{4\cdot 8}{5\cdot 8}+\frac{1\cdot 4}{10\cdot 4}+\frac{3\cdot 5}{8\cdot 5}=\frac{32}{40}+\frac{4}{40}+\frac{15}{40}=\frac{51}{40}$

2.2 Addition Property of Equality

1. Solving the equation:
$$\begin{aligned} x-3&=8 \\ x-3+3&=8+3 \\ x&=11 \end{aligned}$$

3. Solving the equation:
$$\begin{aligned} x+2&=6 \\ x+2+(-2)&=6+(-2) \\ x&=4 \end{aligned}$$

5. Solving the equation:
$$\begin{aligned} a+\frac{1}{2}&=-\frac{1}{4} \\ a+\frac{1}{2}+\left(-\frac{1}{2}\right)&=-\frac{1}{4}+\left(-\frac{1}{2}\right) \\ a&=-\frac{1}{4}+\left(-\frac{2}{4}\right) \\ a&=-\frac{3}{4} \end{aligned}$$

7. Solving the equation:
$$\begin{aligned} x+2.3&=-3.5 \\ x+2.3+(-2.3)&=-3.5+(-2.3) \\ x&=-5.8 \end{aligned}$$

9. Solving the equation:
$$\begin{aligned} y+11&=-6 \\ y+11+(-11)&=-6+(-11) \\ y&=-17 \end{aligned}$$

11. Solving the equation:
$$\begin{aligned} x-\frac{5}{8}&=-\frac{3}{4} \\ x-\frac{5}{8}+\frac{5}{8}&=-\frac{3}{4}+\frac{5}{8} \\ x&=-\frac{6}{8}+\frac{5}{8} \\ x&=-\frac{1}{8} \end{aligned}$$

13. Solving the equation:
$$\begin{aligned} m-6&=-10 \\ m-6+6&=-10+6 \\ m&=-4 \end{aligned}$$

15. Solving the equation:
$$\begin{aligned} 6.9+x&=3.3 \\ -6.9+6.9+x&=-6.9+3.3 \\ x&=-3.6 \end{aligned}$$

17. Solving the equation:

$$\begin{aligned} 5 &= a+4 \\ 5+(-4) &= a+4+(-4) \\ a &= 1 \end{aligned}$$

19. Solving the equation:

$$\begin{aligned} -\frac{5}{9} &= x-\frac{2}{5} \\ -\frac{5}{9}+\frac{2}{5} &= x-\frac{2}{5}+\frac{2}{5} \\ -\frac{25}{45}+\frac{18}{45} &= x \\ x &= -\frac{7}{45} \end{aligned}$$

21. Solving the equation:

$$\begin{aligned} 4x+2-3x &= 4+1 \\ x+2 &= 5 \\ x+2+(-2) &= 5+(-2) \\ x &= 3 \end{aligned}$$

23. Solving the equation:

$$\begin{aligned} 8a-\frac{1}{2}-7a &= \frac{3}{4}+\frac{1}{8} \\ a-\frac{1}{2} &= \frac{6}{8}+\frac{1}{8} \\ a-\frac{1}{2} &- \frac{7}{8} \\ a-\frac{1}{2}+\frac{1}{2} &= \frac{7}{8}+\frac{1}{2} \\ a &= \frac{7}{8}+\frac{4}{8} \\ a &= \frac{11}{8} \end{aligned}$$

25. Solving the equation:

$$\begin{aligned} -3-4x+5x &= 18 \\ -3+x &= 18 \\ 3-3+x &= 3+18 \\ x &= 21 \end{aligned}$$

27. Solving the equation:

$$\begin{aligned} -11x+2+10x+2x &= 9 \\ x+2 &= 9 \\ x+2+(-2) &= 9+(-2) \\ x &= 7 \end{aligned}$$

29. Solving the equation:

$$\begin{aligned} -2.5+4.8 &= 8x-1.2-7x \\ 2.3 &= x-1.2 \\ 2.3+1.2 &= x-1.2+1.2 \\ x &= 3.5 \end{aligned}$$

31. Solving the equation:

$$\begin{aligned} 2y-10+3y-4y &= 18-6 \\ y-10 &= 12 \\ y-10+10 &= 12+10 \\ y &= 22 \end{aligned}$$

33. Solving the equation:

$$\begin{aligned} 15-21 &= 8x+3x-10x \\ x &= -6 \end{aligned}$$

35. Solving the equation:

$$\begin{aligned} 24-3+8a-5a-2a &= 21 \\ 21+a &= 21 \\ -21+21+a &= -21+21 \\ a &= 0 \end{aligned}$$

37. Solving the equation:

$$\begin{aligned} 2(x+3)-x &= 4 \\ 2x+6-x &= 4 \\ x+6 &= 4 \\ x+6+(-6) &= 4+(-6) \\ x &= -2 \end{aligned}$$

39. Solving the equation:

$$\begin{aligned} -3(x-4)+4x &= 3-7 \\ -3x+12+4x &= -4 \\ x+12 &= -4 \\ x+12+(-12) &= -4+(-12) \\ x &= -16 \end{aligned}$$

41. Solving the equation:

$$\begin{aligned} 5(2a+1)-9a &= 8-6 \\ 10a+5-9a &= 2 \\ a+5 &= 2 \\ a+5+(-5) &= 2+(-5) \\ a &= -3 \end{aligned}$$

43. Solving the equation:

$$\begin{aligned} -(x+3)+2x-1 &= 6 \\ -x-3+2x-1 &= 6 \\ x-4 &= 6 \\ x-4+4 &= 6+4 \\ x &= 10 \end{aligned}$$

45. Solving the equation:

$$\begin{aligned} 4y-3(y-6)+2 &= 8 \\ 4y-3y+18+2 &= 8 \\ y+20 &= 8 \\ y+20+(-20) &= 8+(-20) \\ y &= -12 \end{aligned}$$

47. Solving the equation:

$$\begin{aligned} 2(3x+1)-5(x+2) &= 1-10 \\ 6x+2-5x-10 &= -9 \\ x-8 &= -9 \\ x-8+8 &= -9+8 \\ x &= -1 \end{aligned}$$

49. Solving the equation:

$$\begin{aligned} -3(2m-9)+7(m-4) &= 12-9 \\ -6m+27+7m-28 &= 3 \\ m-1 &= 3 \\ m-1+1 &= 3+1 \\ m &= 4 \end{aligned}$$

51. Solving the equation:

$$\begin{aligned} 4x &= 3x+2 \\ 4x+(-3x) &= 3x+(-3x)+2 \\ x &= 2 \end{aligned}$$

53. Solving the equation:

$$\begin{aligned} 8a &= 7a-5 \\ 8a+(-7a) &= 7a+(-7a)-5 \\ a &= -5 \end{aligned}$$

55. Solving the equation:

$$\begin{aligned} 2x &= 3x+1 \\ (-2x)+2x &= (-2x)+3x+1 \\ 0 &= x+1 \\ 0+(-1) &= x+1+(-1) \\ x &= -1 \end{aligned}$$

57. Solving the equation:

$$\begin{aligned} 3y+4 &= 2y+1 \\ 3y+(-2y)+4 &= 2y+(-2y)+1 \\ y+4 &= 1 \\ y+4+(-4) &= 1+(-4) \\ y &= -3 \end{aligned}$$

59. Solving the equation:

$$\begin{aligned} 2m-3 &= m+5 \\ 2m+(-m)-3 &= m+(-m)+5 \\ m-3 &= 5 \\ m-3+3 &= 5+3 \\ m &= 8 \end{aligned}$$

61. Solving the equation:

$$\begin{aligned} 4x-7 &= 5x+1 \\ 4x+(-4x)-7 &= 5x+(-4x)+1 \\ -7 &= x+1 \\ -7+(-1) &= x+1+(-1) \\ x &= -8 \end{aligned}$$

63. Solving the equation:

$$\begin{aligned} 5x-\frac{2}{3} &= 4x+\frac{4}{3} \\ 5x+(-4x)-\frac{2}{3} &= 4x+(-4x)+\frac{4}{3} \\ x-\frac{2}{3} &= \frac{4}{3} \\ x-\frac{2}{3}+\frac{2}{3} &= \frac{4}{3}+\frac{2}{3} \\ x &= \frac{6}{3} = 2 \end{aligned}$$

65. Solving the equation:

$$\begin{aligned} 8a-7.1 &= 7a+3.9 \\ 8a+(-7a)-7.1 &= 7a+(-7a)+3.9 \\ a-7.1 &= 3.9 \\ a-7.1+7.1 &= 3.9+7.1 \\ a &= 11 \end{aligned}$$

67. Applying the associative property: $3(6x)=(3\bullet 6)x=18x$

69. Applying the associative property: $\frac{1}{5}(5x)=\left(\frac{1}{5}\bullet 5\right)x=1x=x$

71. Applying the associative property: $8\left(\frac{1}{8}y\right)=\left(8\bullet\frac{1}{8}\right)y=1y=y$

73. Applying the associative property: $-2\left(-\frac{1}{2}x\right)=\left[-2\bullet\left(-\frac{1}{2}\right)\right]x=1x=x$

75. Applying the associative property: $-\frac{4}{3}\left(-\frac{3}{4}a\right)=\left[-\frac{4}{3}\bullet\left(-\frac{3}{4}\right)\right]a=1a=a$

2.3 Multiplication Property of Equality

1. Solving the equation:
$$5x = 10$$
$$\frac{1}{5}(5x) = \frac{1}{5}(10)$$
$$x = 2$$

3. Solving the equation:
$$7a = 28$$
$$\frac{1}{7}(7a) = \frac{1}{7}(28)$$
$$a = 4$$

5. Solving the equation:
$$-8x = 4$$
$$-\frac{1}{8}(-8x) = -\frac{1}{8}(4)$$
$$x = -\frac{1}{2}$$

7. Solving the equation:
$$8m = -16$$
$$\frac{1}{8}(8m) = \frac{1}{8}(-16)$$
$$m = -2$$

9. Solving the equation:
$$-3x = -9$$
$$-\frac{1}{3}(-3x) = -\frac{1}{3}(-9)$$
$$x = 3$$

11. Solving the equation:
$$-7y = -28$$
$$-\frac{1}{7}(-7y) = -\frac{1}{7}(-28)$$
$$y = 4$$

13. Solving the equation:
$$2x = 0$$
$$\frac{1}{2}(2x) = \frac{1}{2}(0)$$
$$x = 0$$

15. Solving the equation:
$$-5x = 0$$
$$-\frac{1}{5}(-5x) = -\frac{1}{5}(0)$$
$$x = 0$$

17. Solving the equation:
$$\frac{x}{3} = 2$$
$$3\left(\frac{x}{3}\right) = 3(2)$$
$$x = 6$$

19. Solving the equation:
$$-\frac{m}{5} = 10$$
$$-5\left(-\frac{m}{5}\right) = -5(10)$$
$$m = -50$$

21. Solving the equation:

$$-\frac{x}{2}=-\frac{3}{4}$$
$$-2\left(-\frac{x}{2}\right)=-2\left(-\frac{3}{4}\right)$$
$$x=\frac{3}{2}$$

23. Solving the equation:

$$\frac{2}{3}a=8$$
$$\frac{3}{2}\left(\frac{2}{3}a\right)=\frac{3}{2}(8)$$
$$a=12$$

25. Solving the equation:

$$-\frac{3}{5}x=\frac{9}{5}$$
$$-\frac{5}{3}\left(-\frac{3}{5}x\right)=-\frac{5}{3}\left(\frac{9}{5}\right)$$
$$x=-3$$

27. Solving the equation:

$$-\frac{5}{8}y=-20$$
$$-\frac{8}{5}\left(-\frac{5}{8}y\right)=-\frac{8}{5}(-20)$$
$$y=32$$

29. Simplifying and then solving the equation:

$$-4x-2x+3x=24$$
$$-3x=24$$
$$-\frac{1}{3}(-3x)=-\frac{1}{3}(24)$$
$$x=-8$$

31. Simplifying and then solving the equation:

$$4x+8x-2x=15-10$$
$$10x=5$$
$$\frac{1}{10}(10x)=\frac{1}{10}(5)$$
$$x=\frac{1}{2}$$

33. Simplifying and then solving the equation:

$$-3-5=3x+5x-10x$$
$$-8=-2x$$
$$-\frac{1}{2}(-8)=-\frac{1}{2}(-2x)$$
$$x=4$$

35. Using Method 2 to eliminate fractions:

$$18-13=\frac{1}{2}a+\frac{3}{4}a-\frac{5}{8}a$$
$$8(5)=8\left(\frac{1}{2}a+\frac{3}{4}a-\frac{5}{8}a\right)$$
$$40=4a+6a-5a$$
$$40=5a$$
$$\frac{1}{5}(40)=\frac{1}{5}(5a)$$
$$a=8$$

37. Solving by multiplying both sides of the equation by –1:

$$\begin{aligned} -x &= 4 \\ -1(-x) &= -1(4) \\ x &= -4 \end{aligned}$$

39. Solving by multiplying both sides of the equation by –1:

$$\begin{aligned} -x &= -4 \\ -1(-x) &= -1(-4) \\ x &= 4 \end{aligned}$$

41. Solving by multiplying both sides of the equation by –1:

$$\begin{aligned} 15 &= -a \\ -1(15) &= -1(-a) \\ a &= -15 \end{aligned}$$

43. Solving by multiplying both sides of the equation by –1:

$$\begin{aligned} -y &= \frac{1}{2} \\ -1(-y) &= -1\left(\frac{1}{2}\right) \\ y &= -\frac{1}{2} \end{aligned}$$

45. Solving the equation:

$$\begin{aligned} 3x - 2 &= 7 \\ 3x - 2 + 2 &= 7 + 2 \\ 3x &= 9 \\ \frac{1}{3}(3x) &= \frac{1}{3}(9) \\ x &= 3 \end{aligned}$$

47. Solving the equation:

$$\begin{aligned} 2a + 1 &= 3 \\ 2a + 1 + (-1) &= 3 + (-1) \\ 2a &= 2 \\ \frac{1}{2}(2a) &= \frac{1}{2}(2) \\ a &= 1 \end{aligned}$$

49. Using Method 2 to eliminate fractions:

$$\begin{aligned} \frac{1}{8} + \frac{1}{2}x &= \frac{1}{4} \\ 8\left(\frac{1}{8} + \frac{1}{2}x\right) &= 8\left(\frac{1}{4}\right) \\ 1 + 4x &= 2 \\ (-1) + 1 + 4x &= (-1) + 2 \\ 4x &= 1 \\ \frac{1}{4}(4x) &= \frac{1}{4}(1) \\ x &= \frac{1}{4} \end{aligned}$$

51. Solving the equation:

$$\begin{aligned} 6x &= 2x - 12 \\ 6x + (-2x) &= 2x + (-2x) - 12 \\ 4x &= -12 \\ \frac{1}{4}(4x) &= \frac{1}{4}(-12) \\ x &= -3 \end{aligned}$$

53. Solving the equation:

$$\begin{aligned} 2y &= -4y+18 \\ 2y+4y &= -4y+4y+18 \\ 6y &= 18 \\ \frac{1}{6}(6y) &= \frac{1}{6}(18) \\ y &= 3 \end{aligned}$$

55. Solving the equation:

$$\begin{aligned} -7x &= -3x-8 \\ -7x+3x &= -3x+3x-8 \\ -4x &= -8 \\ -\frac{1}{4}(-4x) &= -\frac{1}{4}(-8) \\ x &= 2 \end{aligned}$$

57. Solving the equation:

$$\begin{aligned} 8x+4 &= 2x-5 \\ 8x+(-2x)+4 &= 2x+(-2x)-5 \\ 6x+4 &= -5 \\ 6x+4+(-4) &= -5+(-4) \\ 6x &= -9 \\ \frac{1}{6}(6x) &= \frac{1}{6}(-9) \\ x &= -\frac{3}{2} \end{aligned}$$

59. Using Method 2 to eliminate fractions:

$$\begin{aligned} x+\frac{1}{2} &= \frac{1}{4}x-\frac{5}{8} \\ 8\left(x+\frac{1}{2}\right) &= 8\left(\frac{1}{4}x-\frac{5}{8}\right) \\ 8x+4 &= 2x-5 \\ 8x+(-2x)+4 &= 2x+(-2x)-5 \\ 6x+4 &= -5 \\ 6x+4+(-4) &= -5+(-4) \\ 6x &= -9 \\ \frac{1}{6}(6x) &= \frac{1}{6}(-9) \\ x &= -\frac{3}{2} \end{aligned}$$

61. Solving the equation:

$$\begin{aligned} 6m-3 &= m+2 \\ 6m+(-m)-3 &= m+(-m)+2 \\ 5m-3 &= 2 \\ 5m-3+3 &= 2+3 \\ 5m &= 5 \\ \frac{1}{5}(5m) &= \frac{1}{5}(5) \\ m &= 1 \end{aligned}$$

63. Using Method 2 to eliminate fractions:

$$\begin{aligned}\frac{1}{2}m-\frac{1}{4}&=\frac{1}{12}m+\frac{1}{6}\\12\left(\frac{1}{2}m-\frac{1}{4}\right)&=12\left(\frac{1}{12}m+\frac{1}{6}\right)\\6m-3&=m+2\\6m+(-m)-3&=m+(-m)+2\\5m-3&=2\\5m-3+3&=2+3\\5m&=5\\\frac{1}{5}(5m)&=\frac{1}{5}(5)\\m&=1\end{aligned}$$

65. Solving the equation:

$$\begin{aligned}9y+2&=6y-4\\9y+(-6y)+2&=6y+(-6y)-4\\3y+2&=-4\\3y+2+(-2)&=-4+(-2)\\3y&=-6\\\frac{1}{3}(3y)&=\frac{1}{3}(-6)\\y&=-2\end{aligned}$$

67. Using Method 2 to eliminate fractions:

$$\begin{aligned}\frac{3}{2}y+\frac{1}{3}&=y-\frac{2}{3}\\6\left(\frac{3}{2}y+\frac{1}{3}\right)&=6\left(y-\frac{2}{3}\right)\\9y+2&=6y-4\\9y+(-6y)+2&=6y+(-6y)-4\\3y+2&=-4\\3y+2+(-2)&=-4+(-2)\\3y&=-6\\\frac{1}{3}(3y)&=\frac{1}{3}(-6)\\y&=-2\end{aligned}$$

69. Using the distributive property and combining like terms: $5(2x-8)-3=10x-40-3=10x-43$

71. Using the distributive property and combining like terms:

$$-2(3x+5)+3(x-1)=-6x-10+3x-3=-6x+3x-10-3=-3x-13$$

73. Using the distributive property and combining like terms: $7-3(2y+1)=7-6y-3=-6y+7-3=-6y+4$

75. Using the distributive property and combining like terms: $4x-(9x-3)+4=4x-9x+3+4=-5x+7$

2.4 Solving Linear Equations

1. Solving the equation:

$$\begin{aligned}2(x+3)&=12\\2x+6&=12\\2x+6+(-6)&=12+(-6)\\2x&=6\\\frac{1}{2}(2x)&=\frac{1}{2}(6)\\x&=3\end{aligned}$$

3. Solving the equation:

$$\begin{aligned}6(x-1)&=-18\\6x-6&=-18\\6x-6+6&=-18+6\\6x&=-12\\\frac{1}{6}(6x)&=\frac{1}{6}(-12)\\x&=-2\end{aligned}$$

5. Solving the equation:

$$\begin{aligned}2(4a+1)&=-6\\8a+2&=-6\\8a+2+(-2)&=-6+(-2)\\8a&=-8\\\frac{1}{8}(8a)&=\frac{1}{8}(-8)\\a&=-1\end{aligned}$$

7. Solving the equation:

$$\begin{aligned}14&=2(5x-3)\\14&=10x-6\\14+6&=10x-6+6\\20&=10x\\\frac{1}{10}(20)&=\frac{1}{10}(10x)\\x&=2\end{aligned}$$

9. Solving the equation:

$$\begin{aligned}-2(3y+5)&=14\\-6y-10&=14\\-6y-10+10&=14+10\\-6y&=24\\-\frac{1}{6}(-6y)&=-\frac{1}{6}(24)\\y&=-4\end{aligned}$$

11. Solving the equation:

$$\begin{aligned}-5(2a+4)&=0\\-10a-20&=0\\-10a-20+20&=0+20\\-10a&=20\\-\frac{1}{10}(-10a)&=-\frac{1}{10}(20)\\a&=-2\end{aligned}$$

13. Solving the equation:

$$\begin{aligned}1&=\frac{1}{2}(4x+2)\\1&=2x+1\\1+(-1)&=2x+1+(-1)\\0&=2x\\\frac{1}{2}(0)&=\frac{1}{2}(2x)\\x&=0\end{aligned}$$

15. Solving the equation:

$$\begin{aligned}
3(t-4)+5&=-4\\
3t-12+5&=-4\\
3t-7&=-4\\
3t-7+7&=-4+7\\
3t&=3\\
\frac{1}{3}(3t)&=\frac{1}{3}(3)\\
t&=1
\end{aligned}$$

17. Solving the equation:

$$\begin{aligned}
4(2y+1)-7&=1\\
8y+4-7&=1\\
8y-3&=1\\
8y-3+3&=1+3\\
8y&=4\\
\frac{1}{8}(8y)&=\frac{1}{8}(4)\\
y&=\frac{1}{2}
\end{aligned}$$

19. Solving the equation:

$$\begin{aligned}
\frac{1}{2}(x-3)&=\frac{1}{4}(x+1)\\
\frac{1}{2}x-\frac{3}{2}&=\frac{1}{4}x+\frac{1}{4}\\
4\left(\frac{1}{2}x-\frac{3}{2}\right)&=4\left(\frac{1}{4}x+\frac{1}{4}\right)\\
2x-6&=x+1\\
2x+(-x)-6&=x+(-x)+1\\
x-6&=1\\
x-6+6&=1+6\\
x&=7
\end{aligned}$$

21. Solving the equation:

$$\begin{aligned}
-0.7(2x-7)&=0.3(11-4x)\\
-1.4x+4.9&=3.3-1.2x\\
-1.4x+1.2x+4.9&=3.3-1.2x+1.2x\\
-0.2x+4.9&=3.3\\
-0.2x+4.9+(-4.9)&=3.3+(-4.9)\\
-0.2x&=-1.6\\
\frac{-0.2x}{-0.2}&=\frac{-1.6}{-0.2}\\
x&=8
\end{aligned}$$

23. Solving the equation:

$$\begin{aligned}
-2(3y+1)&=3(1-6y)-9\\
-6y-2&=3-18y-9\\
-6y-2&=-18y-6\\
-6y+18y-2&=-18y+18y-6\\
12y-2&=-6\\
12y-2+2&=-6+2\\
12y&=-4\\
\frac{1}{12}(12y)&=\frac{1}{12}(-4)\\
y&=-\frac{1}{3}
\end{aligned}$$

25. Solving the equation:

$$\begin{aligned}\frac{3}{4}(8x-4)+3&=\frac{2}{5}(5x+10)-1\\6x-3+3&=2x+4-1\\6x&=2x+3\\6x+(-2x)&=2x+(-2x)+3\\4x&=3\\\frac{1}{4}(4x)&=\frac{1}{4}(3)\\x&=\frac{3}{4}\end{aligned}$$

27. Solving the equation:

$$\begin{aligned}0.06x+0.08(100-x)&=6.5\\0.06x+8-0.08x&=6.5\\-0.02x+8&=6.5\\-0.02x+8+(-8)&=6.5+(-8)\\-0.02x&=-1.5\\\frac{-0.02x}{-0.02}&=\frac{-1.5}{-0.02}\\x&=75\end{aligned}$$

29. Solving the equation:

$$\begin{aligned}6-5(2a-3)&=1\\6-10a+15&=1\\-10a+21&=1\\-10a+21+(-21)&=1+(-21)\\-10a&=-20\\-\frac{1}{10}(-10a)&=-\frac{1}{10}(-20)\\a&=2\end{aligned}$$

31. Solving the equation:

$$\begin{aligned}0.2x-0.5&=0.5-0.2(2x-13)\\0.2x-0.5&=0.5-0.4x+2.6\\0.2x-0.5&=-0.4x+3.1\\0.2x+0.4x-0.5&=-0.4x+0.4x+3.1\\0.6x-0.5&=3.1\\0.6x-0.5+0.5&=3.1+0.5\\0.6x&=3.6\\\frac{0.6x}{0.6}&=\frac{3.6}{0.6}\\x&=6\end{aligned}$$

33. Solving the equation:

$$\begin{aligned}2(t-3)+3(t-2)&=28\\2t-6+3t-6&=28\\5t-12&=28\\5t-12+12&=28+12\\5t&=40\\\frac{1}{5}(5t)&=\frac{1}{5}(40)\\t&=8\end{aligned}$$

35. Solving the equation:

$$\begin{aligned} 5(x-2)-(3x+4) &= 3(6x-8)+10 \\ 5x-10-3x-4 &= 18x-24+10 \\ 2x-14 &= 18x-14 \\ 2x+(-18x)-14 &= 18x+(-18x)-14 \\ -16x-14 &= -14 \\ -16x-14+14 &= -14+14 \\ -16x &= 0 \\ -\frac{1}{16}(-16x) &= -\frac{1}{16}(0) \\ x &= 0 \end{aligned}$$

37. Solving the equation:

$$\begin{aligned} 2(5x-3)-(2x-4) &= 5-(6x+1) \\ 10x-6-2x+4 &= 5-6x-1 \\ 8x-2 &= -6x+4 \\ 8x+6x-2 &= -6x+6x+4 \\ 14x-2 &= 4 \\ 14x-2+2 &= 4+2 \\ 14x &= 6 \\ \frac{1}{14}(14x) &= \frac{1}{14}(6) \\ x &= \frac{3}{7} \end{aligned}$$

39. Solving the equation:

$$\begin{aligned} -(3x+1)-(4x-7) &= 4-(3x+2) \\ -3x-1-4x+7 &= 4-3x-2 \\ -7x+6 &= -3x+2 \\ -7x+3x+6 &= -3x+3x+2 \\ -4x+6 &= 2 \\ -4x+6+(-6) &= 2+(-6) \\ -4x &= -4 \\ -\frac{1}{4}(-4x) &= -\frac{1}{4}(-4) \\ x &= 1 \end{aligned}$$

41. Multiplying the fractions: $\frac{1}{2}(3) = \frac{1}{2} \bullet \frac{3}{1} = \frac{3}{2}$

43. Multiplying the fractions: $\frac{2}{3}(6) = \frac{2}{3} \bullet \frac{6}{1} = \frac{12}{3} = 4$

45. Multiplying the fractions: $\frac{5}{9} \bullet \frac{9}{5} = \frac{45}{45} = 1$

47. Applying the distributive property: $2(3x-5) = 2 \bullet 3x - 2 \bullet 5 = 6x-10$

49. Applying the distributive property: $\frac{1}{2}(3x+6) = \frac{1}{2} \bullet 3x + \frac{1}{2} \bullet 6 = \frac{3}{2}x+3$

51. Applying the distributive property: $\frac{1}{3}(-3x+6) = \frac{1}{3} \bullet (-3x) + \frac{1}{3} \bullet 6 = -x+2$

2.5 Formulas

1. Using the perimeter formula:
$$\begin{aligned} P &= 2l + 2w \\ 300 &= 2l + 2(50) \\ 300 &= 2l + 100 \\ 200 &= 2l \\ l &= 100 \end{aligned}$$
The length is 100 feet.

3. Substituting $x = 3$:
$$\begin{aligned} 2(3) + 3y &= 6 \\ 6 + 3y &= 6 \\ 6 + (-6) + 3y &= 6 + (-6) \\ 3y &= 0 \\ y &= 0 \end{aligned}$$

5. Substituting $x = 0$:
$$\begin{aligned} 2(0) + 3y &= 6 \\ 0 + 3y &= 6 \\ 3y &= 6 \\ y &= 2 \end{aligned}$$

7. Substituting $y = 2$:
$$\begin{aligned} 2x - 5(2) &= 20 \\ 2x - 10 &= 20 \\ 2x - 10 + 10 &= 20 + 10 \\ 2x &= 30 \\ x &= 15 \end{aligned}$$

9. Substituting $y = 0$:
$$\begin{aligned} 2x - 5(0) &= 20 \\ 2x - 0 &= 20 \\ 2x &= 20 \\ x &= 10 \end{aligned}$$

11. Substituting $y = 7$:
$$\begin{aligned} 7 &= 2x - 1 \\ 7 + 1 &= 2x - 1 + 1 \\ 2x &= 8 \\ x &= 4 \end{aligned}$$

13. Substituting $y = 3$:
$$\begin{aligned} 3 &= 2x - 1 \\ 3 + 1 &= 2x - 1 + 1 \\ 2x &= 4 \\ x &= 2 \end{aligned}$$

15. Solving for l:
$$\begin{aligned} lw &= A \\ \frac{lw}{w} &= \frac{A}{w} \\ l &= \frac{A}{w} \end{aligned}$$

17. Solving for r:
$$\begin{aligned} rt &= d \\ \frac{rt}{t} &= \frac{d}{t} \\ r &= \frac{d}{t} \end{aligned}$$

19. Solving for h:

$$\begin{aligned} lwh &= V \\ \frac{lwh}{lw} &= \frac{V}{lw} \\ h &= \frac{V}{lw} \end{aligned}$$

21. Solving for P:

$$\begin{aligned} PV &= nRT \\ \frac{PV}{V} &= \frac{nRT}{V} \\ P &= \frac{nRT}{V} \end{aligned}$$

23. Solving for a:

$$\begin{aligned} a+b+c &= P \\ a+b+c-b-c &= P-b-c \\ a &= P-b-c \end{aligned}$$

25. Solving for x:

$$\begin{aligned} x-3y &= -1 \\ x-3y+3y &= -1+3y \\ x &= 3y-1 \end{aligned}$$

27. Solving for y:

$$\begin{aligned} -3x+y &= 6 \\ -3x+3x+y &= 6+3x \\ y &= 3x+6 \end{aligned}$$

29. Solving for y:

$$\begin{aligned} 2x+3y &= 6 \\ -2x+2x+3y &= -2x+6 \\ 3y &= -2x+6 \\ \frac{1}{3}(3y) &= \frac{1}{3}(-2x+6) \\ y &= -\frac{2}{3}x+2 \end{aligned}$$

31. Solving for y:

$$\begin{aligned} 6x+3y &= 12 \\ -6x+6x+3y &= -6x+12 \\ 3y &= -6x+12 \\ \frac{1}{3}(3y) &= \frac{1}{3}(-6x+12) \\ y &= -2x+4 \end{aligned}$$

33. Solving for y:

$$\begin{aligned} 5x-2y &= 3 \\ -5x+5x-2y &= -5x+3 \\ -2y &= -5x+3 \\ -\frac{1}{2}(-2y) &= -\frac{1}{2}(-5x+3) \\ y &= \frac{5}{2}x-\frac{3}{2} \end{aligned}$$

35. Solving for w:

$$\begin{aligned} 2l+2w &= P \\ 2l-2l+2w &= P-2l \\ 2w &= P-2l \\ \frac{2w}{2} &= \frac{P-2l}{2} \\ w &= \frac{P-2l}{2} \end{aligned}$$

37. Solving for v:

$$\begin{aligned} vt+16t^2 &= h \\ vt+16t^2-16t^2 &= h-16t^2 \\ vt &= h-16t^2 \\ \frac{vt}{t} &= \frac{h-16t^2}{t} \\ v &= \frac{h-16t^2}{t} \end{aligned}$$

39. Solving for h:

$$\begin{aligned} \pi r^2+2\pi rh &= A \\ \pi r^2-\pi r^2+2\pi rh &= A-\pi r^2 \\ 2\pi rh &= A-\pi r^2 \\ \frac{2\pi rh}{2\pi r} &= \frac{A-\pi r^2}{2\pi r} \\ h &= \frac{A-\pi r^2}{2\pi r} \end{aligned}$$

41. Solving for y:

$$\begin{aligned} \frac{x}{2}+\frac{y}{3} &= 1 \\ -\frac{x}{2}+\frac{x}{2}+\frac{y}{3} &= -\frac{x}{2}+1 \\ \frac{y}{3} &= -\frac{x}{2}+1 \\ 3\left(\frac{y}{3}\right) &= 3\left(-\frac{x}{2}+1\right) \\ y &= -\frac{3}{2}x+3 \end{aligned}$$

43. Solving for y:

$$\begin{aligned} \frac{x}{7}-\frac{y}{3} &= 1 \\ -\frac{x}{7}+\frac{x}{7}-\frac{y}{3} &= -\frac{x}{7}+1 \\ -\frac{y}{3} &= -\frac{x}{7}+1 \\ -3\left(-\frac{y}{3}\right) &= -3\left(-\frac{x}{7}+1\right) \\ y &= \frac{3}{7}x-3 \end{aligned}$$

45. Solving for y:

$$-\frac{1}{4}x+\frac{1}{8}y=1$$
$$-\frac{1}{4}x+\frac{1}{4}x+\frac{1}{8}y=1+\frac{1}{4}x$$
$$\frac{1}{8}y=\frac{1}{4}x+1$$
$$8\left(\frac{1}{8}y\right)=8\left(\frac{1}{4}x+1\right)$$
$$y=2x+8$$

47. The complement of 30° is 90° – 30° = 60°, and the supplement is 180° – 30° = 150°.
49. The complement of 45° is 90° – 45° = 45°, and the supplement is 180° – 45° = 135°.
51. Translating into an equation and solving:

$$x=0.25\bullet 40$$
$$x=10$$

The number 10 is 25% of 40.

53. Translating into an equation and solving:

$$x=0.12\bullet 2000$$
$$x=240$$

The number 240 is 12% of 2000.

55. Translating into an equation and solving:

$$x\bullet 28=7$$
$$28x=7$$
$$\frac{1}{28}(28x)=\frac{1}{28}(7)$$
$$x=0.25=25\%$$

The number 7 is 25% of 28.

57. Translating into an equation and solving:

$$x\bullet 40=14$$
$$40x=14$$
$$\frac{1}{40}(40x)=\frac{1}{40}(14)$$
$$x=0.35=35\%$$

The number 14 is 35% of 40.

59. Translating into an equation and solving:

$$0.50\bullet x=32$$
$$\frac{0.50x}{0.50}=\frac{32}{0.50}$$
$$x=64$$

The number 32 is 50% of 64.

61. Translating into an equation and solving:

$$0.12\bullet x=240$$
$$\frac{0.12x}{0.12}=\frac{240}{0.12}$$
$$x=2000$$

The number 240 is 12% of 2000.

63. Substituting $F=212$: $C=\frac{5}{9}(212-32)=\frac{5}{9}(180)=100°\text{C}$

This value agrees with the information in Table 1.

65. Substituting $F=68$: $C=\frac{5}{9}(68-32)=\frac{5}{9}(36)=20°\text{C}$

This value agrees with the information in Table 1.

67. Solving for C:

$$\frac{9}{5}C+32=F$$
$$\frac{9}{5}C+32-32=F-32$$
$$\frac{9}{5}C=F-32$$
$$\frac{5}{9}\left(\frac{9}{5}C\right)=\frac{5}{9}(F-32)$$
$$C=\frac{5}{9}(F-32)$$

69. We need to find what percent of 150 is 90:

$$x \bullet 150=90$$
$$\frac{1}{150}(150x)=\frac{1}{150}(90)$$
$$x=0.60=60\%$$

So 60% of the calories in one serving of vanilla ice cream are fat calories.

71. We need to find what percent of 98 is 26:

$$x \bullet 98=26$$
$$\frac{1}{98}(98x)=\frac{1}{98}(26)$$
$$x\approx 0.265=26.5\%$$

So 26.5% of one serving of frozen yogurt are carbohydrates.

73. Solving for r:

$$2\pi r=C$$
$$2\bullet\frac{22}{7}r=44$$
$$\frac{44}{7}r=44$$
$$\frac{7}{44}\left(\frac{44}{7}r\right)=\frac{7}{44}(44)$$
$$r=7 \text{ meters}$$

75. Solving for r:

$$2\pi r=C$$
$$2\bullet 3.14r=9.42$$
$$6.28r=9.42$$
$$\frac{6.28r}{6.28}=\frac{9.42}{6.28}$$
$$r=1.5 \text{ inches}$$

77. Solving for h:

$$\pi r^2h=V$$
$$\frac{22}{7}\left(\frac{7}{22}\right)^2h=42$$
$$\frac{7}{22}h=42$$
$$\frac{22}{7}\left(\frac{7}{22}h\right)=\frac{22}{7}(42)$$
$$h=132 \text{ feet}$$

79. Solving for h:

$$\pi r^2 h = V$$
$$3.14(3)^2 h = 6.28$$
$$28.26h = 6.28$$
$$\frac{28.26h}{28.26} = \frac{6.28}{28.26}$$
$$h = \frac{2}{9} \text{ centimeters}$$

81. The sum of 4 and 1 is 5.
83. The difference of 6 and 2 is 4.
85. An equivalent expression is: $2(6+3) = 2(9) = 18$
87. An equivalent expression is: $2(5)+3 = 10+3 = 13$

2.6 Applications

1. Let x represent the number. The equation is:

$$x + 5 = 13$$
$$x = 8$$

The number is 8.

3. Let x represent the number. The equation is:

$$2x + 4 = 14$$
$$2x = 10$$
$$x = 5$$

The number is 5.

5. Let x represent the number. The equation is:

$$5(x+7) = 30$$
$$5x + 35 = 30$$
$$5x = -5$$
$$x = -1$$

The number is –1.

7. Let x and $x + 2$ represent the two numbers. The equation is:

$$x + x + 2 = 8$$
$$2x + 2 = 8$$
$$2x = 6$$
$$x = 3$$
$$x + 2 = 5$$

The two numbers are 3 and 5.

9. Let x and $3x - 4$ represent the two numbers. The equation is:

$$(x + 3x - 4) + 5 = 25$$
$$4x + 1 = 25$$
$$4x = 24$$
$$x = 6$$
$$3x - 4 = 3(6) - 4 = 14$$

The two numbers are 6 and 14.

11. Completing the table:

	Five Years Ago	Now
Fred	$x + 4 - 5 = x - 1$	$x + 4$
Barney	$x - 5$	x

The equation is:

$$\begin{aligned} x-1+x-5&=48\\ 2x-6&=48\\ 2x&=54\\ x&=27\\ x+4&=31 \end{aligned}$$

Barney is 27 and Fred is 31.

13. Completing the table:

	Now	Three Years from Now
Jack	$2x$	$2x+3$
Lacy	x	$x+3$

The equation is:

$$\begin{aligned} 2x+3+x+3&=54\\ 3x+6&=54\\ 3x&=48\\ x&=16\\ 2x&=32 \end{aligned}$$

Lacy is 16 and Jack is 32.

15. Completing the table:

	Now	Two Years from Now
Pat	$x+20$	$x+20+2=x+22$
Patrick	x	$x+2$

The equation is:

$$\begin{aligned} x+22&=2(x+2)\\ x+22&=2x+4\\ 22&=x+4\\ x&=18\\ x+20&=38 \end{aligned}$$

Patrick is 18 and Pat is 38.

17. Let w represent the width and $w+5$ represent the length. The equation is:

$$\begin{aligned} 2w+2(w+5)&=34\\ 2w+2w+10&=34\\ 4w+10&=34\\ 4w&=24\\ w&=6\\ w+5&=11 \end{aligned}$$

The length is 11 inches and the width is 6 inches.

19. Let s represent the side of the square. The equation is:

$$\begin{aligned} 4s&=48\\ s&=12 \end{aligned}$$

The length of one side is 12 meters.

21. Let w represent the width and $2w-3$ represent the length. The equation is:

$$\begin{aligned} 2w+2(2w-3)&=54\\ 2w+4w-6&=54\\ 6w-6&=54\\ 6w&=60\\ w&=10\\ 2w-3=2(10)-3&=17 \end{aligned}$$

The length is 17 inches and the width is 10 inches.

23. Completing the table:

	Nickels	Dimes
Number	x	$x+9$
Value (cents)	$5(x)$	$10(x+9)$

The equation is:

$$\begin{aligned} 5(x)+10(x+9) &= 210 \\ 5x+10x+90 &= 210 \\ 15x+90 &= 210 \\ 15x &= 120 \\ x &= 8 \\ x+9 &= 17 \end{aligned}$$

Sue has 8 nickels and 17 dimes.

25. Completing the table:

	Dimes	Quarters
Number	x	$2x$
Value (cents)	$10(x)$	$25(2x)$

The equation is:

$$\begin{aligned} 10(x)+25(2x) &= 900 \\ 10x+50x &= 900 \\ 60x &= 900 \\ x &= 15 \\ 2x &= 30 \end{aligned}$$

You have 15 dimes and 30 quarters.

27. Completing the table:

	Nickels	Dimes	Quarters
Number	x	$x+3$	$x+5$
Value (cents)	$5(x)$	$10(x+3)$	$25(x+5)$

The equation is:

$$\begin{aligned} 5(x)+10(x+3)+25(x+5) &= 435 \\ 5x+10x+30+25x+125 &= 435 \\ 40x+155 &= 435 \\ 40x &= 280 \\ x &= 7 \\ x+3 &= 10 \\ x+5 &= 12 \end{aligned}$$

Katie has 7 nickels, 10 dimes, and 12 quarters.

29. The statement is: 4 is less than 10
31. The statement is: 9 is greater than or equal to –5
33. The correct statement is: $12 < 20$
35. The correct statement is: $-8 < -6$
37. Simplifying: $|8-3|-|5-2| = |5|-|3| = 5-3 = 2$
39. Simplifying: $15-|9-3(7-5)| = 15-|9-3(2)| = 15-|9-6| = 15-|3| = 15-3 = 12$

2.7 More Applications

1. Completing the table:

	Dollars Invested at 8%	Dollars Invested at 9%
Number of	x	$x+2000$
Interest on	$0.08(x)$	$0.09(x+2000)$

The equation is:

$$\begin{aligned} 0.08(x)+0.09(x+2000) &= 860 \\ 0.08x+0.09x+180 &= 860 \\ 0.17x+180 &= 860 \\ 0.17x &= 680 \\ x &= 4000 \\ x+2000 &= 6000 \end{aligned}$$

You have $4,000 invested at 8% and $6,000 invested at 9%.

3. Completing the table:

	Dollars Invested at 10%	Dollars Invested at 12%
Number of	x	$x+500$
Interest on	$0.10(x)$	$0.12(x+500)$

The equation is:

$$\begin{aligned} 0.10(x)+0.12(x+500) &= 214 \\ 0.10x+0.12x+60 &= 214 \\ 0.22x+60 &= 214 \\ 0.22x &= 154 \\ x &= 700 \\ x+500 &= 1200 \end{aligned}$$

Tyler has $700 invested at 10% and $1,200 invested at 12%.

5. Completing the table:

	Dollars Invested at 8%	Dollars Invested at 9%	Dollars Invested at 10%
Number of	x	$2x$	$3x$
Interest on	$0.08(x)$	$0.09(2x)$	$0.10(3x)$

The equation is:

$$\begin{aligned} 0.08(x)+0.09(2x)+0.10(3x) &= 280 \\ 0.08x+0.18x+0.30x &= 280 \\ 0.56x &= 280 \\ x &= 500 \\ 2x &= 1000 \\ 3x &= 1500 \end{aligned}$$

She has $500 invested at 8%, $1,000 invested at 9%, and $1,500 invested at 10%.

7. Let x represent the measure of the two equal angles, so $x+x=2x$ represents the measure of the third angle. Since the sum of the three angles is 180°, the equation is:

$$\begin{aligned} x+x+2x &= 180^\circ \\ 4x &= 180^\circ \\ x &= 45^\circ \\ 2x &= 90^\circ \end{aligned}$$

The measures of the three angles are 45°, 45°, and 90°.

9. Let x represent the measure of the largest angle. Then $\frac{1}{5}x$ represents the measure of the smallest angle, and $2\left(\frac{1}{5}x\right)=\frac{2}{5}x$ represents the measure of the other angle. Since the sum of the three angles is 180°, the equation is:

$$\begin{aligned} x+\frac{1}{5}x+\frac{2}{5}x &= 180° \\ \frac{5}{5}x+\frac{1}{5}x+\frac{2}{5}x &= 180° \\ \frac{8}{5}x &= 180° \\ x &= 112.5° \\ \frac{1}{5}x &= 22.5° \\ \frac{2}{5}x &= 45° \end{aligned}$$

The measures of the three angles are 22.5°, 45°, and 112.5°.

11. Let x represent the measure of the other acute angle, and 90° is the measure of the right angle. Since the sum of the three angles is 180°, the equation is:

$$\begin{aligned} x+37°+90° &= 180° \\ x+127° &= 180° \\ x &= 53° \end{aligned}$$

The other two angles are 53° and 90°.

13. Let x represent the total minutes for the call. Then \$0.41 is charged for the first minute, and \$0.32 is charged for the additional $x-1$ minutes. The equation is:

$$\begin{aligned} 0.41(1)+0.32(x-1) &= 5.21 \\ 0.41+0.32x-0.32 &= 5.21 \\ 0.32x+0.09 &= 5.21 \\ 0.32x &= 5.12 \\ x &= 16 \end{aligned}$$

The call was 16 minutes long.

15. Let x represent the hours JoAnn worked that week. Then \$12/hour is paid for the first 35 hours and \$18/hour is paid for the additional $x-35$ hours. The equation is:

$$\begin{aligned} 12(35)+18(x-35) &= 492 \\ 420+18x-630 &= 492 \\ 18x-210 &= 492 \\ 18x &= 702 \\ x &= 39 \end{aligned}$$

JoAnn worked 39 hours that week.

17. Let x represent the number of children's tickets Stacey sold, so $2x$ represents the number of adult tickets sold. The equation is:

$$\begin{aligned} 6.00(2x)+4.50(x) &= 115.50 \\ 12x+4.5x &= 115.5 \\ 16.5x &= 115.5 \\ x &= 7 \\ 2x &= 14 \end{aligned}$$

Stacey sold 7 children's tickets and 14 adult tickets.

19. For Jeff, the total time traveled is $\dfrac{425 \text{ miles}}{55 \text{ miles / hour}} \approx 7.72 \text{ hours} \approx 463 \text{ minutes}$. Since he left at 11:00 AM, he will arrive at 6:43 PM. For Carla, the total time traveled is $\dfrac{425 \text{ miles}}{65 \text{ miles / hour}} \approx 6.54 \text{ hours} \approx 392 \text{ minutes}$. Since she left at 1:00 PM, she will arrive at 7:32 PM. Thus Jeff will arrive in Lake Tahoe first.

21. Since $\frac{1}{5}$ mile $= 0.2$ mile, the taxi charge is \$1.25 for the first $\frac{1}{5}$ mile and \$0.25 per fifth mile for the remaining 7.3 miles. Since 7.3 miles $= \frac{7.3}{0.2} = 36.5$ fifths, the total charge is: $\$1.25 + \$0.25(36.5) \approx \$10.38$

23. The first $\frac{1}{5}$ mile is \$1.25, and the remaining $12.4 - 0.2 = 12.2$ miles will be charged at \$0.25 per fifth mile. Since 12.2 miles $= \frac{12.2}{0.2} = 61$ fifths, the total charge is: $\$1.25 + \$0.25(61) = \$16.50$
Yes, the meter is working correctly.

25. If all 36 people are Elk's Lodge members (which would be the least amount), the cost of the lessons would be $\$3(36) = \$108$. Since half of the money is paid to Ike and Nancy, the least amount they could make is $\frac{1}{2}(\$108) = \$54$.

27. Yes. The total receipts were \$160, which is possible if there were 10 Elk's members and 26 nonmembers. Computing the total receipts: $10(\$3) + 26(\$5) = \$30 + \$130 = \$160$

29. The pattern is to add –4, so the next number is: $-4 + (-4) = -8$

31. The pattern is to multiply by $-\frac{1}{2}$, so the next number is: $-\frac{3}{2}\left(-\frac{1}{2}\right) = \frac{3}{4}$

33. Each number is the square of a number, with alternating signs. For example, $1^2 = 1$, $-2^2 = -4$, $3^2 = 9$, and $-4^2 = -16$. Based on this pattern, the next number is: $5^2 = 25$

35. The pattern is to add $-\frac{1}{2}$, so the next number is: $\frac{1}{2} + \left(-\frac{1}{2}\right) = 0$

2.8 Linear Inequalities

1. Solving the inequality:

$$x - 5 < 7$$
$$x - 5 + 5 < 7 + 5$$
$$x < 12$$

Graphing the solution set:

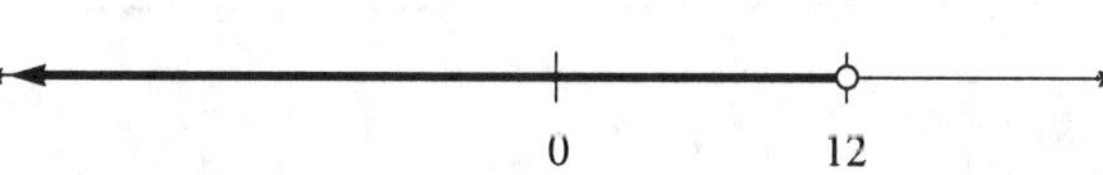

3. Solving the inequality:

$$a - 4 \le 8$$
$$a - 4 + 4 \le 8 + 4$$
$$a \le 12$$

Graphing the solution set:

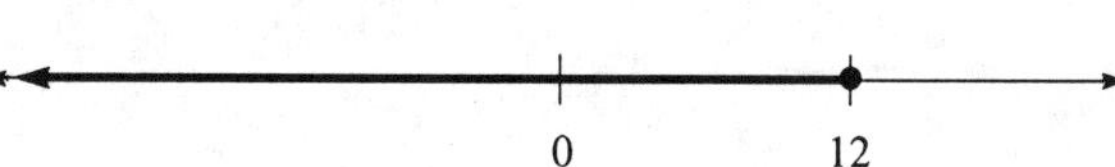

5. Solving the inequality:

$$x - 4.3 > 8.7$$
$$x - 4.3 + 4.3 > 8.7 + 4.3$$
$$x > 13$$

Graphing the solution set:

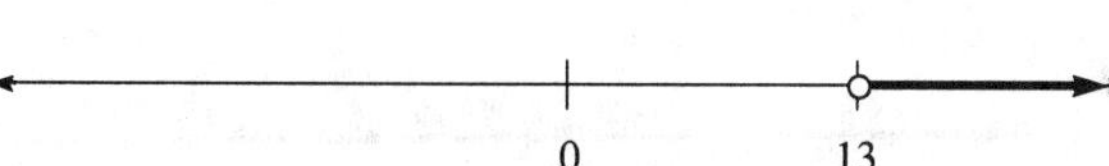

7. Solving the inequality:

$$y + 6 \ge 10$$
$$y + 6 + (-6) \ge 10 + (-6)$$
$$y \ge 4$$

Graphing the solution set:

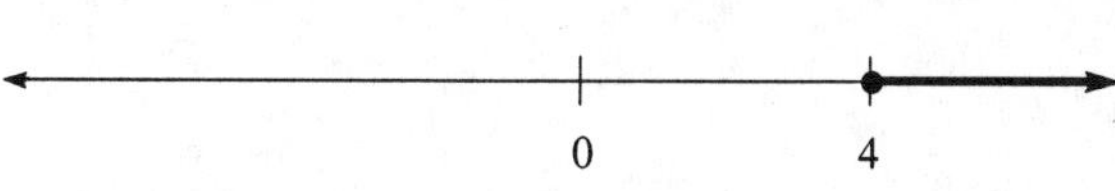

9. Solving the inequality:

$$2 < x - 7$$
$$2 + 7 < x - 7 + 7$$
$$9 < x$$
$$x > 9$$

Graphing the solution set:

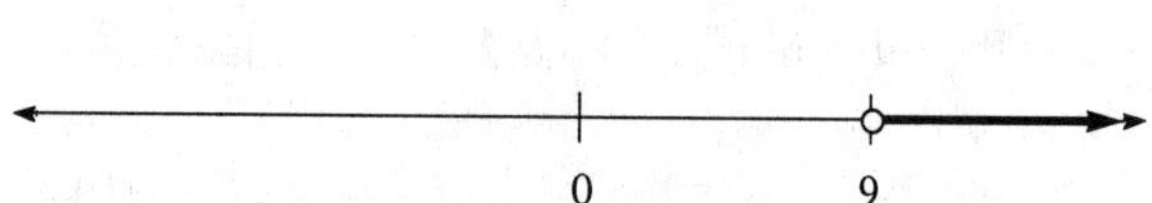

11. Solving the inequality:

$$3x < 6$$
$$\frac{1}{3}(3x) < \frac{1}{3}(6)$$
$$x < 2$$

Graphing the solution set:

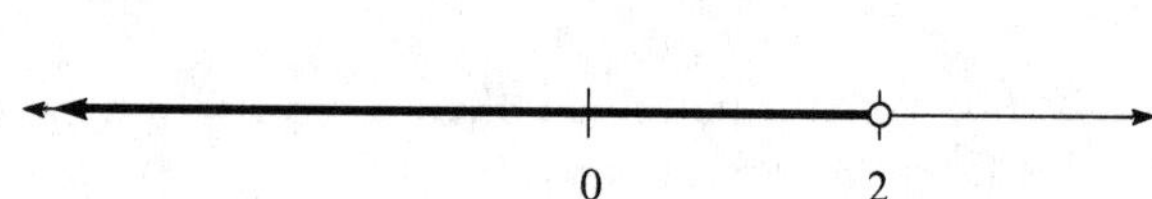

13. Solving the inequality:

$$5a \le 25$$
$$\frac{1}{5}(5a) \le \frac{1}{5}(25)$$
$$a \le 5$$

Graphing the solution set:

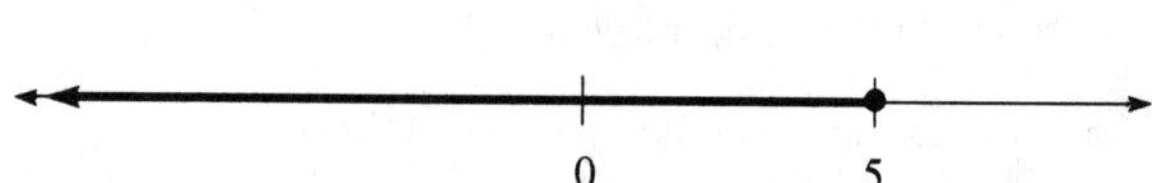

15. Solving the inequality:

$$\frac{x}{3} > 5$$
$$3\left(\frac{x}{3}\right) > 3(5)$$
$$x > 15$$

Graphing the solution set:

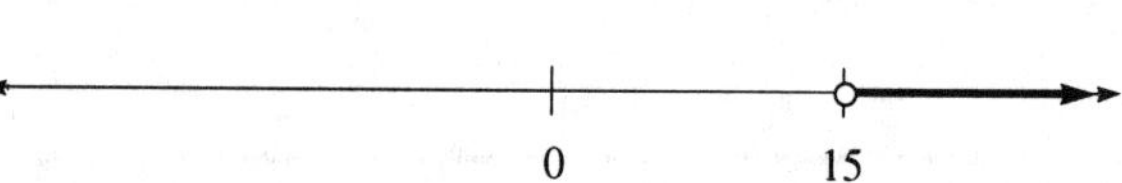

17. Solving the inequality:

$$-2x > 6$$
$$-\frac{1}{2}(-2x) < -\frac{1}{2}(6)$$
$$x < -3$$

Graphing the solution set:

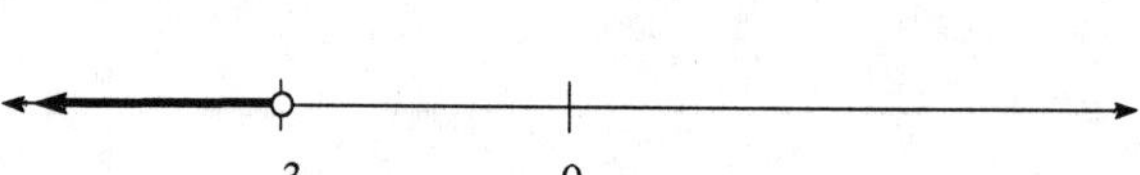

19. Solving the inequality:

$$-3x \ge -18$$
$$-\frac{1}{3}(-3x) \le -\frac{1}{3}(-18)$$
$$x \le 6$$

Graphing the solution set:

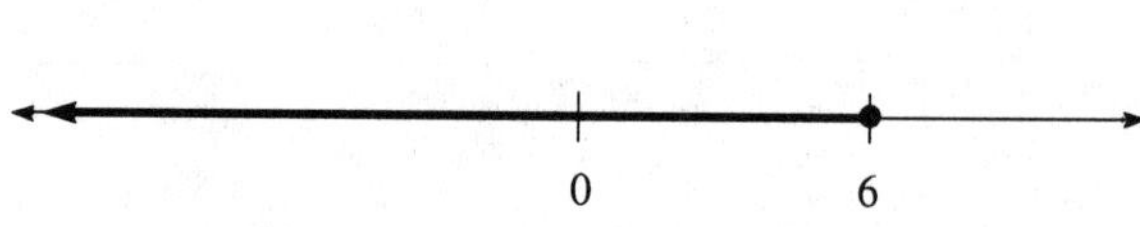

21. Solving the inequality:

$$-\frac{x}{5} \le 10$$
$$-5\left(-\frac{x}{5}\right) \ge -5(10)$$
$$x \ge -50$$

Graphing the solution set:

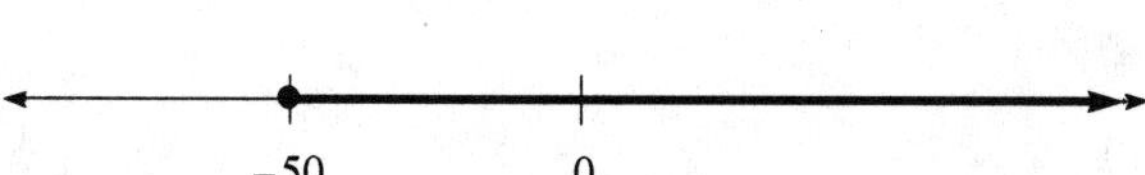

23. Solving the inequality:

$$-\frac{2}{3}y > 4$$
$$-\frac{3}{2}\left(-\frac{2}{3}y\right) < -\frac{3}{2}(4)$$
$$y < -6$$

Graphing the solution set:

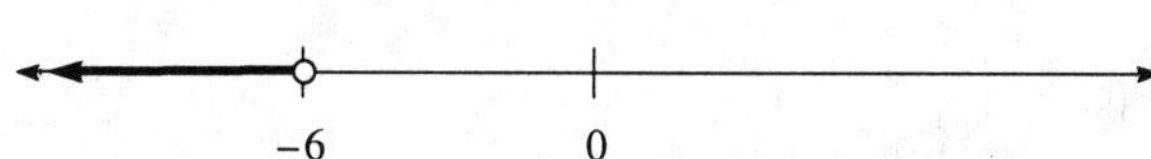

25. Solving the inequality:

$$2x - 3 < 9$$
$$2x - 3 + 3 < 9 + 3$$
$$2x < 12$$
$$\frac{1}{2}(2x) < \frac{1}{2}(12)$$
$$x < 6$$

Graphing the solution set:

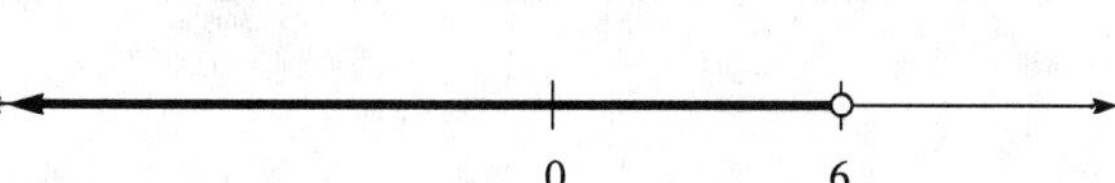

27. Solving the inequality:

$$-\frac{1}{5}y - \frac{1}{3} \le \frac{2}{3}$$
$$-\frac{1}{5}y - \frac{1}{3} + \frac{1}{3} \le \frac{2}{3} + \frac{1}{3}$$
$$-\frac{1}{5}y \le 1$$
$$-5\left(-\frac{1}{5}y\right) \ge -5(1)$$
$$y \ge -5$$

Graphing the solution set:

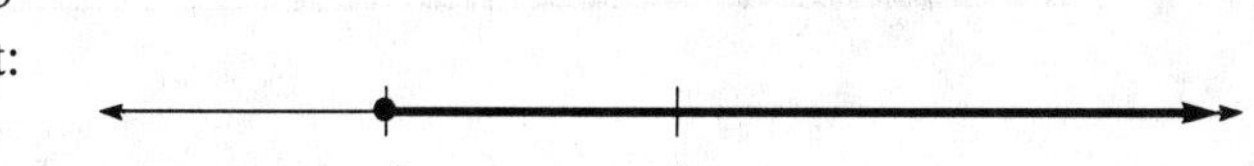

29. Solving the inequality:

$$-4x + 1 > -11$$
$$-4x + 1 + (-1) > -11 + (-1)$$
$$-4x > -12$$
$$-\frac{1}{4}(-4x) < -\frac{1}{4}(-12)$$
$$x < 3$$

Graphing the solution set:

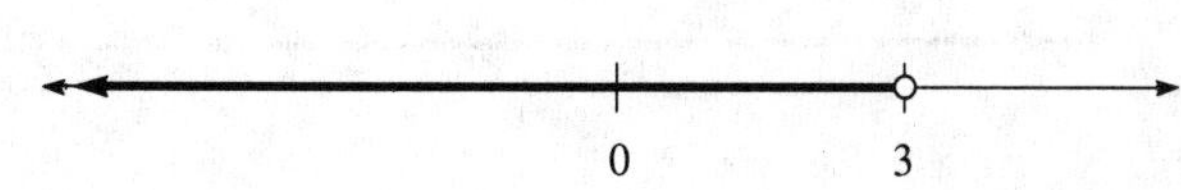

31. Solving the inequality:

$$\frac{2}{3}x-5\le 7$$
$$\frac{2}{3}x-5+5\le 7+5$$
$$\frac{2}{3}x\le 12$$
$$\frac{3}{2}\left(\frac{2}{3}x\right)\le \frac{3}{2}(12)$$
$$x\le 18$$

Graphing the solution set:

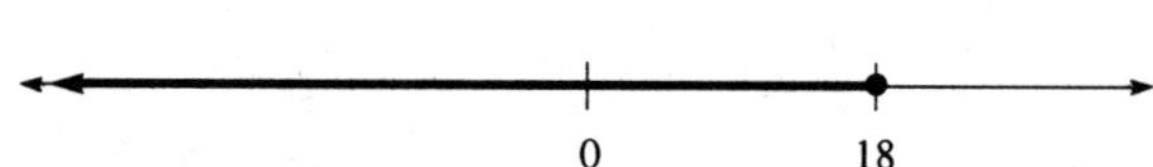

33. Solving the inequality:

$$-\frac{2}{5}a-3>5$$
$$-\frac{2}{5}a-3+3>5+3$$
$$-\frac{2}{5}a>8$$
$$-\frac{5}{2}\left(-\frac{2}{5}a\right)<-\frac{5}{2}(8)$$
$$a<-20$$

Graphing the solution set:

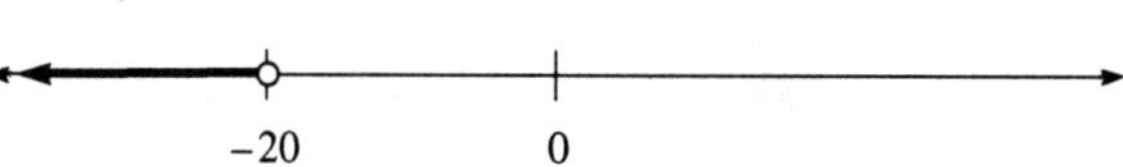

35. Solving the inequality:

$$5-\frac{3}{5}y>-10$$
$$-5+5-\frac{3}{5}y>-5+(-10)$$
$$-\frac{3}{5}y>-15$$
$$-\frac{5}{3}\left(-\frac{3}{5}y\right)<-\frac{5}{3}(-15)$$
$$y<25$$

Graphing the solution set:

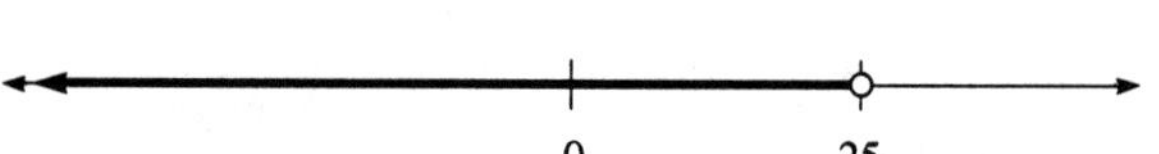

37. Solving the inequality:

$$0.3(a+1)\le 1.2$$
$$0.3a+0.3\le 1.2$$
$$0.3a+0.3+(-0.3)\le 1.2+(-0.3)$$
$$0.3a\le 0.9$$
$$\frac{0.3a}{0.3}\le\frac{0.9}{0.3}$$
$$a\le 3$$

Graphing the solution set:

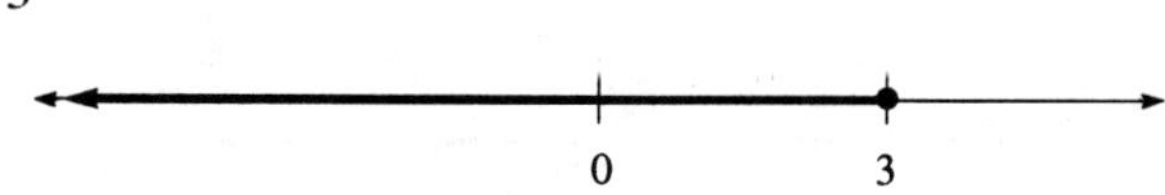

39. Solving the inequality:

$$\begin{aligned} 2(5-2x) &\le -20 \\ 10-4x &\le -20 \\ -10+10-4x &\le -10+(-20) \\ -4x &\le -30 \\ -\frac{1}{4}(-4x) &\ge -\frac{1}{4}(-30) \\ x &\ge \frac{15}{2} \end{aligned}$$

Graphing the solution set:

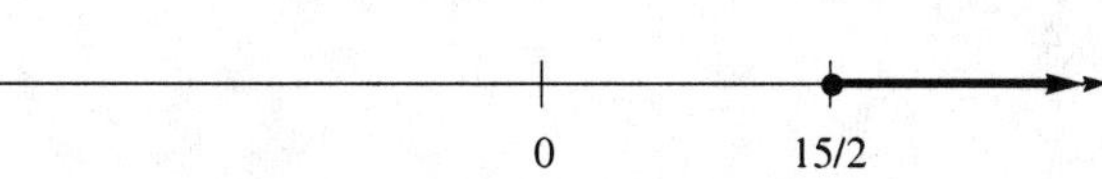

41. Solving the inequality:

$$\begin{aligned} 3x-5 &> 8x \\ -3x+3x-5 &> -3x+8x \\ -5 &> 5x \\ \frac{1}{5}(-5) &> \frac{1}{5}(5x) \\ -1 &> x \\ x &< -1 \end{aligned}$$

Graphing the solution set:

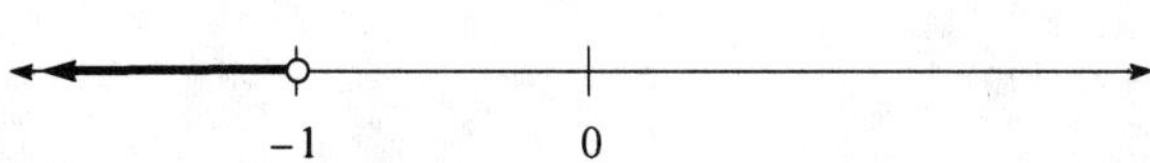

43. First multiply by 6 to clear the inequality of fractions:

$$\begin{aligned} \frac{1}{3}y-\frac{1}{2} &\le \frac{5}{6}y+\frac{1}{2} \\ 6\left(\frac{1}{3}y-\frac{1}{2}\right) &\le 6\left(\frac{5}{6}y+\frac{1}{2}\right) \\ 2y-3 &\le 5y+3 \\ -5y+2y-3 &\le -5y+5y+3 \\ -3y-3 &\le 3 \\ -3y-3+3 &\le 3+3 \\ -3y &\le 6 \\ -\frac{1}{3}(-3y) &\ge -\frac{1}{3}(6) \\ y &\ge -2 \end{aligned}$$

Graphing the solution set:

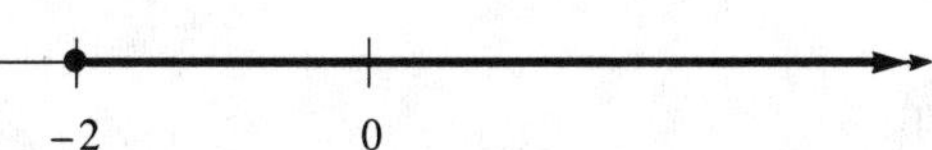

45. Solving the inequality:

$$\begin{aligned} -0.4x+1.2 &< -2x-0.4 \\ 2x-0.4x+1.2 &< 2x-2x-0.4 \\ 1.6x+1.2 &< -0.4 \\ 1.6x+1.2+(-1.2) &< -0.4+(-1.2) \\ 1.6x &< -1.6 \\ \frac{1.6x}{1.6} &< \frac{-1.6}{1.6} \\ x &< -1 \end{aligned}$$

Graphing the solution set:

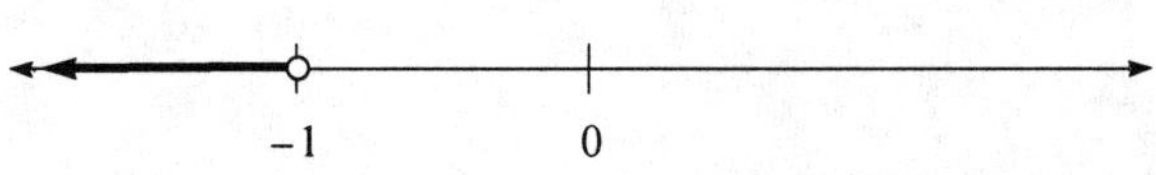

47. Solving the inequality:

$$
\begin{aligned}
3(m-2)-4 &\geq 7m+14 \\
3m-6-4 &\geq 7m+14 \\
3m-10 &\geq 7m+14 \\
-7m+3m-10 &\geq -7m+7m+14 \\
-4m-10 &\geq 14 \\
-4m-10+10 &\geq 14+10 \\
-4m &\geq 24 \\
-\frac{1}{4}(-4m) &\leq -\frac{1}{4}(24) \\
m &\leq -6
\end{aligned}
$$

Graphing the solution set:

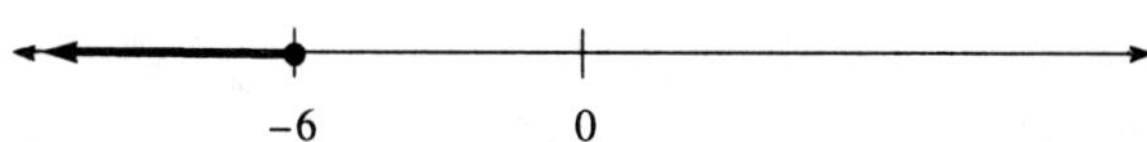

49. Solving the inequality:

$$
\begin{aligned}
3-4(x-2) &\leq -5x+6 \\
3-4x+8 &\leq -5x+6 \\
-4x+11 &\leq -5x+6 \\
-4x+5x+11 &\leq -5x+5x+6 \\
x+11 &\leq 6 \\
x+11+(-11) &\leq 6+(-11) \\
x &\leq -5
\end{aligned}
$$

Graphing the solution set:

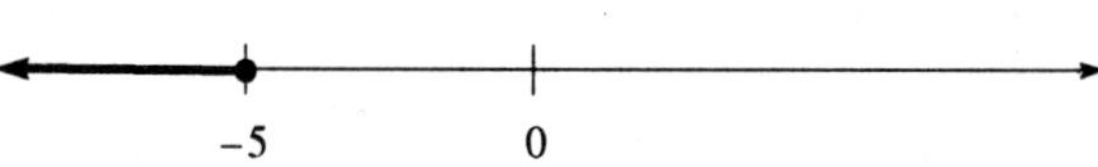

51. Solving for y:

$$
\begin{aligned}
3x+2y &< 6 \\
2y &< -3x+6 \\
y &< -\frac{3}{2}x+3
\end{aligned}
$$

53. Solving for y:

$$
\begin{aligned}
2x-5y &> 10 \\
-5y &> -2x+10 \\
y &< \frac{2}{5}x-2
\end{aligned}
$$

55. Solving for y:

$$
\begin{aligned}
-3x+7y &\leq 21 \\
7y &\leq 3x+21 \\
y &\leq \frac{3}{7}x+3
\end{aligned}
$$

57. Solving for y:

$$
\begin{aligned}
2x-4y &\geq -4 \\
-4y &\geq -2x-4 \\
y &\leq \frac{1}{2}x+1
\end{aligned}
$$

59. Let x represent the number. Solving the inequality:

$$
\begin{aligned}
2x+6 &< 10 \\
2x &< 4 \\
x &< 2
\end{aligned}
$$

61. Let x represent the number. Solving the inequality:
$$4x > x - 8$$
$$3x > -8$$
$$x > -\frac{8}{3}$$
63. Let x represent the number. Solving the inequality:
$$2(x+5) \le 12$$
$$2x + 10 \le 12$$
$$2x \le 2$$
$$x \le 1$$
65. Let x represent the number. Solving the inequality:
$$3x - 5 < x + 7$$
$$2x - 5 < 7$$
$$2x < 12$$
$$x < 6$$
67. Let w represent the width, so $3w$ represents the length. Using the formula for perimeter:
$$2(w) + 2(3w) \ge 48$$
$$2w + 6w \ge 48$$
$$8w \ge 48$$
$$w \ge 6$$
The width is at least 6 meters.
69. Let x, $x + 2$, and $x + 4$ represent the sides of the triangle. The inequality is:
$$x + (x+2) + (x+4) > 24$$
$$3x + 6 > 24$$
$$3x > 18$$
$$x > 6$$
The shortest side is an even number greater than 6 inches (greater than or equal to 8 inches).
71. b (commutative property of addition)
73. a (distributive property)
75. b and c (commutative and associative properties of addition)

2.9 Compound Inequalities

1. Graphing the solution set:

−1 5

3. Graphing the solution set:

−3 0

5. Graphing the solution set:

7. Graphing the solution set:

2 4

9. Graphing the solution set:

11. Graphing the solution set:

−1 5

13. Graphing the solution set:

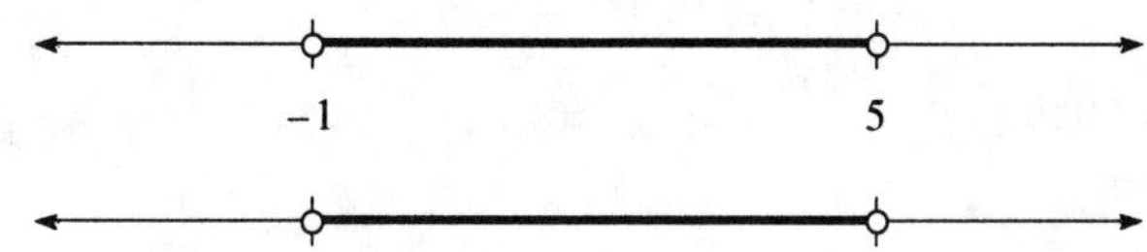

15. Graphing the solution set:

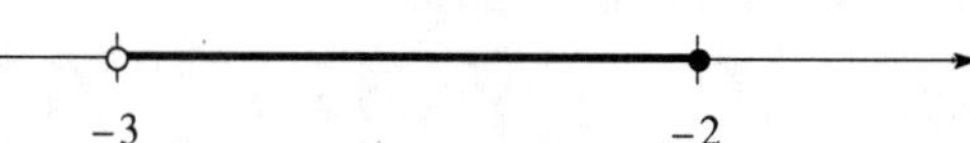

17. Solving the compound inequality:

$$\begin{aligned} 3x-1&<5 \quad &\text{or}\quad 5x-5&>10\\ 3x&<6 & 5x&>15\\ x&<2 & x&>3 \end{aligned}$$

Graphing the solution set:

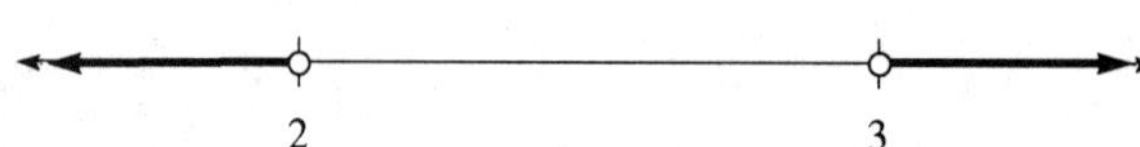

19. Solving the compound inequality:

$$\begin{aligned} x-2&>-5 \quad &\text{and}\quad x+7&<13\\ x&>-3 & x&<6 \end{aligned}$$

Graphing the solution set:

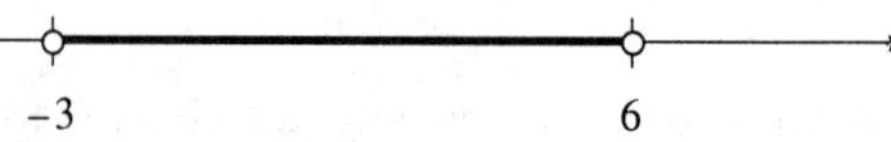

21. Solving the compound inequality:

$$\begin{aligned} 11x&<22 \quad &\text{or}\quad 12x&>36\\ x&<2 & x&>3 \end{aligned}$$

Graphing the solution set:

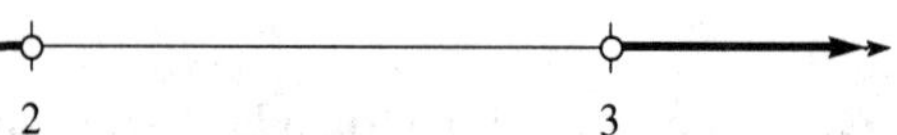

23. Solving the compound inequality:

$$\begin{aligned} 3x-5&<10 \quad &\text{and}\quad 2x+1&>-5\\ 3x&<15 & 2x&>-6\\ x&<5 & x&>-3 \end{aligned}$$

Graphing the solution set:

−3 5

25. Solving the compound inequality:

$$\begin{aligned} 2x-3&<8 \quad &\text{and}\quad 3x+1&>-10\\ 2x&<11 & 3x&>-11\\ x&<\frac{11}{2} & x&>-\frac{11}{3} \end{aligned}$$

Graphing the solution set:

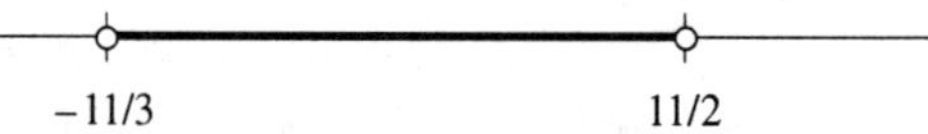

27. Solving the compound inequality:

$$\begin{aligned} 2x-1&<3 \quad &\text{and}\quad 3x-2&>1\\ 2x&<4 & 3x&>3\\ x&<2 & x&>1 \end{aligned}$$

Graphing the solution set:

29. Solving the compound inequality:

$$-1\le x-5\le 2$$
$$4\le x\le 7$$

Graphing the solution set:

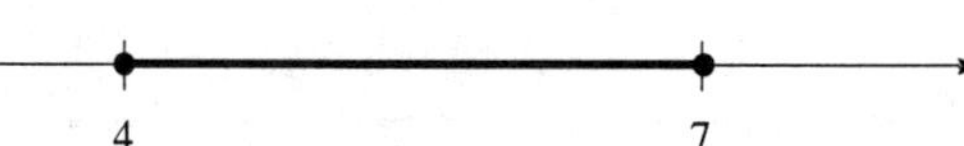

31. Solving the compound inequality:

$$-4\le 2x\le 6$$
$$-2\le x\le 3$$

Graphing the solution set:

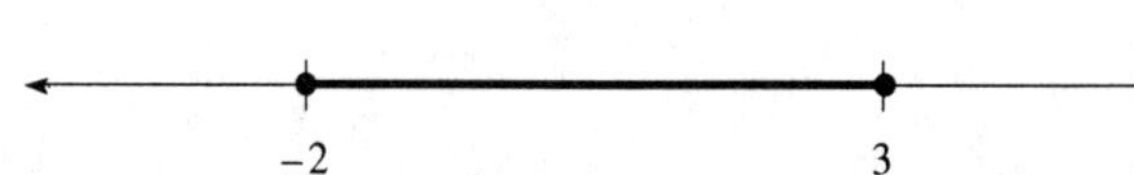

33. Solving the compound inequality:

$$-3 < 2x + 1 < 5$$
$$-4 < 2x < 4$$
$$-2 < x < 2$$

Graphing the solution set:

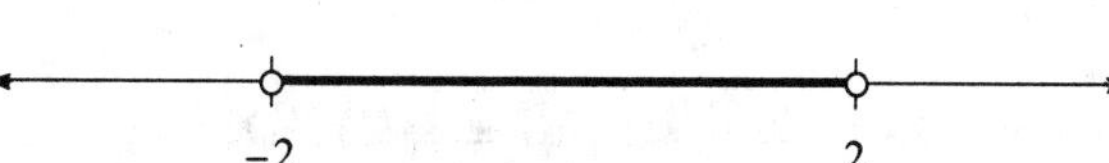

35. Solving the compound inequality:

$$0 \le 3x + 2 \le 7$$
$$-2 \le 3x \le 5$$
$$-\frac{2}{3} \le x \le \frac{5}{3}$$

Graphing the solution set:

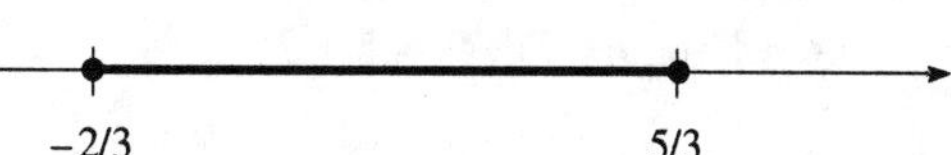

37. Solving the compound inequality:

$$-7 < 2x + 3 < 11$$
$$-10 < 2x < 8$$
$$-5 < x < 4$$

Graphing the solution set:

39. Solving the compound inequality:

$$-1 \le 4x + 5 \le 9$$
$$-6 \le 4x \le 4$$
$$-\frac{3}{2} \le x \le 1$$

Graphing the solution set:

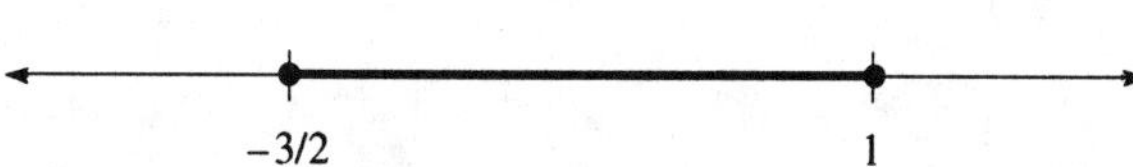

41. Let x represent the number. Solving the inequality:

$$10 < x + 5 < 20$$
$$5 < x < 15$$

The number is between 5 and 15.

43. Let x represent the number. Solving the inequality:

$$5 < 2x - 3 < 7$$
$$8 < 2x < 10$$
$$4 < x < 5$$

The number is between 4 and 5.

45. Let w represent the width and $w + 4$ represent the length. Using the perimeter formula:

$$20 < 2w + 2(w + 4) < 30$$
$$20 < 2w + 2w + 8 < 30$$
$$20 < 4w + 8 < 30$$
$$12 < 4w < 22$$
$$3 < w < \frac{11}{2}$$

The width is between 3 inches and $\frac{11}{2} = 5\frac{1}{2}$ inches.

47. Simplifying the expression: $-|-5| = -(5) = -5$

49. Simplifying the expression: $-3 - 4(-2) = -3 + 8 = 5$

51. Simplifying the expression: $5|3-8| - 6|2-5| = 5|-5| - 6|-3| = 5(5) - 6(3) = 25 - 18 = 7$

53. Simplifying the expression: $5 - 2[-3(5-7) - 8] = 5 - 2[-3(-2) - 8] = 5 - 2(6-8) = 5 - 2(-2) = 5 + 4 = 9$

55. The expression is: $-3-(-9)=-3+9=6$

57. Applying the distributive property: $\frac{1}{2}(4x-6)=\frac{1}{2}\bullet 4x-\frac{1}{2}\bullet 6=2x-3$

59. The integers are: $-3,0,2$

Chapter 2 Review

1. Simplifying the expression: $5x-8x=(5-8)x=-3x$
3. Simplifying the expression: $-a+2+5a-9=-a+5a+2-9=4a-7$
5. Simplifying the expression: $6-2(3y+1)-4=6-6y-2-4=-6y+6-2-4=-6y$
7. Evaluating when $x=3$: $7x-2=7(3)-2=21-2=19$
9. Evaluating when $x=3$: $-x-2x-3x=-6x=-6(3)=-18$
11. Evaluating when $x=-2$: $-3x+2=-3(-2)+2=6+2=8$
13. Solving the equation:
$$\begin{aligned} x+2&=-6\\ x+2+(-2)&=-6+(-2)\\ x&=-8 \end{aligned}$$
15. Solving the equation:
$$\begin{aligned} 10-3y+4y&=12\\ 10+y&=12\\ -10+10+y&=-10+12\\ y&=2 \end{aligned}$$
17. Solving the equation:
$$\begin{aligned} 2x&=-10\\ \frac{1}{2}(2x)&=\frac{1}{2}(-10)\\ x&=-5 \end{aligned}$$
19. Solving the equation:
$$\begin{aligned} \frac{x}{3}&=4\\ 3\left(\frac{x}{3}\right)&=3(4)\\ x&=12 \end{aligned}$$
21. Solving the equation:
$$\begin{aligned} 3a-2&=5a\\ -3a+3a-2&=-3a+5a\\ -2&=2a\\ \frac{1}{2}(-2)&=\frac{1}{2}(2a)\\ a&=-1 \end{aligned}$$
23. Solving the equation:
$$\begin{aligned} 3x+2&=5x-8\\ 3x+(-5x)+2&=5x+(-5x)-8\\ -2x+2&=-8\\ -2x+2+(-2)&=-8+(-2)\\ -2x&=-10\\ -\frac{1}{2}(-2x)&=-\frac{1}{2}(-10)\\ x&=5 \end{aligned}$$

25. Solving the equation:

$$\begin{aligned} 0.7x - 0.1 &= 0.5x - 0.1 \\ 0.7x + (-0.5x) - 0.1 &= 0.5x + (-0.5x) - 0.1 \\ 0.2x - 0.1 &= -0.1 \\ 0.2x - 0.1 + 0.1 &= -0.1 + 0.1 \\ 0.2x &= 0 \\ \frac{0.2x}{0.2} &= \frac{0}{0.2} \\ x &= 0 \end{aligned}$$

27. Simplifying and then solving the equation:

$$\begin{aligned} 2(x-5) &= 10 \\ 2x - 10 &= 10 \\ 2x - 10 + 10 &= 10 + 10 \\ 2x &= 20 \\ \frac{1}{2}(2x) &= \frac{1}{2}(20) \\ x &= 10 \end{aligned}$$

29. Simplifying and then solving the equation:

$$\begin{aligned} \frac{1}{2}(3t-2) + \frac{1}{2} &= \frac{5}{2} \\ \frac{3}{2}t - 1 + \frac{1}{2} &= \frac{5}{2} \\ \frac{3}{2}t - \frac{1}{2} &= \frac{5}{2} \\ \frac{3}{2}t - \frac{1}{2} + \frac{1}{2} &= \frac{5}{2} + \frac{1}{2} \\ \frac{3}{2}t &= 3 \\ \frac{2}{3}\left(\frac{3}{2}t\right) &= \frac{2}{3}(3) \\ t &= 2 \end{aligned}$$

31. Simplifying and then solving the equation:

$$\begin{aligned} 2(3x+7) &= 4(5x-1) + 18 \\ 6x + 14 &= 20x - 4 + 18 \\ 6x + 14 &= 20x + 14 \\ 6x + (-20x) + 14 &= 20x + (-20x) + 14 \\ -14x + 14 &= 14 \\ -14x + 14 + (-14) &= 14 + (-14) \\ -14x &= 0 \\ -\frac{1}{14}(-14x) &= -\frac{1}{14}(0) \\ x &= 0 \end{aligned}$$

33. Substituting $x = 5$:

$$\begin{aligned} 4(5) - 5y &= 20 \\ 20 - 5y &= 20 \\ -5y &= 0 \\ y &= 0 \end{aligned}$$

35. Substituting $x = -5$:

$$\begin{aligned} 4(-5) - 5y &= 20 \\ -20 - 5y &= 20 \\ -5y &= 40 \\ y &= -8 \end{aligned}$$

37. Solving for y:

$$\begin{aligned} 2x - 5y &= 10 \\ -5y &= -2x + 10 \\ y &= \frac{2}{5}x - 2 \end{aligned}$$

39. Solving for h:

$$\begin{aligned} \pi r^2 h &= V \\ \frac{\pi r^2 h}{\pi r^2} &= \frac{V}{\pi r^2} \\ h &= \frac{V}{\pi r^2} \end{aligned}$$

41. Computing the amount:

$$\begin{aligned} 0.86(240) &= x \\ x &= 206.4 \end{aligned}$$

86% of 240 is 206.4

43. Let x represent the number. The equation is:

$$\begin{aligned} 2x + 6 &= 28 \\ 2x &= 22 \\ x &= 11 \end{aligned}$$

The number is 11.

45. Completing the table:

	Dollars Invested at 9%	Dollars Invested at 10%
Number of	x	$x+300$
Interest on	$0.09(x)$	$0.10(x+300)$

The equation is:

$$\begin{aligned} 0.09(x) + 0.10(x+300) &= 125 \\ 0.09x + 0.10x + 30 &= 125 \\ 0.19x + 30 &= 125 \\ 0.19x &= 95 \\ x &= 500 \\ x + 300 &= 800 \end{aligned}$$

The man invested \$500 at 9% and \$800 at 10%.

47. Solving the inequality:

$$\begin{aligned} -2x &< 4 \\ -\frac{1}{2}(-2x) &> -\frac{1}{2}(4) \\ x &> -2 \end{aligned}$$

49. Solving the inequality:

$$\begin{aligned} -\frac{a}{2} &\le -3 \\ -2\left(-\frac{a}{2}\right) &\ge -2(-3) \\ a &\ge 6 \end{aligned}$$

51. Solving the inequality:

$$\begin{aligned} -4x + 5 &> 37 \\ -4x &> 32 \\ x &< -8 \end{aligned}$$

Graphing the solution set:

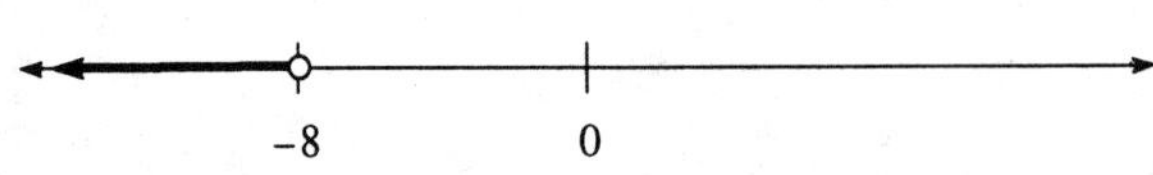

53. Solving the inequality:

$$\begin{aligned} 2(3t+1)+6 &\geq 5(2t+4) \\ 6t+2+6 &\geq 10t+20 \\ 6t+8 &\geq 10t+20 \\ -4t+8 &\geq 20 \\ -4t &\geq 12 \\ t &\leq -3 \end{aligned}$$

Graphing the solution set:

55. Solving the compound inequality:

$$\begin{aligned} -5x &\geq 25 \quad \text{or} \quad & 2x-3 &\geq 9 \\ x &\leq -5 & 2x &\geq 12 \\ x &\leq -5 & x &\geq 6 \end{aligned}$$

Graphing the solution set:

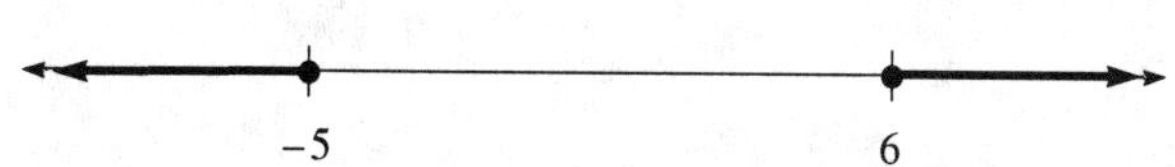

Cumulative Review: Chapters 1-2

1. Simplifying: $6+3(6+2)=6+3(8)=6+24=30$
3. Simplifying: $7-9-12=7+(-9)+(-12)=7+(-21)=-14$
5. Using the associative property: $\frac{1}{5}(10x)=\left(\frac{1}{5}\cdot 10\right)x=2x$
7. Simplifying: $\left(-\frac{2}{3}\right)^3=\left(-\frac{2}{3}\right)\left(-\frac{2}{3}\right)\left(-\frac{2}{3}\right)=-\frac{8}{27}$
9. Simplifying: $-\frac{3}{4}\div\frac{15}{16}=-\frac{3}{4}\cdot\frac{16}{15}=-\frac{48}{60}=-\frac{4}{5}$
11. Simplifying: $\frac{-4(-6)}{-9}=\frac{24}{-9}=-\frac{8}{3}$
13. Simplifying: $\frac{(5-3)^2}{5^2-3^2}=\frac{2^2}{25-9}=\frac{4}{16}=\frac{1}{4}$
15. First factor the denominators to find the LCM:

$$\begin{aligned} 21 &= 3\cdot 7 \\ 35 &= 5\cdot 7 \\ \text{LCM} &= 3\cdot 5\cdot 7=105 \end{aligned}$$

Simplifying: $\frac{4}{21}-\frac{9}{35}=\frac{4\cdot 5}{21\cdot 5}-\frac{9\cdot 3}{35\cdot 3}=\frac{20}{105}-\frac{27}{105}=-\frac{7}{105}=-\frac{7}{3\cdot 5\cdot 7}=-\frac{1}{3\cdot 5}=-\frac{1}{15}$

17. Solving the equation:

$$\begin{aligned} 7x &= 6x+4 \\ 7x+(-6x) &= 6x+(-6x)+4 \\ x &= 4 \end{aligned}$$

19. Solving the equation:

$$\begin{aligned} -\frac{3}{5}x &= 30 \\ -\frac{5}{3}\left(-\frac{3}{5}x\right) &= -\frac{5}{3}(30) \\ x &= -50 \end{aligned}$$

21. Solving the equation:

$$\begin{aligned}5x-7&=x-1\\5x+(-x)-7&=x+(-x)-1\\4x-7&=-1\\4x-7+7&=-1+7\\4x&=6\\\frac{1}{4}(4x)&=\frac{1}{4}(6)\\x&=\frac{3}{2}\end{aligned}$$

23. Solving the equation:

$$\begin{aligned}15-3(2t+4)&=1\\15-6t-12&=1\\-6t+3&=1\\-6t+3+(-3)&=1+(-3)\\-6t&=-2\\-\frac{1}{6}(-6t)&=-\frac{1}{6}(-2)\\t&=\frac{1}{3}\end{aligned}$$

25. Solving the equation:

$$\begin{aligned}\frac{1}{3}(x-6)&=\frac{1}{4}(x+8)\\\frac{1}{3}x-2&=\frac{1}{4}x+2\\12\left(\frac{1}{3}x-2\right)&=12\left(\frac{1}{4}x+2\right)\\4x-24&=3x+24\\4x+(-3x)-24&=3x+(-3x)+24\\x-24&=24\\x-24+24&=24+24\\x&=48\end{aligned}$$

27. Solving for y:

$$\begin{aligned}3x+4y&=12\\3x+(-3x)+4y&=12+(-3x)\\4y&=-3x+12\\\frac{1}{4}(4y)&=\frac{1}{4}(-3x+12)\\y&=-\frac{3}{4}x+3\end{aligned}$$

29. Solving the inequality:

$$\begin{aligned}-5x+9&<-6\\-5x+9+(-9)&<-6+(-9)\\-5x&<-15\\-\frac{1}{5}(-5x)&>-\frac{1}{5}(-15)\\x&>3\end{aligned}$$

Graphing the solution set:

0 3

31. Solving the compound inequality:
$$-2 < x+1 < 5$$
$$-3 < x < 4$$
Graphing the solution set:

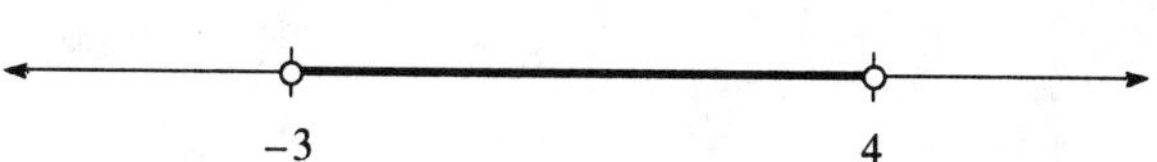

33. The opposite of $-\frac{2}{3}$ is $\frac{2}{3}$, the reciprocal is $-\frac{3}{2}$, and the absolute value is $\left|-\frac{2}{3}\right| = \frac{2}{3}$.

35. The pattern is to add –3, so the next term is: $-5+(-3)=-8$

37. Using the distributive property: $\frac{1}{4}(8x-4)=\frac{1}{4}\bullet 8x-\frac{1}{4}\bullet 4=2x-1$

39. Reducing the fraction: $\frac{234}{312}=\frac{2\bullet 3\bullet 3\bullet 13}{2\bullet 2\bullet 2\bullet 3\bullet 13}=\frac{3}{2\bullet 2}=\frac{3}{4}$

41. Evaluating when $a=3$ and $b=-2$: $a^2-2ab+b^2=(3)^2-2(3)(-2)+(-2)^2=9+12+4=25$

43. Let x represent the number. The equation is:
$$2x+7=31$$
$$2x=24$$
$$x=12$$
The number is 12.

45. Let x represent the acute angle and 90° is the right angle. Since the sum of the three angles is 180°, the equation is:
$$x+42°+90°=180°$$
$$x+132°=180°$$
$$x=48°$$
The other two angles are 48° and 90°.

47. The other angle must be $90°-25°=65°$.

49. Completing the table:

	Dollars Invested at 5%	Dollars Invested at 6%
Number of	x	$x+200$
Interest on	$0.05(x)$	$0.06(x+200)$

The equation is:
$$0.05(x)+0.06(x+200)=56$$
$$0.05x+0.06x+12=56$$
$$0.11x+12=56$$
$$0.11x=44$$
$$x=400$$
$$x+200=600$$
You have \$400 invested at 5% and \$600 invested at 6%.

Chapter 2 Test

1. Simplifying: $3x+2-7x+3=3x-7x+2+3=-4x+5$
2. Simplifying: $4a-5-a+1=4a-a-5+1=3a-4$
3. Simplifying: $7-3(y+5)-4=7-3y-15-4=-3y+7-15-4=-3y-12$
4. Simplifying: $8(2x+1)-5(x-4)=16x+8-5x+20=16x-5x+8+20=11x+28$
5. Evaluating when $x=-5$: $2x-3-7x=-5x-3=-5(-5)-3=25-3=22$
6. Evaluating when $x=2$ and $y=3$: $x^2+2xy+y^2=(2)^2+2(2)(3)+(3)^2=4+12+9=25$
7. Solving the equation:
$$2x-5=7$$
$$2x-5+5=7+5$$
$$2x=12$$
$$\frac{1}{2}(2x)=\frac{1}{2}(12)$$
$$x=6$$

8. Solving the equation:

$$\begin{aligned} 2y+4&=5y \\ -2y+2y+4&=-2y+5y \\ 4&=3y \\ \frac{1}{3}(4)&=\frac{1}{3}(3y) \\ y&=\frac{4}{3} \end{aligned}$$

9. First clear the equation of fractions by multiplying by 10:

$$\begin{aligned} \frac{1}{2}x-\frac{1}{10}&=\frac{1}{5}x+\frac{1}{2} \\ 10\left(\frac{1}{2}x-\frac{1}{10}\right)&=10\left(\frac{1}{5}x+\frac{1}{2}\right) \\ 5x-1&=2x+5 \\ 5x+(-2x)-1&=2x+(-2x)+5 \\ 3x-1&=5 \\ 3x-1+1&=5+1 \\ 3x&=6 \\ \frac{1}{3}(3x)&=\frac{1}{3}(6) \\ x&=2 \end{aligned}$$

10. Solving the equation:

$$\begin{aligned} \frac{2}{5}(5x-10)&=-5 \\ 2x-4&=-5 \\ 2x-4+4&=-5+4 \\ 2x&=-1 \\ \frac{1}{2}(2x)&=\frac{1}{2}(-1) \\ x&=-\frac{1}{2} \end{aligned}$$

11. Solving the equation:

$$\begin{aligned} -5(2x+1)-6&=19 \\ -10x-5-6&=19 \\ -10x-11&=19 \\ -10x-11+11&=19+11 \\ -10x&=30 \\ -\frac{1}{10}(-10x)&=-\frac{1}{10}(30) \\ x&=-3 \end{aligned}$$

12. Solving the equation:

$$\begin{aligned} 0.04x+0.06(100-x)&=4.6 \\ 0.04x+6-0.06x&=4.6 \\ -0.02x+6&=4.6 \\ -0.02x+6+(-6)&=4.6+(-6) \\ -0.02x&=-1.4 \\ \frac{-0.02x}{-0.02}&=\frac{-1.4}{-0.02} \\ x&=70 \end{aligned}$$

13. Solving the equation:

$$\begin{aligned}
2(t-4)+3(t+5)&=2t-2\\
2t-8+3t+15&=2t-2\\
5t+7&=2t-2\\
5t+(-2t)+7&=2t+(-2t)-2\\
3t+7&=-2\\
3t+7+(-7)&=-2+(-7)\\
3t&=-9\\
\frac{1}{3}(3t)&=\frac{1}{3}(-9)\\
t&=-3
\end{aligned}$$

14. Solving the equation:

$$\begin{aligned}
2x-4(5x+1)&=3x+17\\
2x-20x-4&=3x+17\\
-18x-4&=3x+17\\
-18x+(-3x)-4&=3x+(-3x)+17\\
-21x-4&=17\\
-21x-4+4&=17+4\\
-21x&=21\\
-\frac{1}{21}(21x)&=-\frac{1}{21}(21)\\
x&=-1
\end{aligned}$$

15. Finding the amount:

$$\begin{aligned}
0.15(38)&=x\\
5.7&=x
\end{aligned}$$

15% of 38 is 5.7.

16. Finding the base:

$$\begin{aligned}
0.12x&=240\\
\frac{0.12x}{0.12}&=\frac{240}{0.12}\\
x&=2000
\end{aligned}$$

12% of 2,000 is 240.

17. Substituting $y=-2$:

$$\begin{aligned}
2x-3(-2)&=12\\
2x+6&=12\\
2x&=6\\
x&=3
\end{aligned}$$

18. Substituting $V=88$, $\pi=\frac{22}{7}$, and $r=3$:

$$\begin{aligned}
\frac{1}{3}\bullet\frac{22}{7}\bullet(3)^2h&=88\\
\frac{66}{7}h&=88\\
\frac{7}{66}\left(\frac{66}{7}h\right)&=\frac{7}{66}(88)\\
h&=\frac{28}{3}\text{ inches}
\end{aligned}$$

19. Solving for y:

$$\begin{aligned}
2x+5y&=20\\
5y&=-2x+20\\
y&=-\frac{2}{5}x+4
\end{aligned}$$

20. Solving for v:

$$x + vt + 16t^2 = h$$
$$vt = h - x - 16t^2$$
$$v = \frac{h - x - 16t^2}{t}$$

21. Completing the table:

	Ten Years Ago	Now
Dave	$2x - 10$	$2x$
Rick	$x - 10$	x

The equation is:

$$2x - 10 + x - 10 = 40$$
$$3x - 20 = 40$$
$$3x = 60$$
$$x = 20$$
$$2x = 40$$

Rick is 20 years old and Dave is 40 years old.

22. Let w represent the width and $2w$ represent the length. Using the perimeter formula:

$$2(w) + 2(2w) = 60$$
$$2w + 4w = 60$$
$$6w = 60$$
$$w = 10$$
$$2w = 20$$

The width is 10 inches and the length is 20 inches.

23. Completing the table:

	Dimes	Quarters
Number	$x + 7$	x
Value (cents)	$10(x + 7)$	$25(x)$

The equation is:

$$10(x + 7) + 25(x) = 350$$
$$10x + 70 + 25x = 350$$
$$35x + 70 = 350$$
$$35x = 280$$
$$x = 8$$
$$x + 7 = 15$$

He has 8 quarters and 15 dimes in his collection.

24. Completing the table:

	Dollars Invested at 7%	Dollars Invested at 9%
Number of	x	$x + 600$
Interest on	$0.07(x)$	$0.09(x + 600)$

The equation is:

$$0.07(x) + 0.09(x + 600) = 182$$
$$0.07x + 0.09x + 54 = 182$$
$$0.16x + 54 = 182$$
$$0.16x = 128$$
$$x = 800$$
$$x + 600 = 1400$$

She has \$800 invested at 7% and \$1,400 invested at 9%.

25. Solving the inequality:

$$\begin{aligned} 2x+3&<5 \\ 2x+3+(-3)&<5+(-3) \\ 2x&<2 \\ \frac{1}{2}(2x)&<\frac{1}{2}(2) \\ x&<1 \end{aligned}$$

Graphing the solution set:

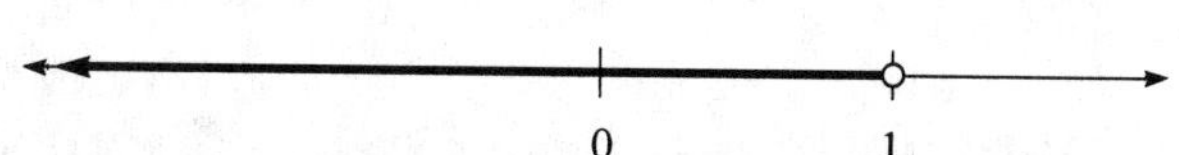

26. Solving the inequality:

$$\begin{aligned} -5a&>20 \\ -\frac{1}{5}(-5a)&<-\frac{1}{5}(20) \\ a&<-4 \end{aligned}$$

Graphing the solution set:

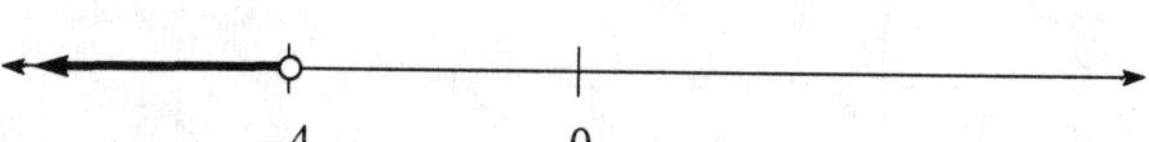

27. Solving the inequality:

$$\begin{aligned} 0.4-0.2x&\ge 1 \\ -0.4+0.4-0.2x&\ge -0.4+1 \\ -0.2x&\ge 0.6 \\ \frac{-0.2x}{-0.2}&\le\frac{0.6}{-0.2} \\ x&\le -3 \end{aligned}$$

Graphing the solution set:

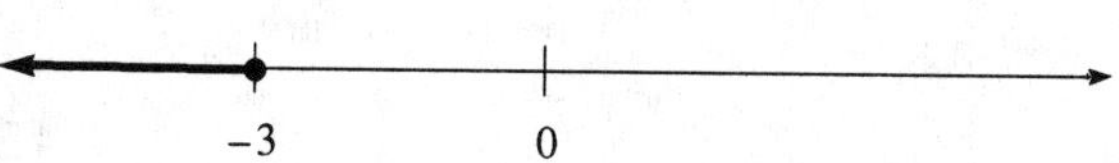

28. Solving the inequality:

$$\begin{aligned} 4-5(m+1)&\le 9 \\ 4-5m-5&\le 9 \\ -5m-1&\le 9 \\ -5m-1+1&\le 9+1 \\ -5m&\le 10 \\ -\frac{1}{5}(-5m)&\ge-\frac{1}{5}(10) \\ m&\ge -2 \end{aligned}$$

Graphing the solution set:

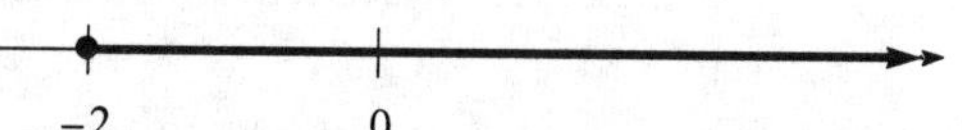

29. Solving the compound inequality:

$$\begin{aligned} 3-4x&\ge -5 \quad \text{or} \quad & 2x&\ge 10 \\ -4x&\ge -8 & x&\ge 5 \\ x&\le 2 & x&\ge 5 \end{aligned}$$

Graphing the solution set:

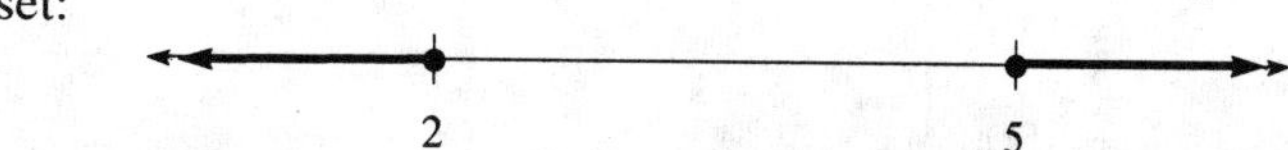

30. Solving the compound inequality:

$$\begin{aligned} -7<2x-1<9 \\ -6<2x<10 \\ -3<x<5 \end{aligned}$$

Graphing the solution set:

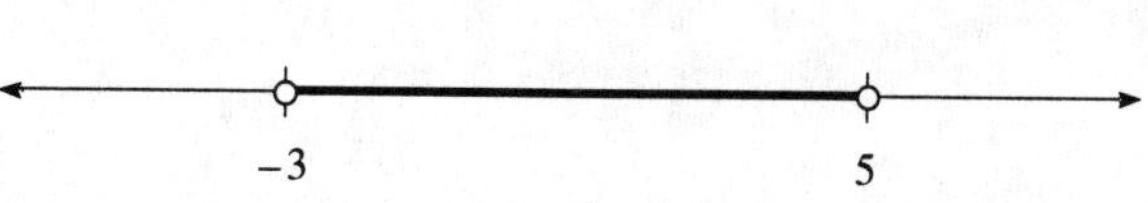

Chapter 3
Linear Equations and Inequalities in Two Variables

3.1 Paired Data and Graphing Ordered Pairs

1. Constructing a histogram:

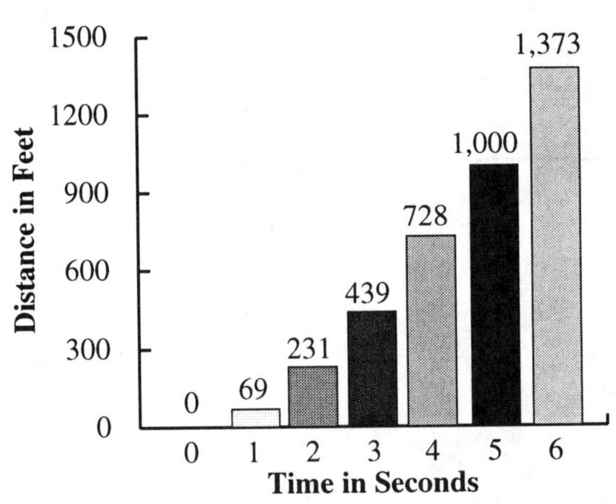

3. Constructing a histogram:

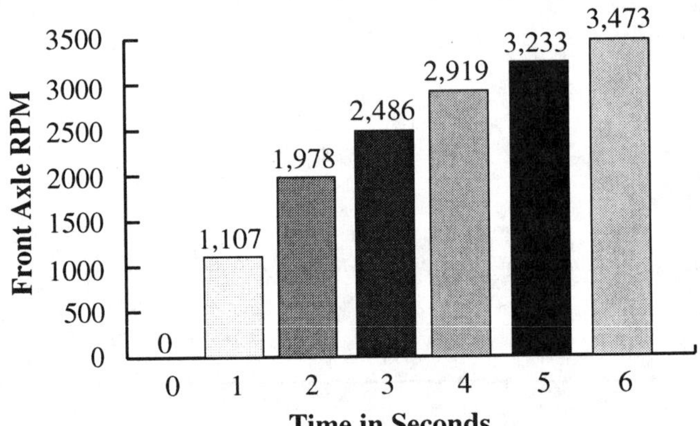

5. Constructing a scatter diagram:

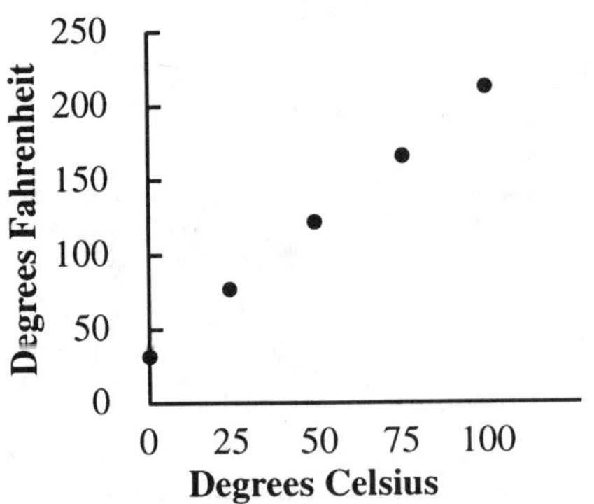

7. Yes, since the points appear to lie along a line.
9. Constructing a scatter diagram:

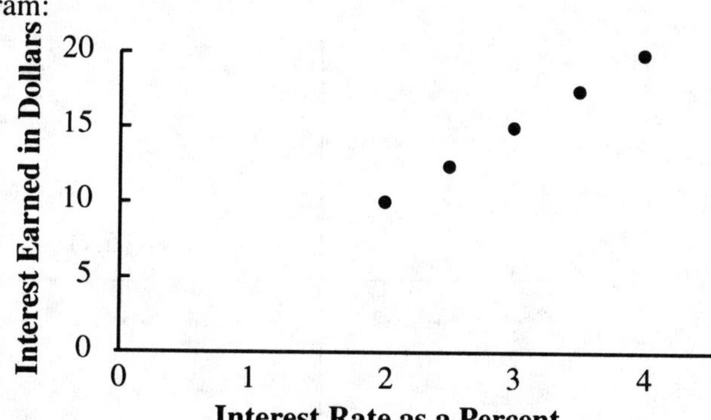

11. Yes, since the points appear to lie along a line.
13. Graphing the ordered pair:

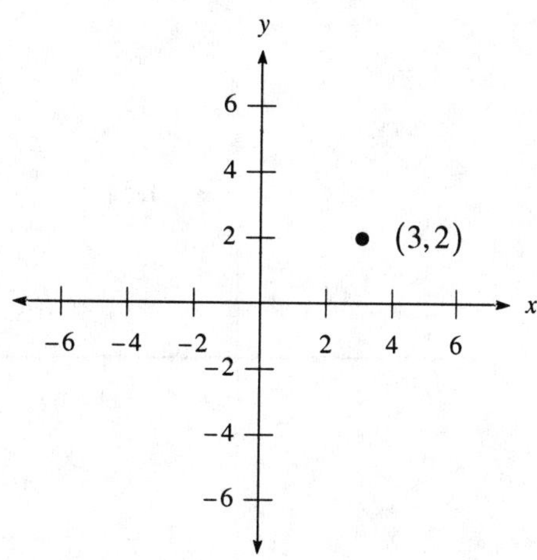

15. Graphing the ordered pair:

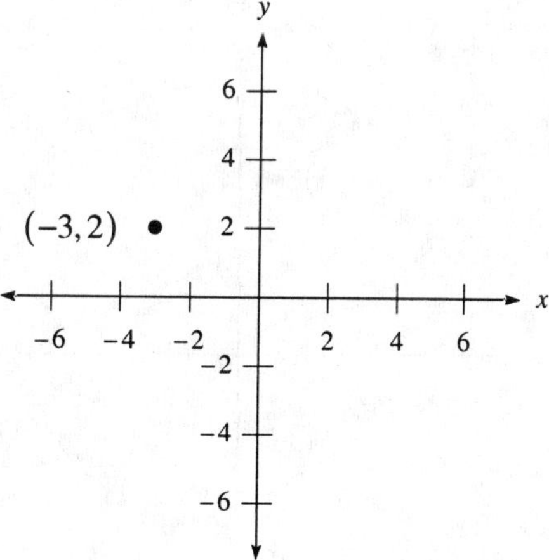

17. Graphing the ordered pair:

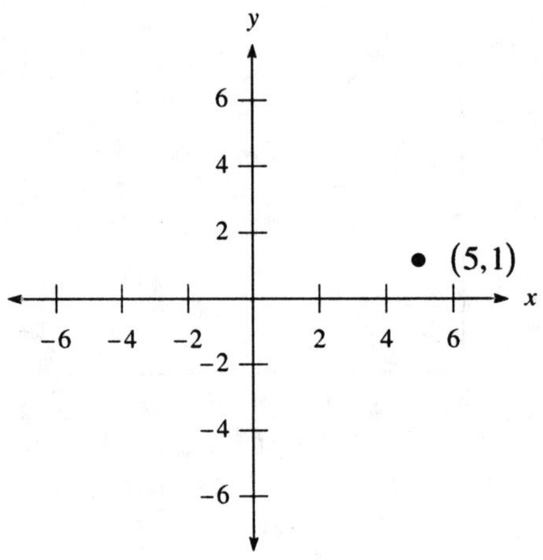

19. Graphing the ordered pair:

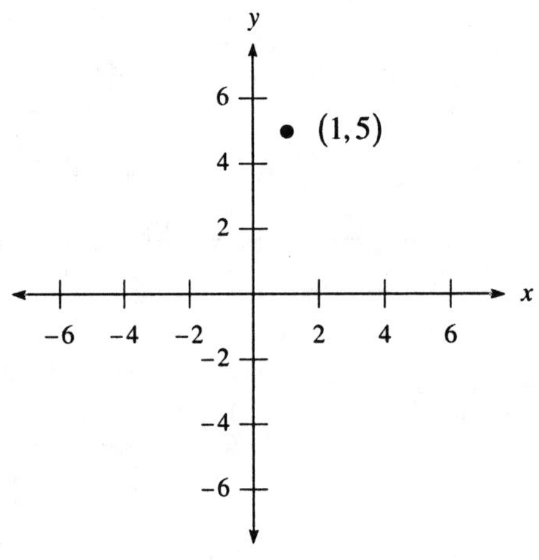

21. Graphing the ordered pair:

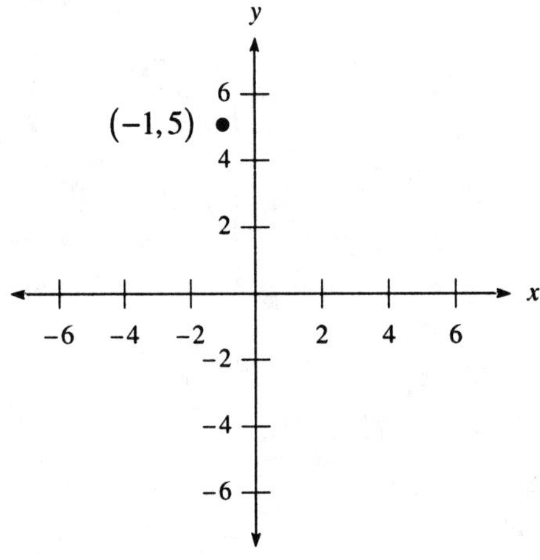

23. Graphing the ordered pair:

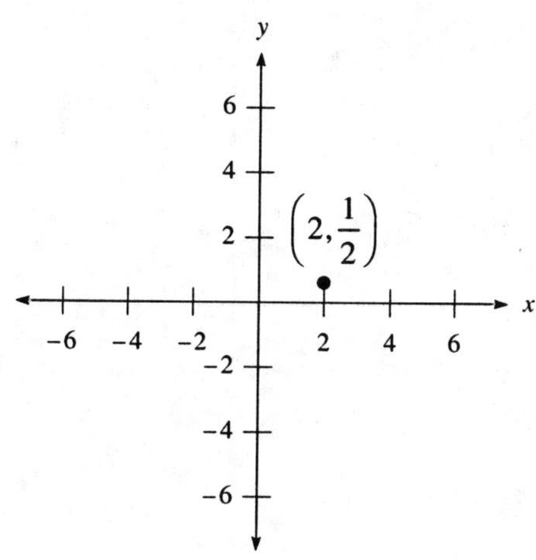

25. Graphing the ordered pair:

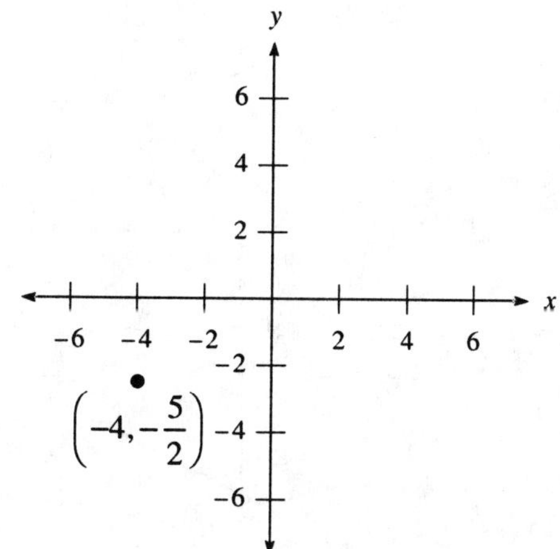

27. Graphing the ordered pair:

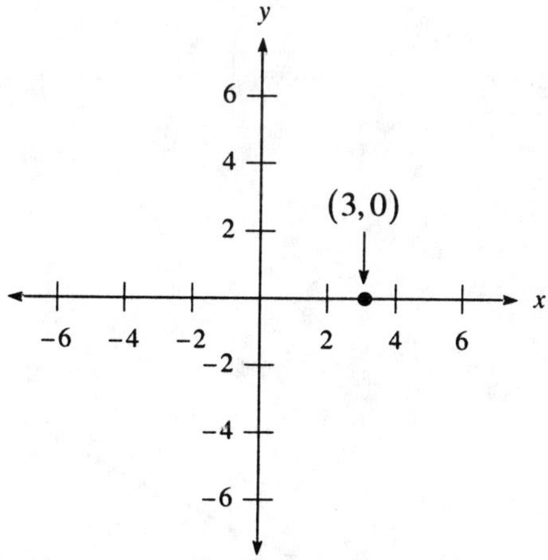

29. Graphing the ordered pair:

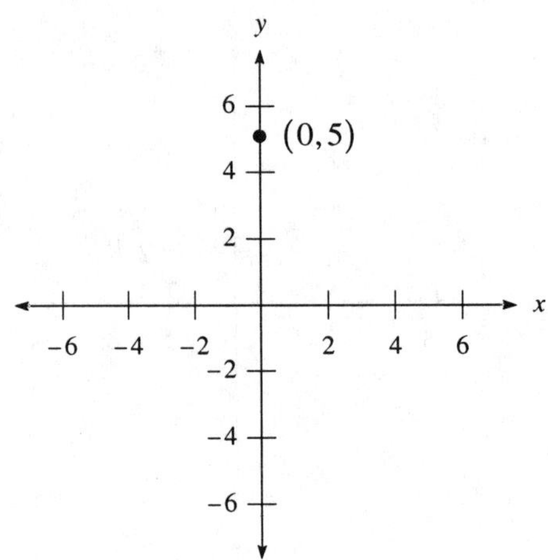

31. The coordinates are $(-4,4)$.
33. The coordinates are $(-4,2)$.
35. The coordinates are $(-3,0)$.
37. The coordinates are $(2,-2)$.
39. The coordinates are $(-5,-5)$.
41. Graphing the line:

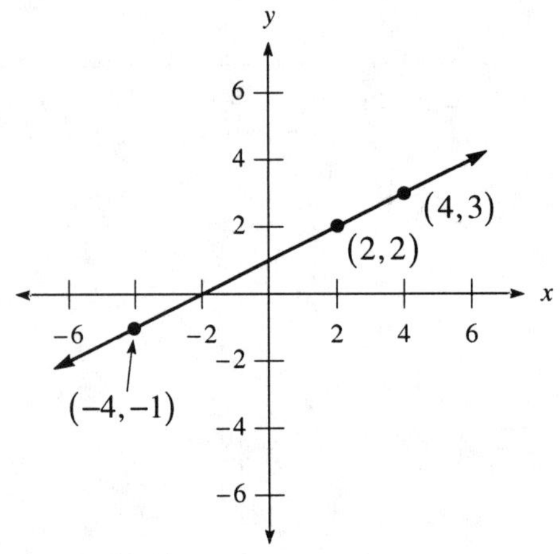

Yes, the point $(2,2)$ lies on the line.

43. Graphing the line:

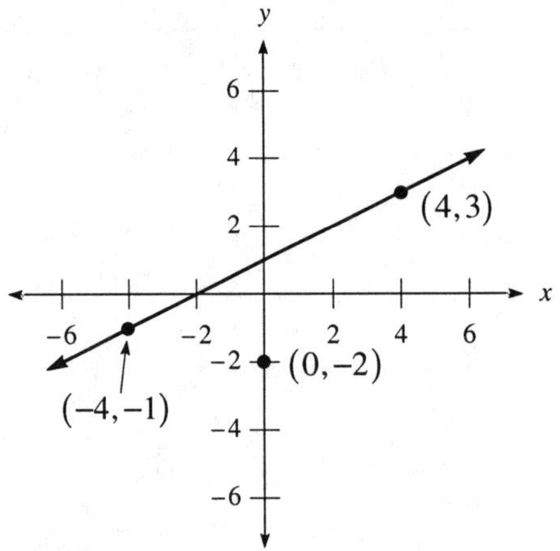

No, the point $(0,-2)$ does not lie on the line.

45. Graphing the line:

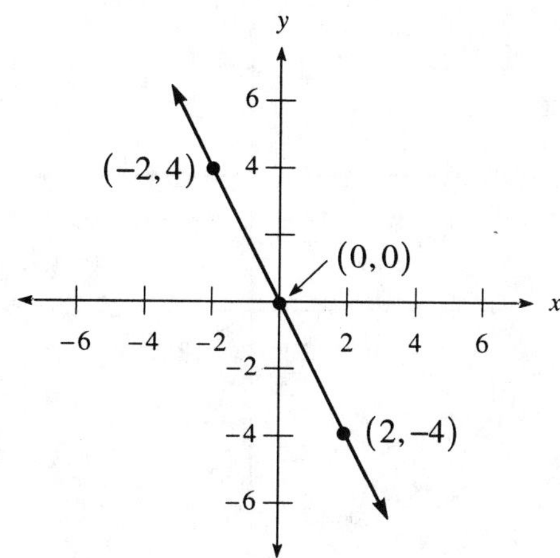

Yes, the point $(0,0)$ lies on the line.

47. Graphing the line:

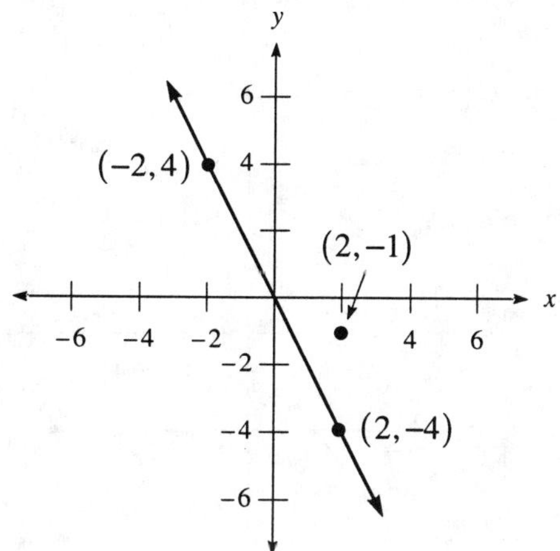

No, the point $(2,-1)$ does not lie on the line.

49. Graphing the line:

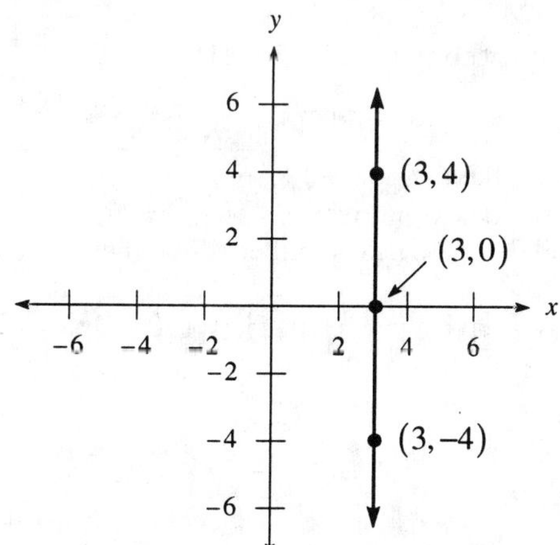

Yes, the point $(3,0)$ lies on the line.

51. No, the x-coordinate of every point on this line is 3.

53. Graphing the line:

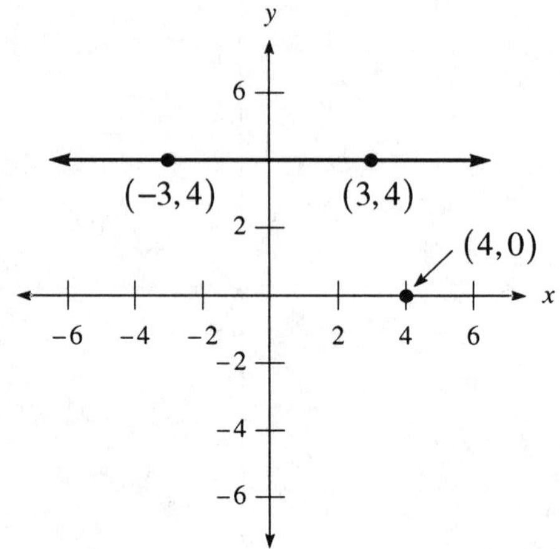

No, the point $(4,0)$ does not lie on the line.

55. No, the y-coordinate of every point on this line is 4.

57. Graphing the line:

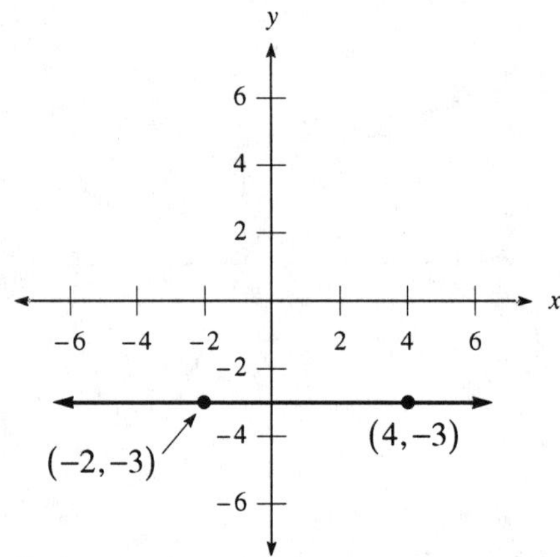

All points on this line have a y-coordinate equal to –3.

59. Points of the form $(0,y)$ must lie on the y-axis.

61. The pattern is to add 7, so the next number is: $24+7=31$

63. The pattern is to multiply by $\frac{1}{3}$, so the next number is: $\frac{1}{9}\cdot\frac{1}{3}=\frac{1}{27}$

65. The pattern is to add –3, so the next number is: $-2+(-3)=-5$

67. The pattern is to multiply by 3, so the next number is: $189\cdot 3=567$

69. The pattern is to add 1, add 2, add 3, so the next number is obtained by adding 4: $11+4=15$

3.2 Solutions to Linear Equations in Two Variables

1. Substituting $x=0$, $x=3$, and $y=-6$:

$$2(0)+y=6 \qquad 2(3)+y=6 \qquad 2x+(-6)=6$$
$$0+y=6 \qquad 6+y=6 \qquad 2x=12$$
$$y=6 \qquad y=0 \qquad x=6$$

The ordered pairs are $(0,6)$, $(3,0)$, and $(6,-6)$.

3. Substituting $x=0$, $y=0$, and $x=-4$:

$$3(0)+4y=12 \qquad 3x+4(0)=12 \qquad 3(-4)+4y=12$$
$$0+4y=12 \qquad 3x+0=12 \qquad -12+4y=12$$
$$4y=12 \qquad 3x=12 \qquad 4y=24$$
$$y=3 \qquad x=4 \qquad y=6$$

The ordered pairs are $(0,3)$, $(4,0)$, and $(-4,6)$.

5. Substituting $x = 1$, $y = 0$, and $x = 5$:

$$y = 4(1) - 3 \qquad 0 = 4x - 3 \qquad y = 4(5) - 3$$
$$y = 4 - 3 \qquad 3 = 4x \qquad y = 20 - 3$$
$$y = 1 \qquad x = \frac{3}{4} \qquad y = 17$$

The ordered pairs are $(1,1)$, $\left(\frac{3}{4},0\right)$, and $(5,17)$.

7. Substituting $x = 2$, $y = 6$, and $x = 0$:

$$y = 7(2) - 1 \qquad 6 = 7x - 1 \qquad y = 7(0) - 1$$
$$y = 14 - 1 \qquad 7 = 7x \qquad y = 0 - 1$$
$$y = 13 \qquad x = 1 \qquad y = -1$$

The ordered pairs are $(2,13)$, $(1,6)$, and $(0,-1)$.

9. Substituting $y = 4$, $y = -3$, and $y = 0$ results (in each case) in $x = -5$. The ordered pairs are $(-5,4)$, $(-5,-3)$, and $(-5,0)$.

11. Completing the table:

x	y
1	3
–3	–9
4	12
6	18

13. Completing the table:

x	y
0	0
–1/2	–2
–3	–12
3	12

15. Completing the table:

x	y
2	3
3	2
5	0
9	–4

17. Completing the table:

x	y
2	0
3	2
1	–2
–3	–10

19. Completing the table:

x	y
0	–1
–1	–7
–3	–19
3/2	8

21. Substituting each ordered pair into the equation:

$$(2,3):\quad 2(2)-5(3)=4-15=-11\neq 10$$
$$(0,-2):\quad 2(0)-5(-2)=0+10=10$$
$$\left(\frac{5}{2},1\right):\quad 2\left(\frac{5}{2}\right)-5(1)=5-5=0\neq 10$$

Only the ordered pair $(0,-2)$ is a solution.

23. Substituting each ordered pair into the equation:

$$(1,5):\quad 7(1)-2=7-2=5$$
$$(0,-2):\quad 7(0)-2=0-2=-2$$
$$(-2,-16):\quad 7(-2)-2=-14-2=-16$$

All the ordered pairs $(1,5)$, $(0,-2)$ and $(-2,-16)$ are solutions.

25. Substituting each ordered pair into the equation:

$$(1,6):\quad 6(1)=6$$
$$(-2,-12):\quad 6(-2)=-12$$
$$(0,0):\quad 6(0)=0$$

All the ordered pairs $(1,6)$, $(-2,-12)$ and $(0,0)$ are solutions.

27. Substituting each ordered pair into the equation:

$$(1,1):\quad 1+1=2\neq 0$$
$$(2,-2):\quad 2+(-2)=0$$
$$(3,3):\quad 3+3=6\neq 0$$

Only the ordered pair $(2,-2)$ is a solution.

29. Since $x = 3$, the ordered pair $(5,3)$ cannot be a solution. The ordered pairs $(3,0)$ and $(3,-3)$ are solutions.

31. Substituting $w = 3$:

$$\begin{aligned} 2l+2(3)&=30\\ 2l+6&=30\\ 2l&=24\\ l&=12 \end{aligned}$$

The length is 12 inches.

33. Since $y=2x$, $y=2(5)=10$. The ordered pair is $(5,10)$.

35. Substituting $x = 3$, $x = -3$, $x = 5$, and $x = -5$ results in the ordered pairs $(3,3)$, $(-3,3)$, $(5,5)$, and $(-5,5)$.

37. The equation $y=13+1.5x$ will calculate the total monthly charge y for picking up x trash bags. Completing the table:

Bags Collected	Cost (in Dollars)
5	\$20.50
7	\$23.50
9	\$26.50
11	\$29.50
13	\$32.50

Constructing a histogram:

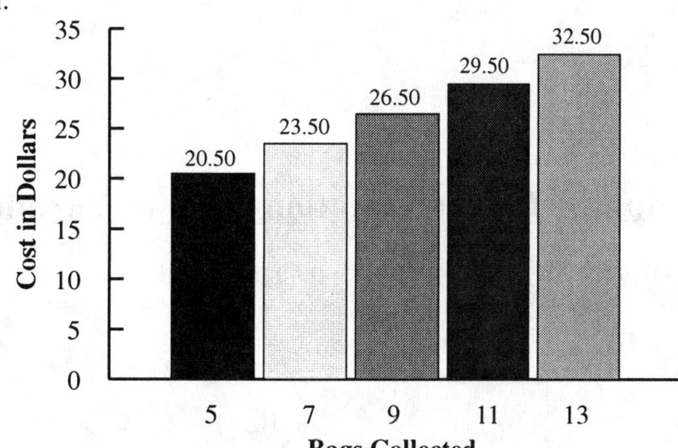

39. The equation $y = 7 + 1.1x$ will calculate the total monthly charge y for drinking x gallons of water. Completing the table:

Gallons per Month	Monthly Cost (in Dollars)
15	\$23.50
20	\$29.00
25	\$34.50
30	\$40.00
35	\$45.50

Constructing a histogram:

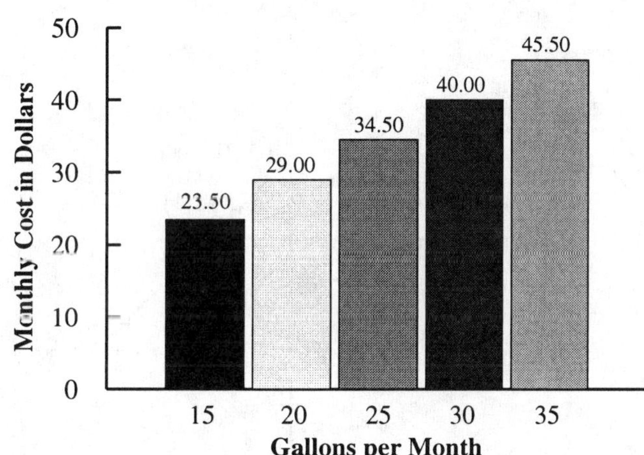

41. Substituting $x = 4$:

$$\begin{aligned} 3(4) + 2y &= 6 \\ 12 + 2y &= 6 \\ 2y &= -6 \\ y &= -3 \end{aligned}$$

43. Substituting $x = 0$: $y = -\frac{1}{3}(0) + 2 = 0 + 2 = 2$

45. Substituting $x = 2$: $y = \frac{3}{2}(2) - 3 = 3 - 3 = 0$

47. Solving for y:

$$\begin{aligned} 5x + y &= 4 \\ y &= -5x + 4 \end{aligned}$$

49. Solving for y:

$$3x - 2y = 6$$
$$-2y = -3x + 6$$
$$y = \frac{3}{2}x - 3$$

3.3 Graphing Linear Equations in Two Variables

1. The ordered pairs are $(0,4)$, $(2,2)$, and $(4,0)$. Graphing the equation:

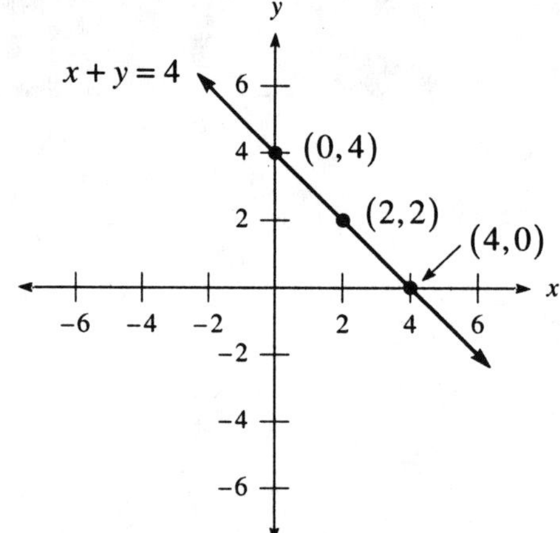

3. The ordered pairs are $(0,3)$, $(2,1)$, and $(4,-1)$. Graphing the equation:

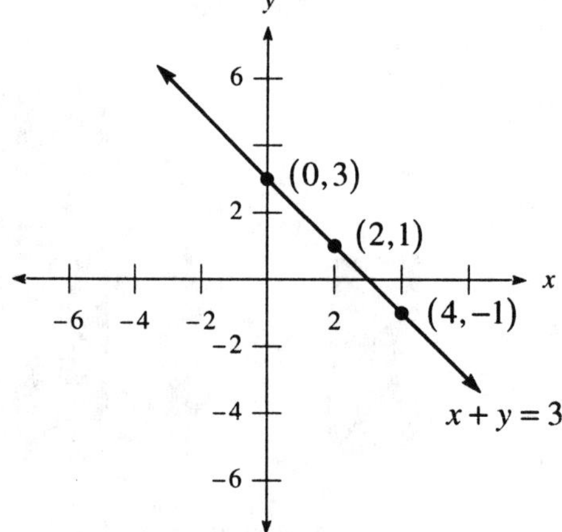

5. The ordered pairs are $(0,0)$, $(-2,-4)$, and $(2,4)$. Graphing the equation:

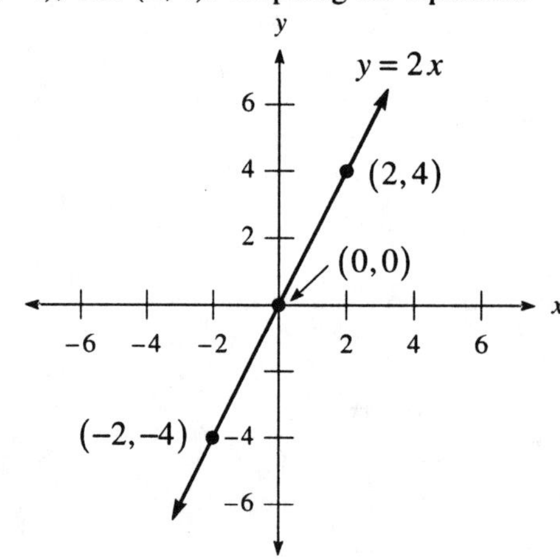

7. The ordered pairs are $(-3,-1)$, $(0,0)$, and $(3,1)$. Graphing the equation:

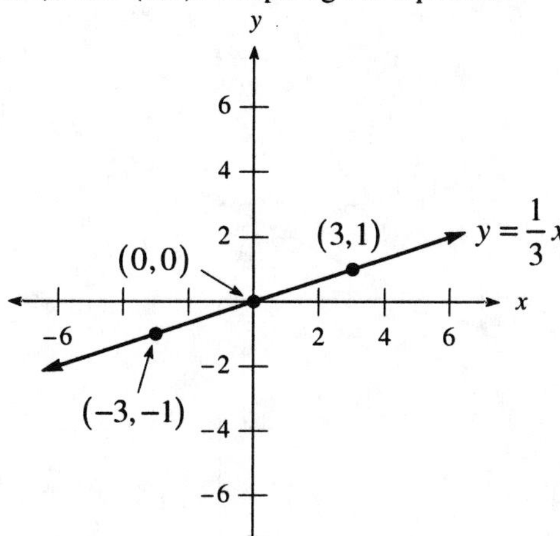

9. The ordered pairs are $(0,1)$, $(-1,-1)$, and $(1,3)$. Graphing the equation:

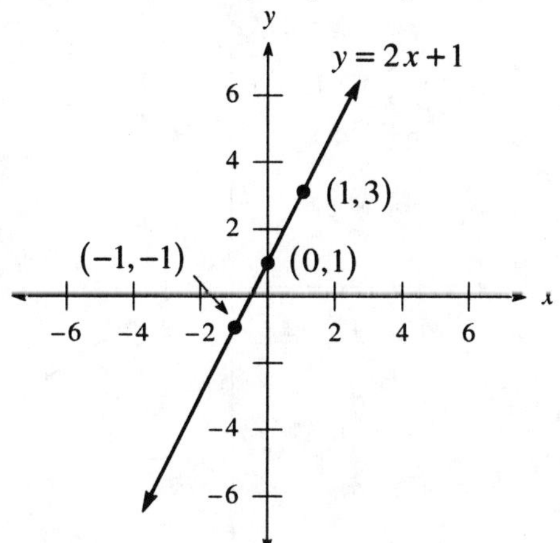

11. The ordered pairs are $(0,4)$, $(-1,4)$, and $(2,4)$. Graphing the equation:

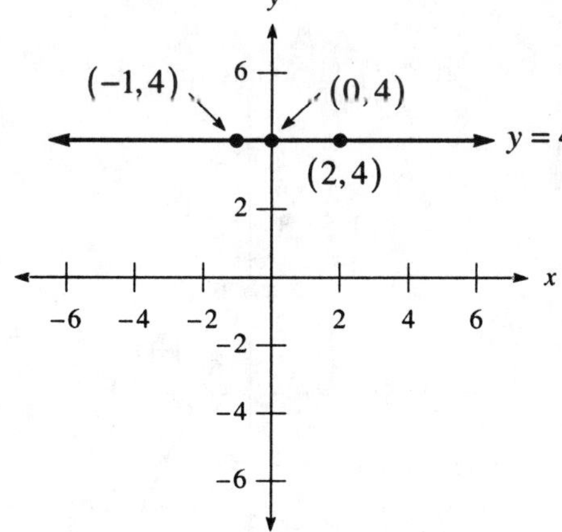

13. The ordered pairs are $(-2,2)$, $(0,3)$, and $(2,4)$. Graphing the equation:

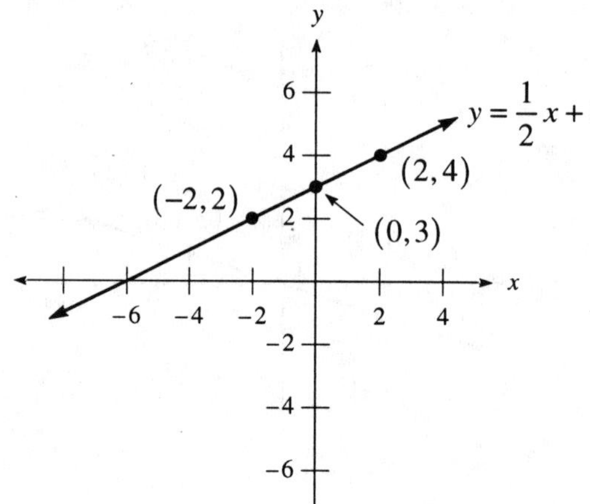

15. The ordered pairs are $(-3,3)$, $(0,1)$, and $(3,-1)$. Graphing the equation:

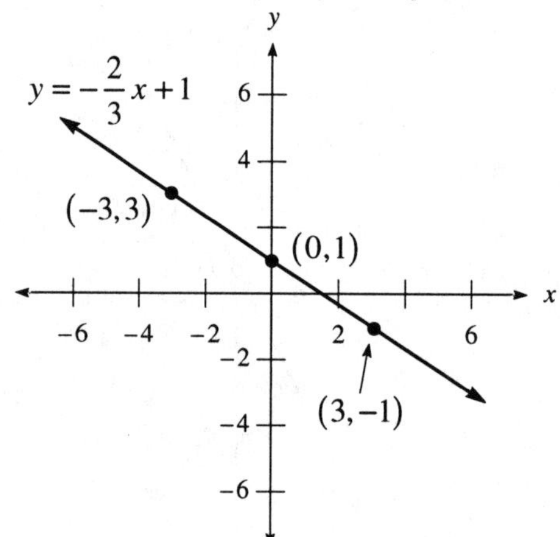

17. Solving for y:

$$2x+y=3$$
$$y=-2x+3$$

The ordered pairs are $(-1,5)$, $(0,3)$, and $(1,1)$. Graphing the equation:

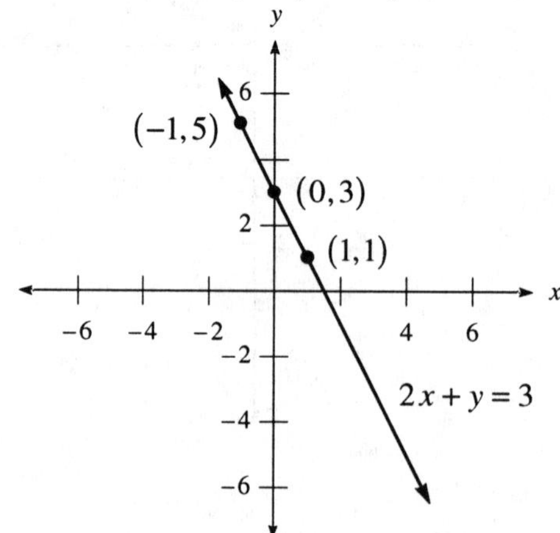

19. Solving for y:

$$3x+2y=6$$
$$2y=-3x+6$$
$$y=-\frac{3}{2}x+3$$

The ordered pairs are $(0,3)$, $(2,0)$, and $(4,-3)$. Graphing the equation:

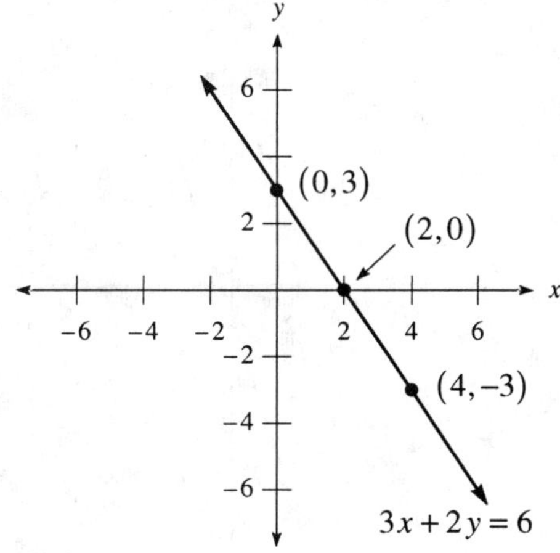

21. Solving for y:

$$\begin{aligned} -x+2y&=6 \\ 2y&=x+6 \\ y&=\frac{1}{2}x+3 \end{aligned}$$

The ordered pairs are $(-2,2)$, $(0,3)$, and $(2,4)$. Graphing the equation:

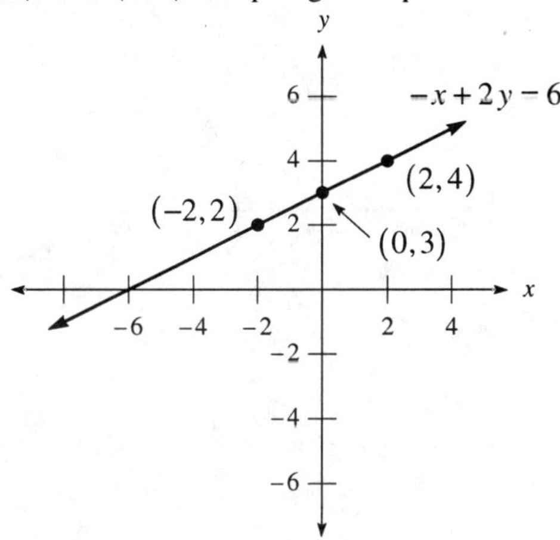

23. Three solutions are $(-4,2)$, $(0,0)$, and $(4,-2)$. Graphing the equation:

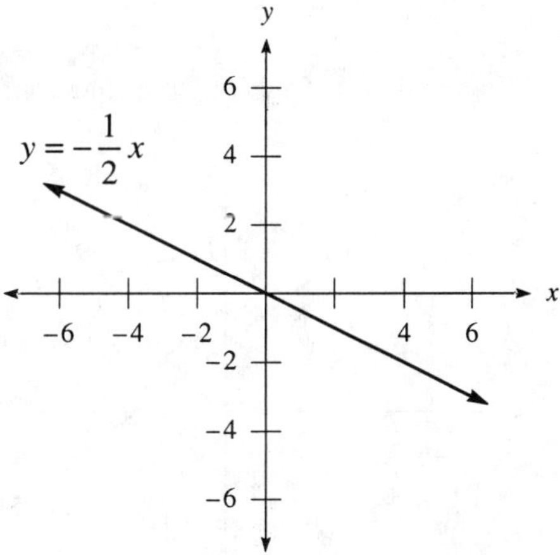

25. Three solutions are $(-1,-4)$, $(0,-1)$, and $(1,2)$. Graphing the equation:

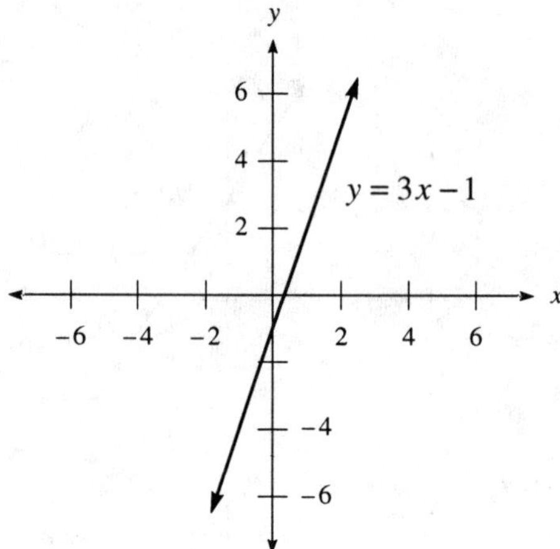

27. Solving for *y*:

$$-2x + y = 1$$
$$y = 2x + 1$$

Three solutions are $(-2,-3)$, $(0,1)$, and $(2,5)$. Graphing the equation:

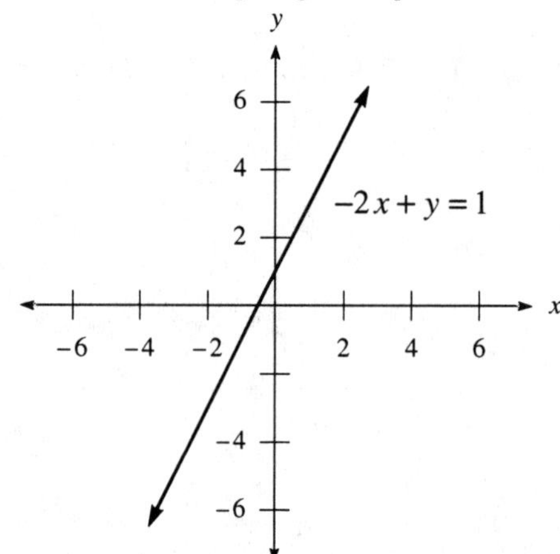

29. Solving for *y*:

$$3x + 4y = 8$$
$$4y = -3x + 8$$
$$y = -\frac{3}{4}x + 2$$

Three solutions are $(-4,5)$, $(0,2)$, and $(4,-1)$. Graphing the equation:

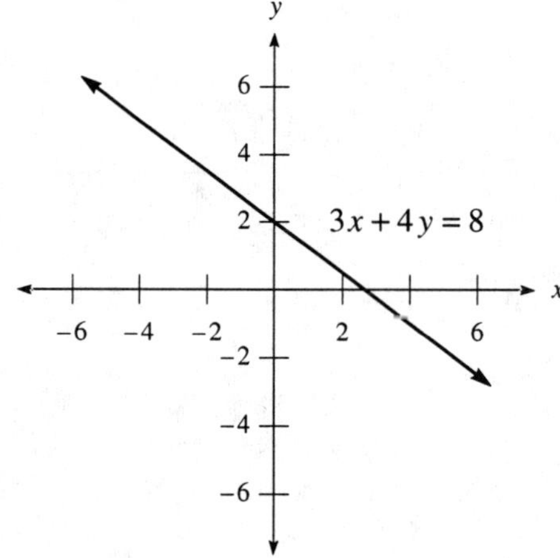

31. Three solutions are $(-2,-4)$, $(-2,0)$, and $(-2,4)$. Graphing the equation:

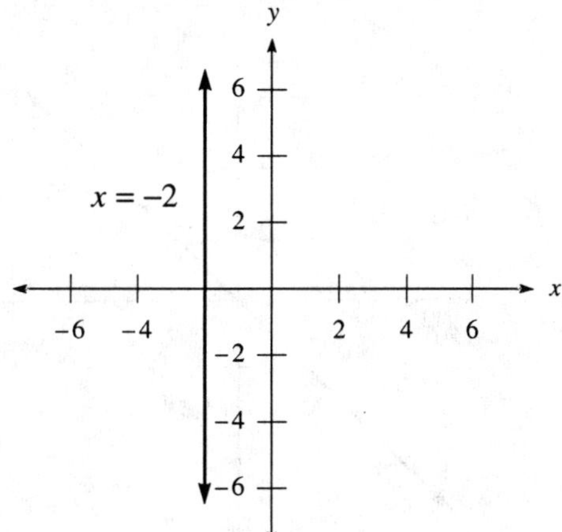

33. Three solutions are $(-4,2)$, $(0,2)$, and $(4,2)$. Graphing the equation:

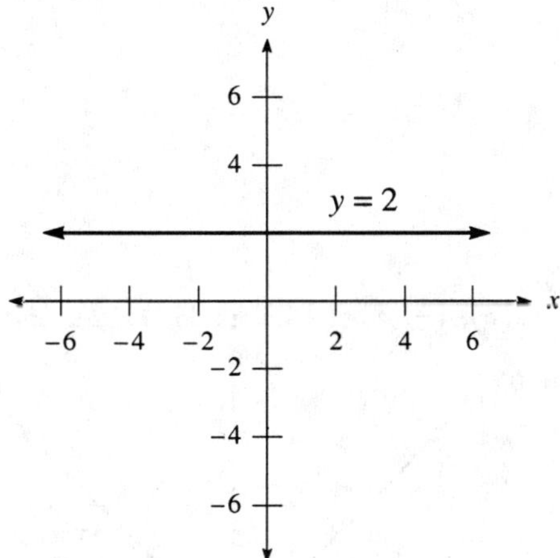

35. The ordered pairs are $(1,4)$, $(2,3)$, and $(3,2)$. Graphing the equation:

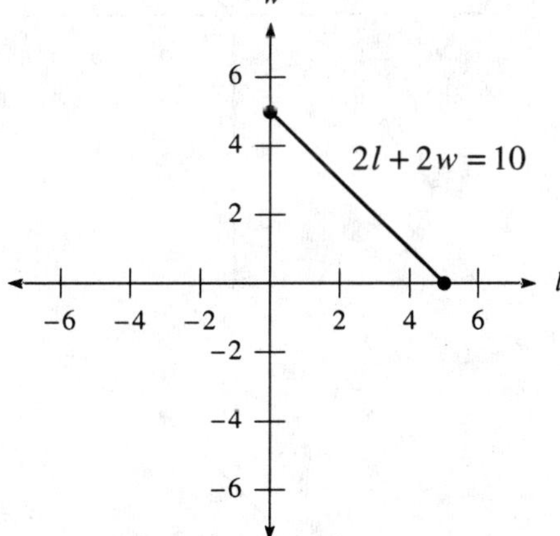

37. Graphing the line $y = x$:

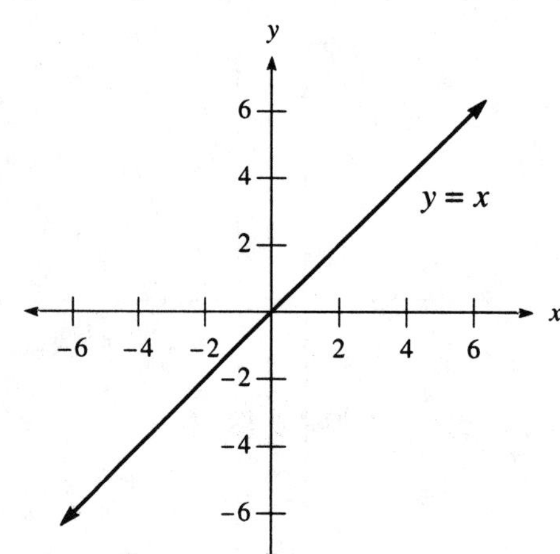

39. Completing the table:

x	y
−4	−3
−2	−2
0	−1
2	0
6	2

41. The ordered pairs are $(-3,3)$, $(-2,2)$, $(-1,1)$, $(0,0)$, $(1,1)$, $(2,2)$, and $(3,3)$. Graphing the equation:

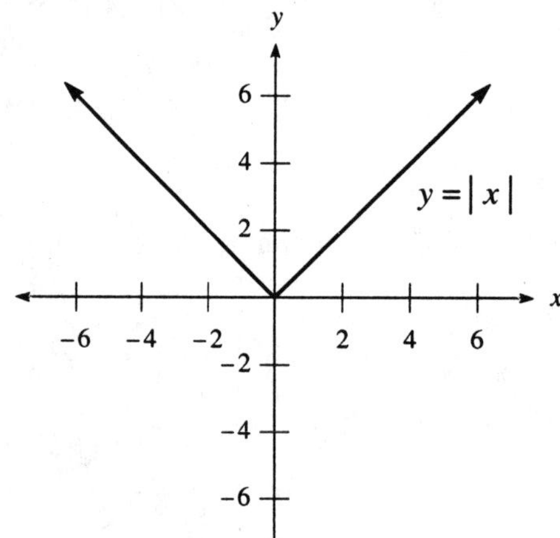

43. Graphing the three lines:

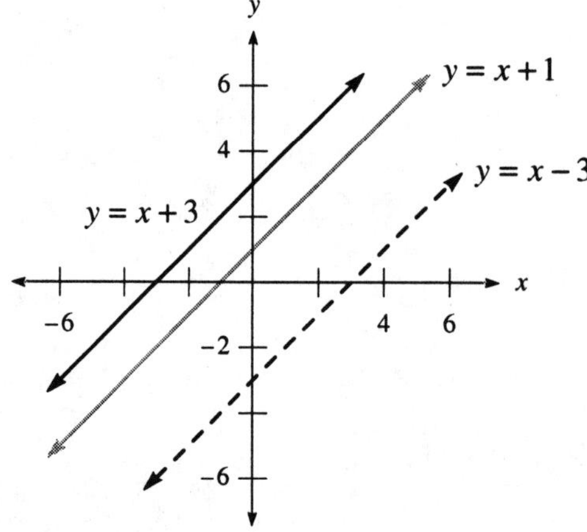

45. Solving the equation:

$$\begin{aligned} 3(x-2) &= 9 \\ 3x-6 &= 9 \\ 3x &= 15 \\ x &= 5 \end{aligned}$$

47. Solving the equation:

$$\begin{aligned} 2(3x-1)+4 &= -10 \\ 6x-2+4 &= -10 \\ 6x+2 &= -10 \\ 6x &= -12 \\ x &= -2 \end{aligned}$$

49. Solving the equation:

$$\begin{aligned} 6-2(4x-7) &= -4 \\ 6-8x+14 &= -4 \\ -8x+20 &= -4 \\ -8x &= -24 \\ x &= 3 \end{aligned}$$

51. First multiply by 6 to clear the equation of fractions:

$$\begin{aligned} 6\left(\frac{1}{2}x+4\right) &= 6\left(\frac{2}{3}x+5\right) \\ 3x+24 &= 4x+30 \\ -x+24 &= 30 \\ -x &= 6 \\ x &= -6 \end{aligned}$$

3.4 More on Graphing: Intercepts

1. To find the x-intercept, let $y = 0$:

$$\begin{aligned} 2x+0 &= 4 \\ 2x &= 4 \\ x &= 2 \end{aligned}$$

To find the y-intercept, let $x = 0$:

$$\begin{aligned} 2(0)+y &= 4 \\ 0+y &= 4 \\ y &= 4 \end{aligned}$$

Graphing the line:

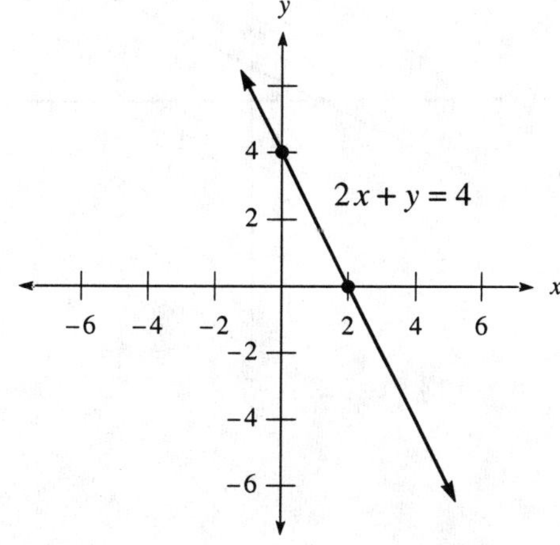

3. To find the x-intercept, let $y = 0$:

$$-x + 0 = 3$$
$$-x = 3$$
$$x = -3$$

To find the y-intercept, let $x = 0$:

$$-0 + y = 3$$
$$y = 3$$

Graphing the line:

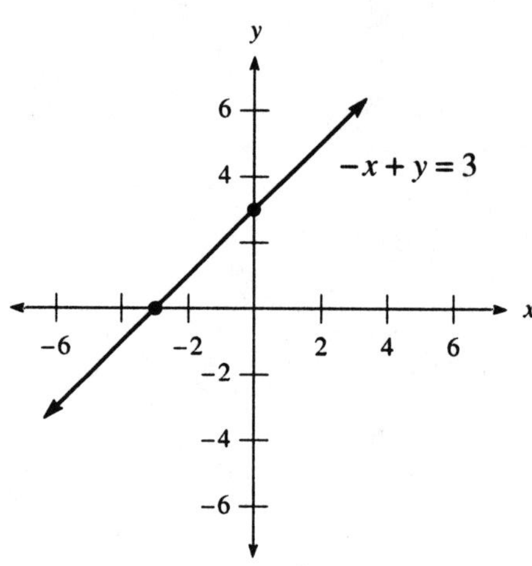

5. To find the x-intercept, let $y = 0$:

$$-x + 2(0) = 2$$
$$-x = 2$$
$$x = -2$$

To find the y-intercept, let $x = 0$:

$$-0 + 2y = 2$$
$$2y = 2$$
$$y = 1$$

Graphing the line:

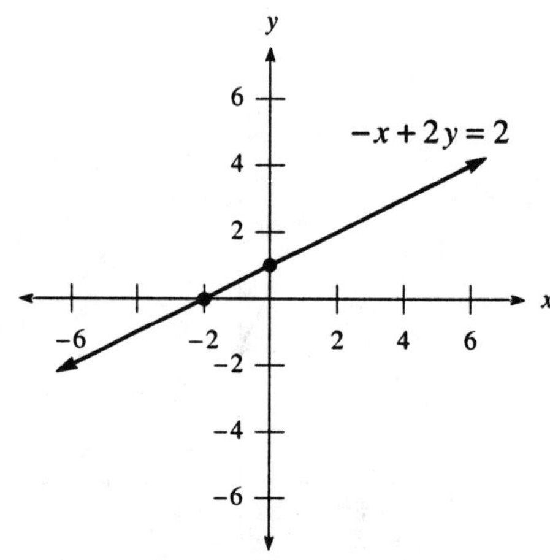

7. To find the x-intercept, let $y = 0$:

$$5x + 2(0) = 10$$
$$5x = 10$$
$$x = 2$$

To find the y-intercept, let $x = 0$:

$$5(0) + 2y = 10$$
$$2y = 10$$
$$y = 5$$

Graphing the line:

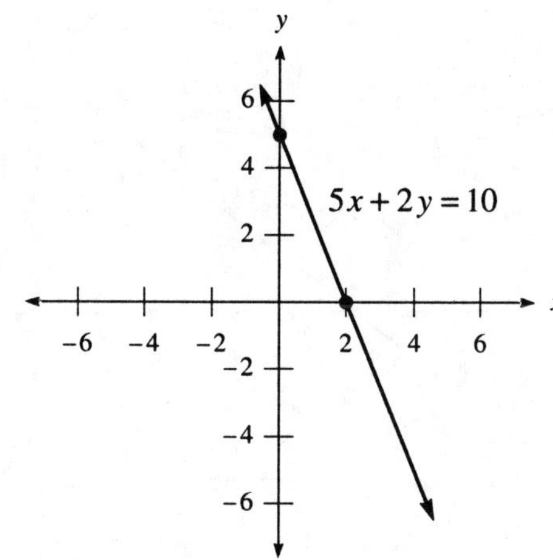

9. To find the x-intercept, let $y = 0$:

$$4x - 2(0) = 8$$
$$4x = 8$$
$$x = 2$$

To find the y-intercept, let $x = 0$:

$$4(0) - 2y = 8$$
$$-2y = 8$$
$$y = -4$$

Graphing the line:

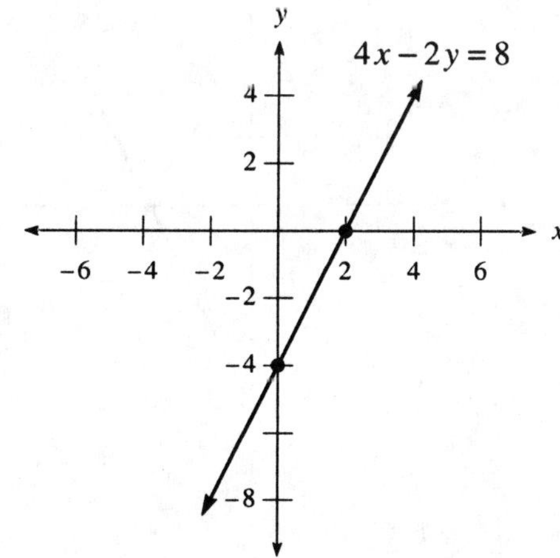

11. To find the x-intercept, let $y = 0$:

$$-4x + 5(0) = 20$$
$$-4x = 20$$
$$x = -5$$

To find the y-intercept, let $x = 0$:

$$-4(0) + 5y = 20$$
$$5y = 20$$
$$y = 4$$

Graphing the line:

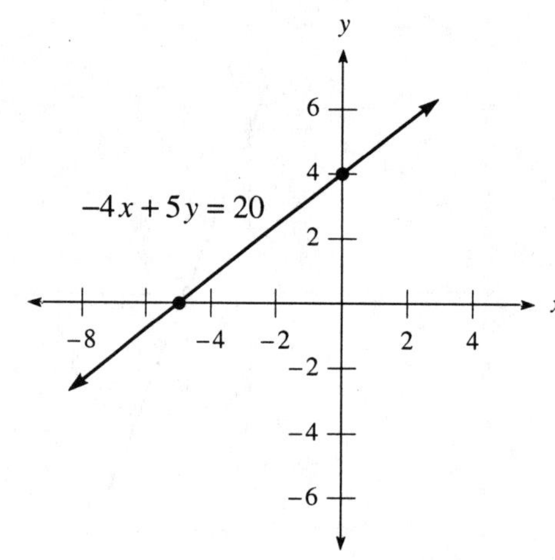

13. To find the x-intercept, let $y = 0$:

$$2x - 6 = 0$$
$$2x = 6$$
$$x = 3$$

To find the y-intercept, let $x = 0$:

$$y = 2(0) - 6$$
$$y = -6$$

Graphing the line:

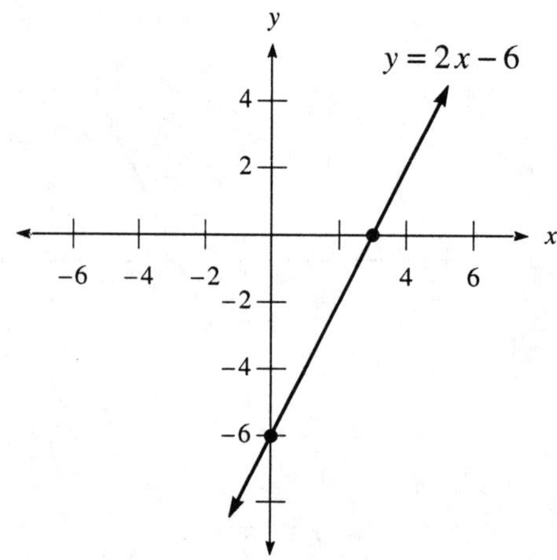

15. To find the x-intercept, let $y = 0$:

$$2x + 2 = 0$$
$$2x = -2$$
$$x = -1$$

To find the y-intercept, let $x = 0$:

$$y = 2(0) + 2$$
$$y = 2$$

Graphing the line:

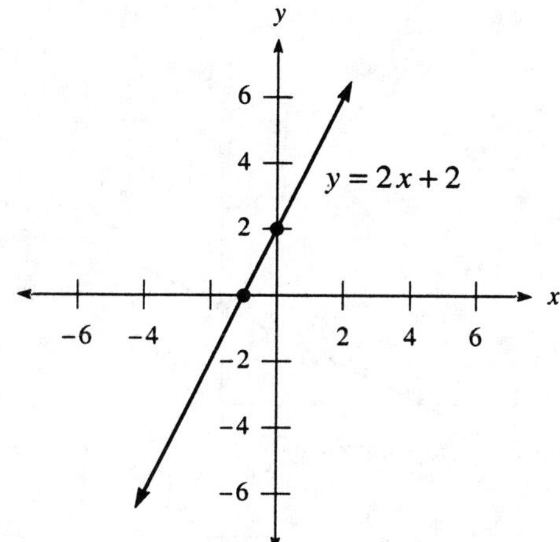

17. To find the x-intercept, let $y = 0$:

$$2x - 1 = 0$$
$$2x = 1$$
$$x = \frac{1}{2}$$

To find the y-intercept, let $x = 0$:

$$y = 2(0) - 1$$
$$y = -1$$

Graphing the line:

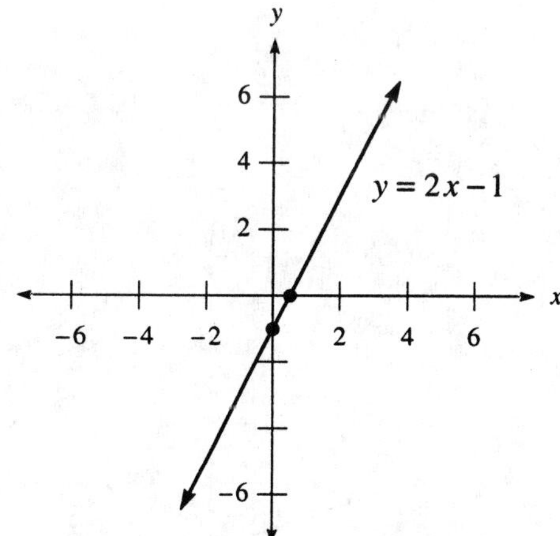

19. To find the x-intercept, let $y = 0$:

$$\frac{1}{2}x + 3 = 0$$
$$\frac{1}{2}x = -3$$
$$x = -6$$

To find the y-intercept, let $x = 0$:

$$y = \frac{1}{2}(0) + 3$$
$$y = 3$$

Graphing the line:

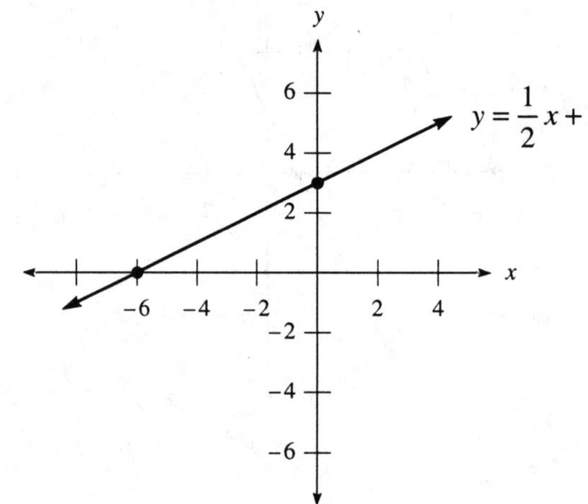

21. To find the x-intercept, let $y = 0$:

$$-\frac{1}{3}x - 2 = 0$$
$$-\frac{1}{3}x = 2$$
$$x = -6$$

To find the y-intercept, let $x = 0$:

$$y = -\frac{1}{3}(0) - 2$$
$$y = -2$$

Graphing the line:

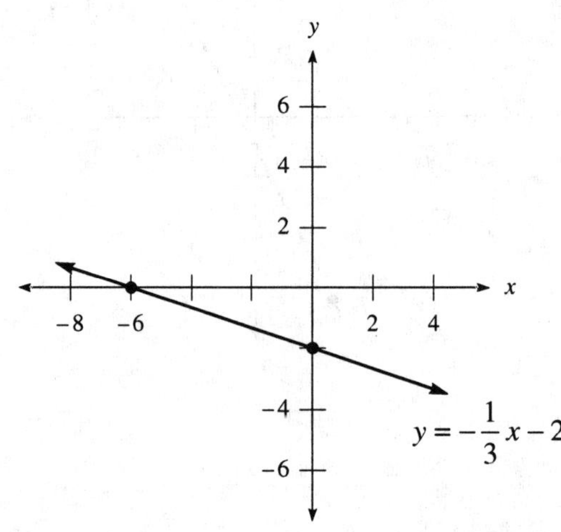

23. Another point on the line is $(2,-4)$. Graphing the line:

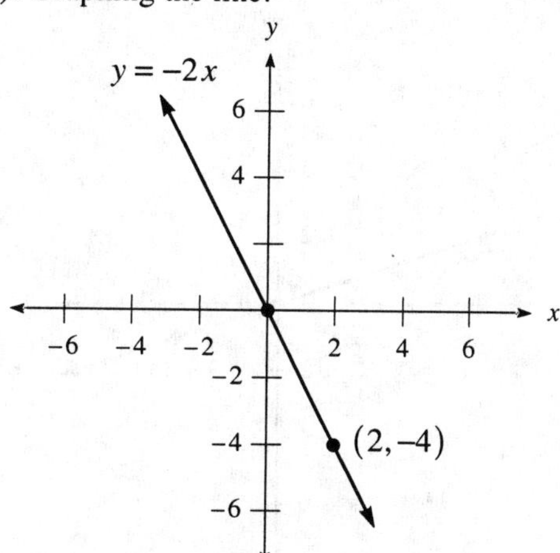

25. Another point on the line is $(2,4)$. Graphing the line:

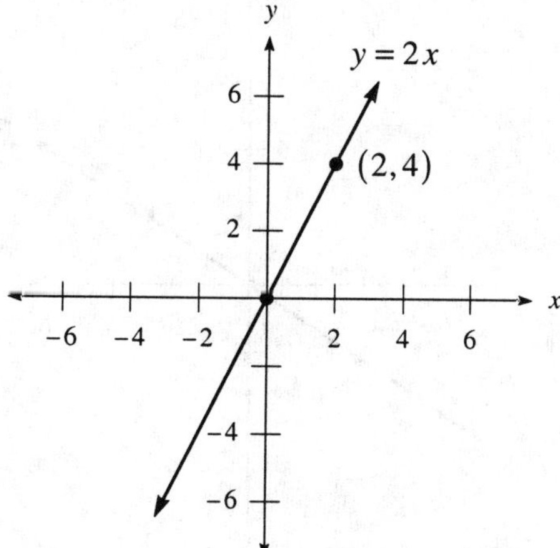

27. Another point on the line is $(3,1)$. Graphing the line:

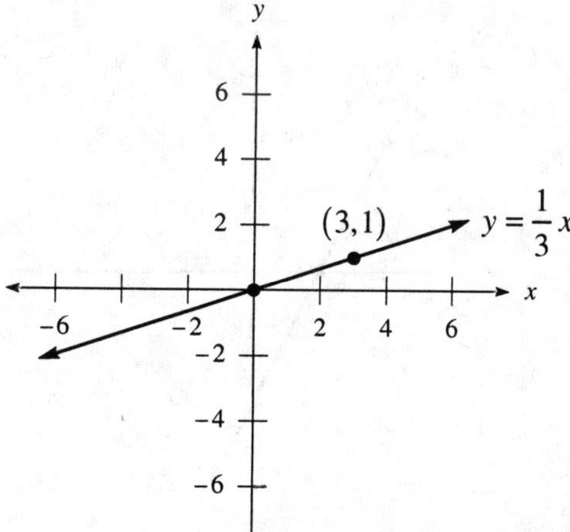

29. Another point on the line is $(3,-1)$. Graphing the line:

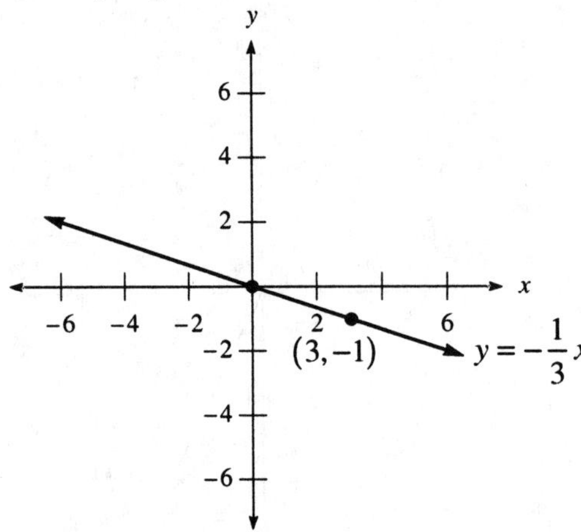

31. Another point on the line is $(3,2)$. Graphing the line:

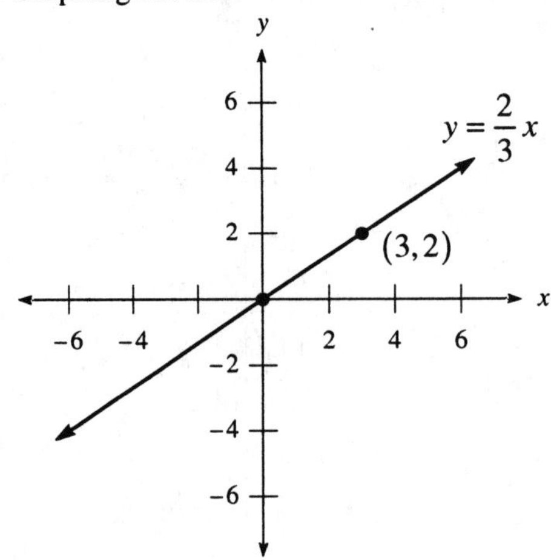

33. Graphing the line:

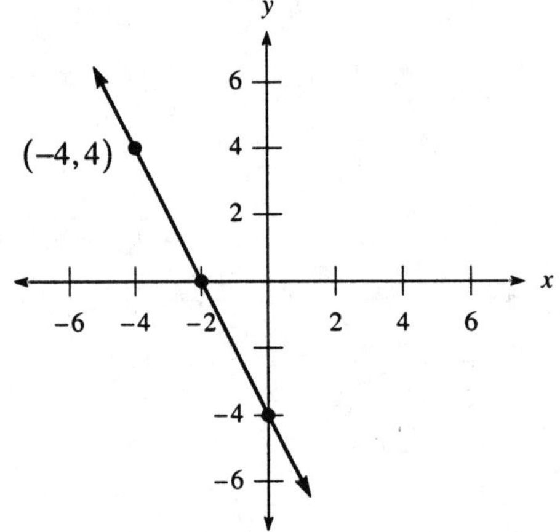

The y-intercept is –4.

35. Graphing the line:

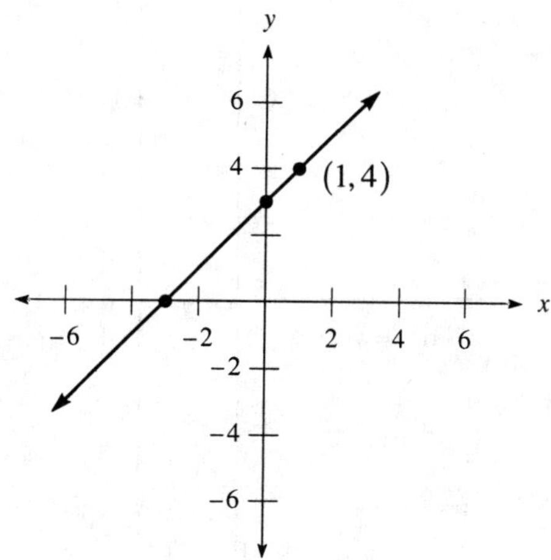

The x-intercept is –3.

37. Graphing the line:

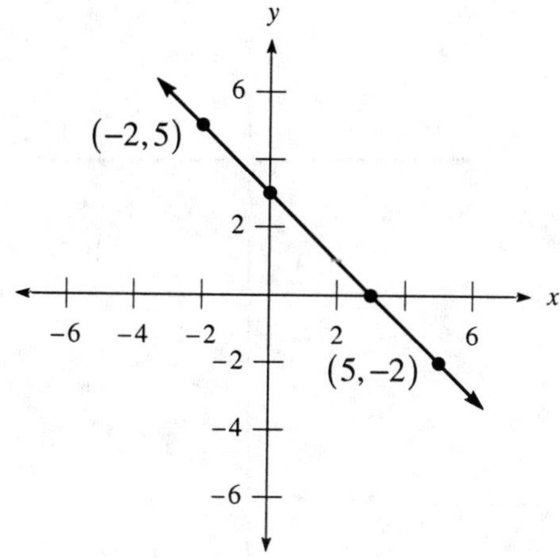

The x- and y-intercepts are both 3.

39. Completing the table:

x	y
–2	1
0	–1
–1	0
1	–2

41. Graphing the line:

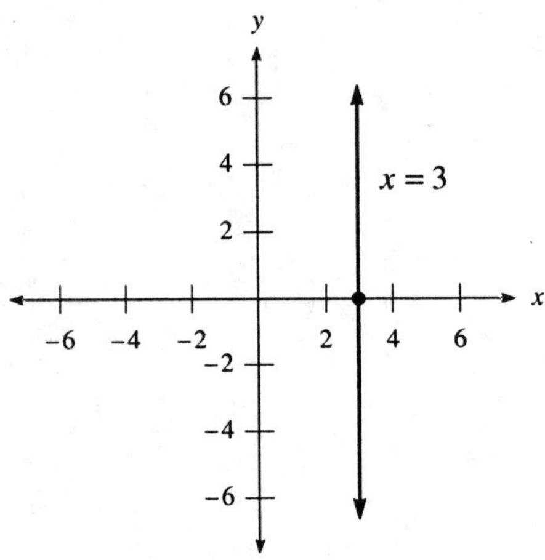

The x-intercept is 3.

43. Graphing the line:

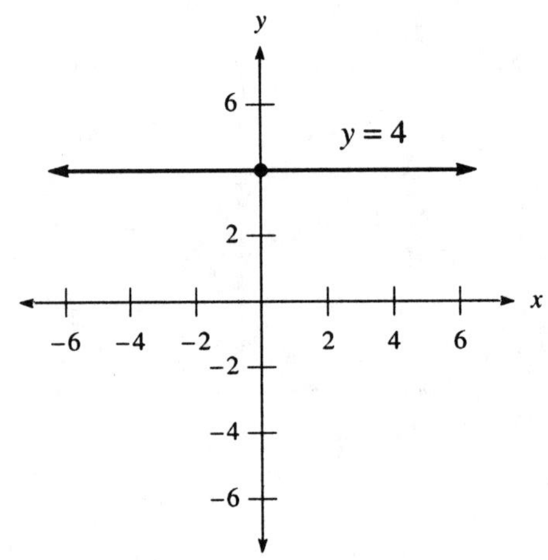

The y-intercept is 4.

45. Solving the inequality:

$$x-3<2$$
$$x-3+3<2+3$$
$$x<5$$

47. Solving the inequality:

$$-3x \geq 12$$
$$-\frac{1}{3}(-3x) \leq -\frac{1}{3}(12)$$
$$x \leq -4$$

49. Solving the inequality:

$$-\frac{x}{3} \leq -1$$
$$-3\left(-\frac{x}{3}\right) \geq -3(-1)$$
$$x \geq 3$$

51. Solving the inequality:

$$-4x+1<17$$
$$-4x+1-1<17-1$$
$$-4x<16$$
$$-\frac{1}{4}(-4x) > -\frac{1}{4}(16)$$
$$x>-4$$

3.5 The Slope of a Line

1. The slope is given by: $m=\frac{4-1}{4-2}=\frac{3}{2}$
Graphing the line:

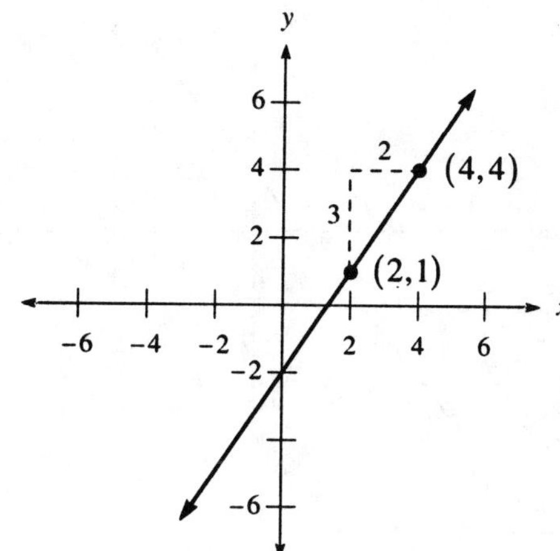

3. The slope is given by: $m=\frac{2-4}{5-1}=\frac{-2}{4}=-\frac{1}{2}$
Graphing the line:

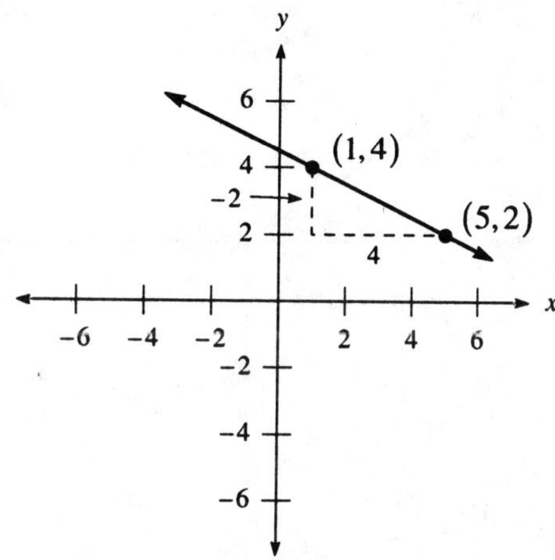

5. The slope is given by: $m=\frac{2-(-3)}{4-1}=\frac{5}{3}$
Graphing the line:

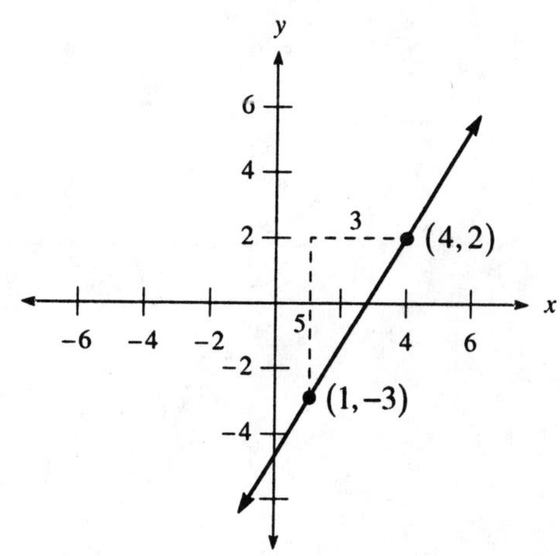

7. The slope is given by: $m=\dfrac{3-(-2)}{1-(-3)}=\dfrac{5}{4}$

Graphing the line:

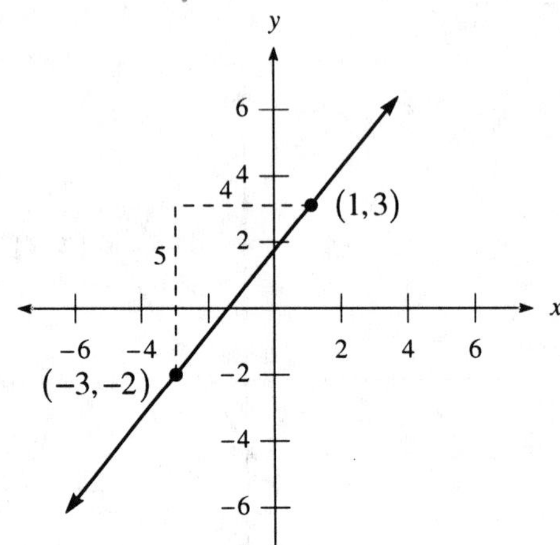

9. The slope is given by: $m=\dfrac{-2-2}{3-(-3)}=\dfrac{-4}{6}=-\dfrac{2}{3}$

Graphing the line:

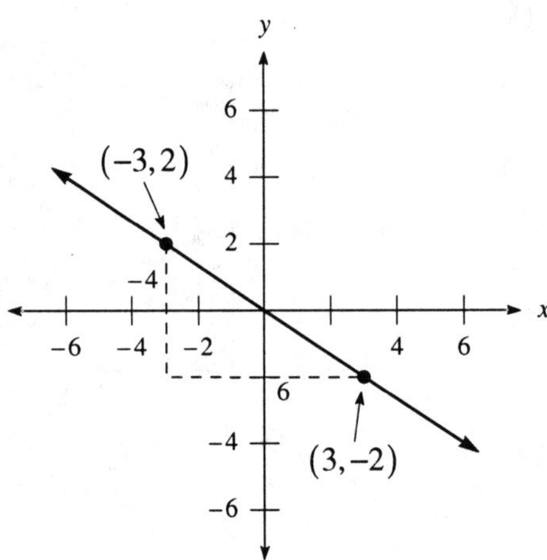

11. The slope is given by: $m=\dfrac{-2-(-5)}{3-2}=\dfrac{3}{1}=3$

Graphing the line:

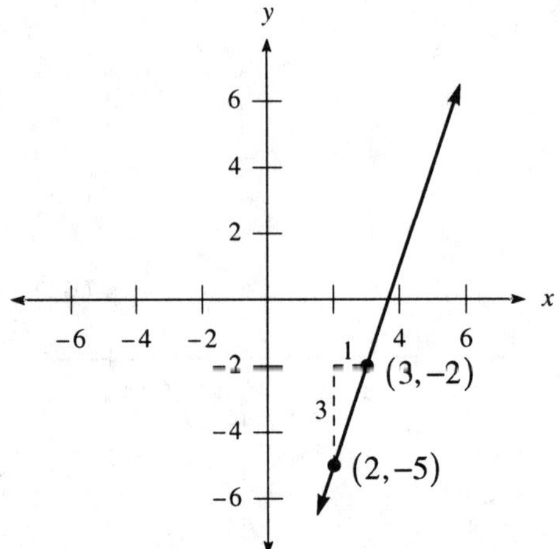

13. Graphing the line:

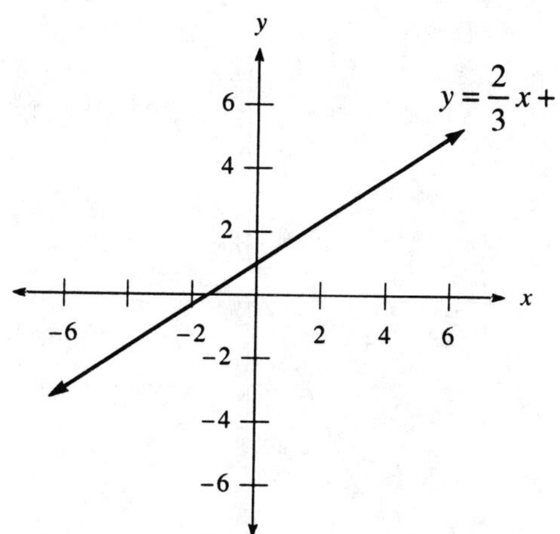

15. Graphing the line:

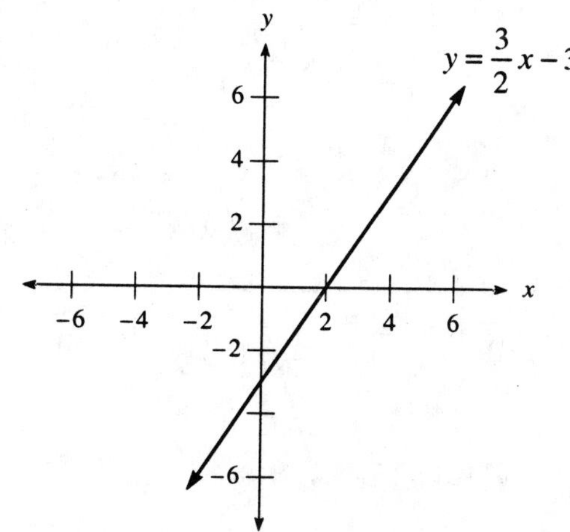

17. Graphing the line:

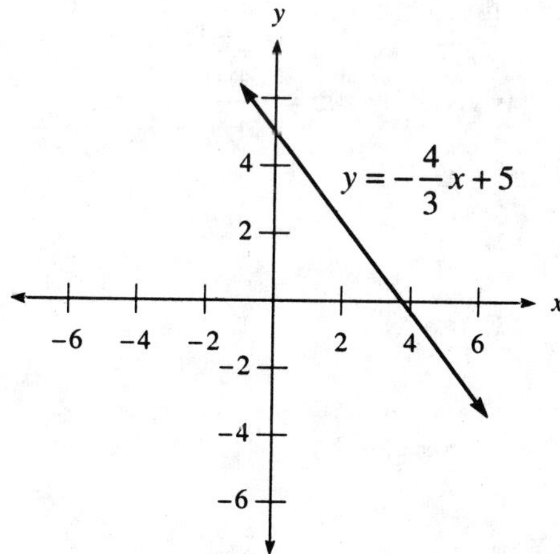

19. Graphing the line:

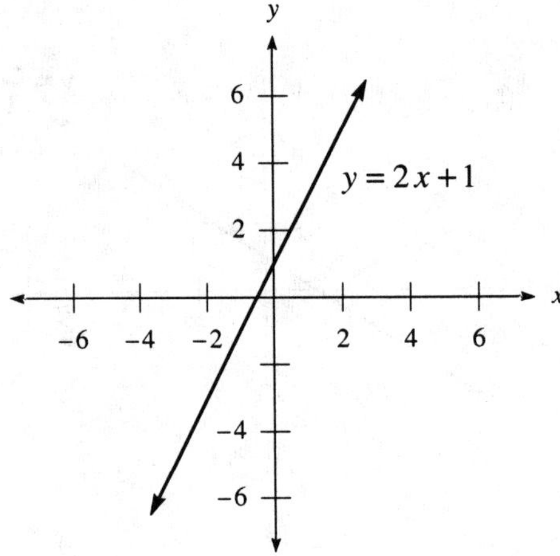

21. Graphing the line:

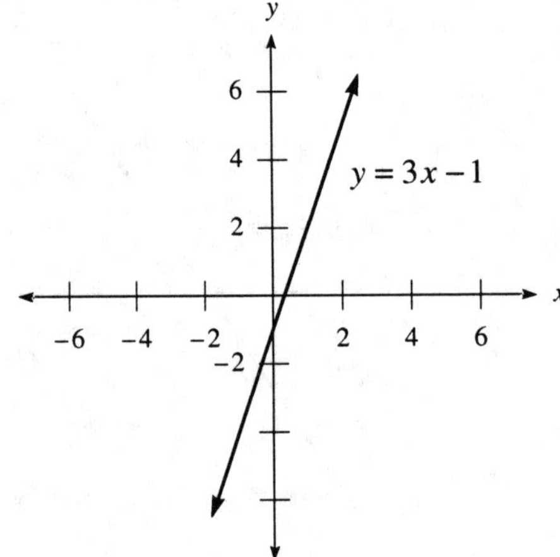

23. The y-intercept is 2, and the slope is given by: $m=\frac{5-(-1)}{1-(-1)}=\frac{6}{2}=3$

25. The y-intercept is –2, and the slope is given by: $m=\frac{2-0}{2-1}=\frac{2}{1}=2$

27. The slope is given by: $m=\frac{0-(-2)}{3-0}=\frac{2}{3}$

Graphing the line:

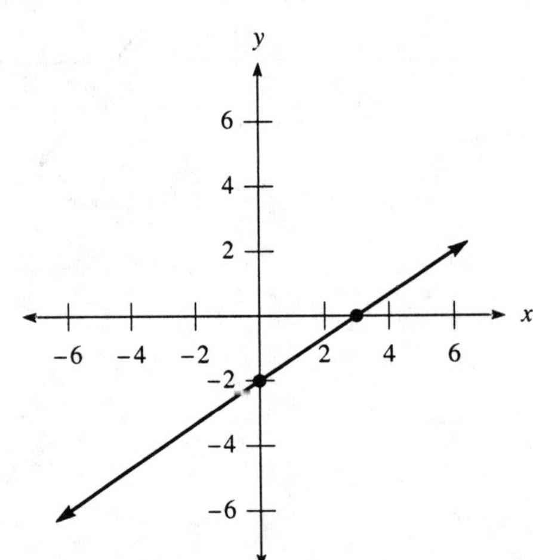

29. The slope is given by: $m = \dfrac{0-2}{4-0} = \dfrac{-2}{4} = -\dfrac{1}{2}$

Graphing the line:

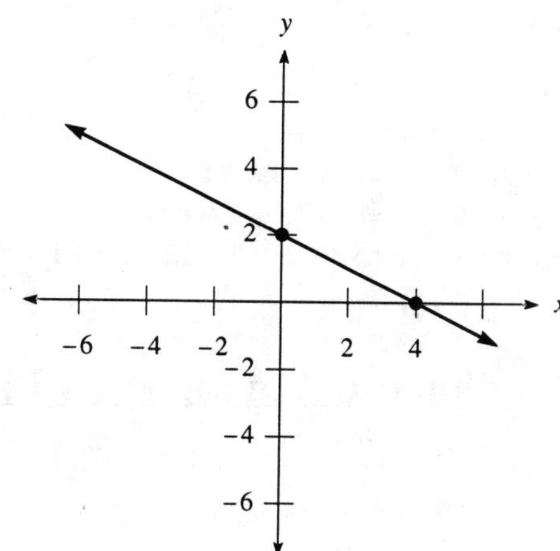

31. Graphing the line:

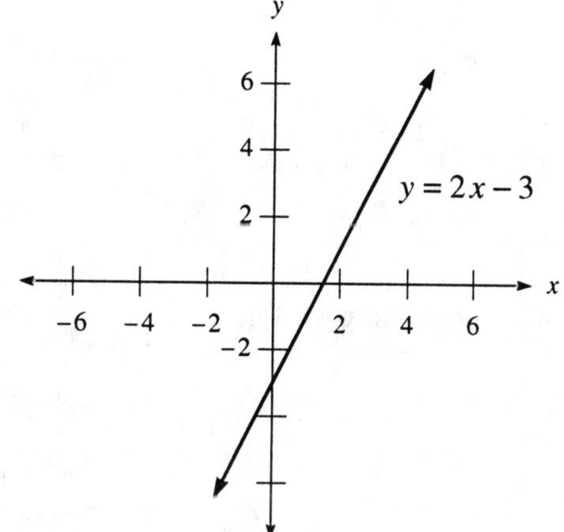

The slope is 2 and the y-intercept is -3.

33. Graphing the line:

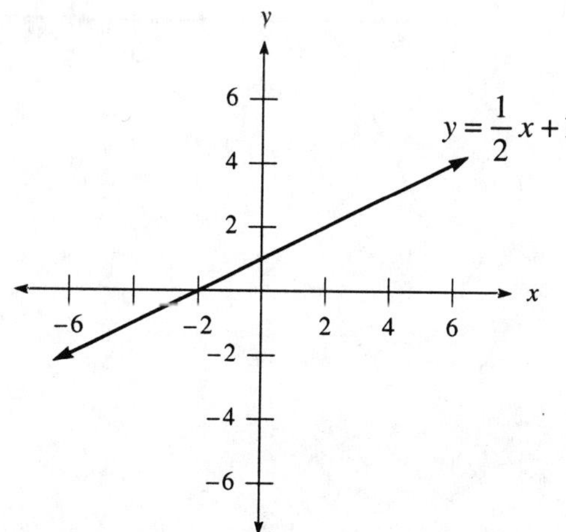

The slope is $\dfrac{1}{2}$ and the y-intercept is 1.

35. Using the slope formula:

$$\frac{y-2}{6-4}=2$$
$$\frac{y-2}{2}=2$$
$$y-2=4$$
$$y=6$$

37. Evaluating when $x=-3$: $2x-9=2(-3)-9=-6-9=-15$
39. Evaluating when $x=-3$: $9-6x=9-6(-3)=9+18=27$
41. Simplifying, then evaluating when $x=-3$: $4(3x+2)+1=12x+8+1=12x+9=12(-3)+9=-36+9=-27$
43. Evaluating when $x=-3$: $2x^2+3x+4=2(-3)^2+3(-3)+4=2(9)-9+4=18-9+4=13$

3.6 Finding the Equation of a Line

1. The slope-intercept form is $y=\frac{2}{3}x+1$.
3. The slope-intercept form is $y=\frac{3}{2}x-1$.
5. The slope-intercept form is $y=-\frac{2}{5}x+3$.
7. The slope-intercept form is $y=2x-4$.
9. The slope-intercept form is $y=-3x+2$.
11. Solving for y:

$$-2x+y=4$$
$$y=2x+4$$

The slope is 2 and the y-intercept is 4. Graphing the line:

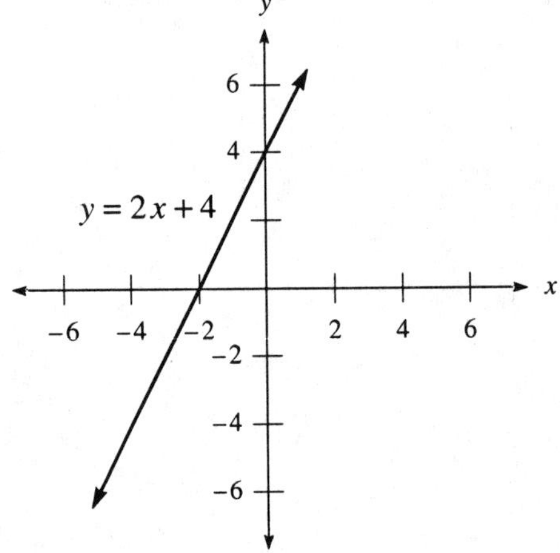

13. Solving for y:

$$3x + y = 3$$
$$y = -3x + 3$$

The slope is –3 and the y-intercept is 3. Graphing the line:

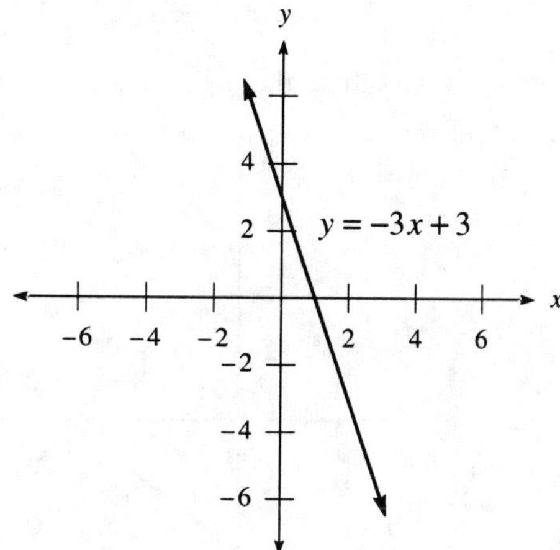

15. Solving for y:

$$3x + 2y = 6$$
$$2y = -3x + 6$$
$$y = -\frac{3}{2}x + 3$$

The slope is $-\frac{3}{2}$ and the y-intercept is 3. Graphing the line:

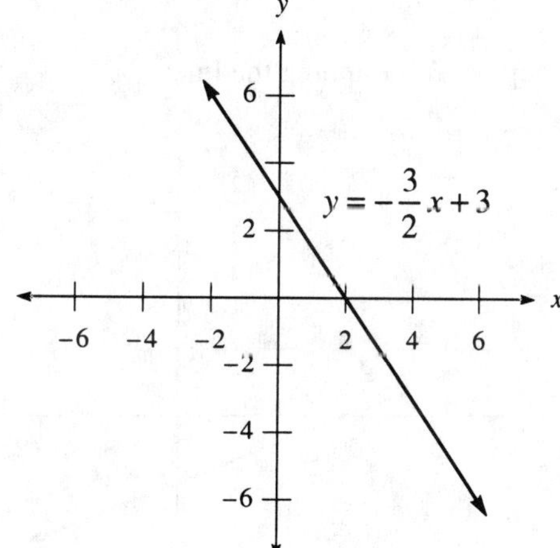

17. Solving for y:

$$\begin{aligned} 4x-5y&=20\\ -5y&=-4x+20\\ y&=\frac{4}{5}x-4 \end{aligned}$$

The slope is $\frac{4}{5}$ and the y-intercept is –4. Graphing the line:

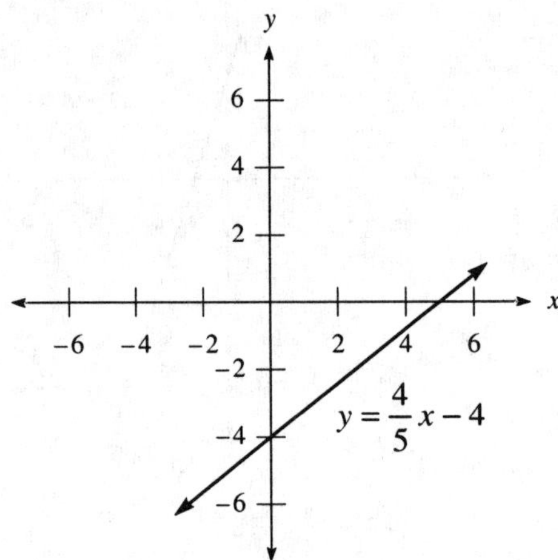

19. Solving for y:

$$\begin{aligned} -2x-5y&=10\\ -5y&=2x+10\\ y&=-\frac{2}{5}x-2 \end{aligned}$$

The slope is $-\frac{2}{5}$ and the y-intercept is –2. Graphing the line:

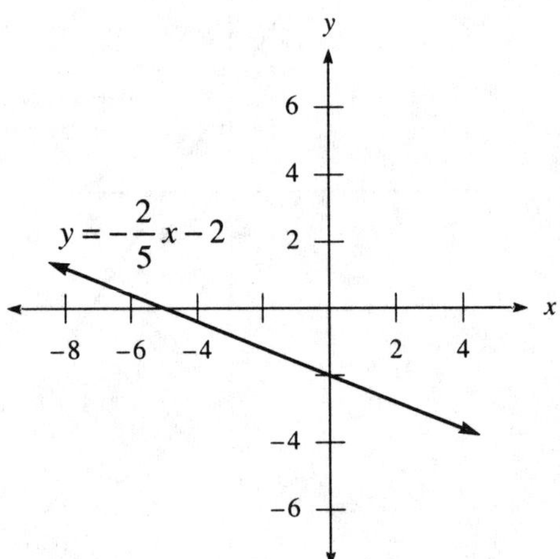

21. Using the point-slope formula:

$$\begin{aligned} y-(-5)&=2(x-(-2))\\ y+5&=2(x+2)\\ y+5&=2x+4\\ y&=2x-1 \end{aligned}$$

23. Using the point-slope formula:

$$y-1=-\frac{1}{2}(x-(-4))$$
$$y-1=-\frac{1}{2}(x+4)$$
$$y-1=-\frac{1}{2}x-2$$
$$y=-\frac{1}{2}x-1$$

25. Using the point-slope formula:

$$y-(-3)=\frac{3}{2}(x-2)$$
$$y+3=\frac{3}{2}x-3$$
$$y=\frac{3}{2}x-6$$

27. Using the point-slope formula:

$$y-4=-3(x-(-1))$$
$$y-4=-3(x+1)$$
$$y-4=-3x-3$$
$$y=-3x+1$$

29. Using the point-slope formula:

$$y-4=1(x-2)$$
$$y-4=x-2$$
$$y=x+2$$

31. Finding the slope: $m=\frac{-1-(-4)}{1-(-2)}=\frac{-1+4}{1+2}=\frac{3}{3}=1$

Using the point-slope formula:

$$y-(-4)=1(x-(-2))$$
$$y+4=x+2$$
$$y=x-2$$

33. Finding the slope: $m=\frac{1-(-5)}{2-(-1)}=\frac{1+5}{2+1}=\frac{6}{3}=2$

Using the point-slope formula:

$$y-1=2(x-2)$$
$$y-1=2x-4$$
$$y=2x-3$$

35. Finding the slope: $m=\frac{6-(-2)}{3-(-3)}=\frac{6+2}{3+3}=\frac{8}{6}=\frac{4}{3}$

Using the point-slope formula:

$$y-6=\frac{4}{3}(x-3)$$
$$y-6=\frac{4}{3}x-4$$
$$y=\frac{4}{3}x+2$$

37. Finding the slope: $m=\frac{-5-(-1)}{3-(-3)}=\frac{-5+1}{3+3}=\frac{-4}{6}=-\frac{2}{3}$

Using the point-slope formula:

$$\begin{aligned} y-(-5) &= -\frac{2}{3}(x-3) \\ y+5 &= -\frac{2}{3}x+2 \\ y &= -\frac{2}{3}x-3 \end{aligned}$$

39. Finding the slope: $m=\frac{-2-1}{1-(-2)}=\frac{-2-1}{1+2}=\frac{-3}{3}=-1$

Using the point-slope formula:

$$\begin{aligned} y-(-2) &= -1(x-1) \\ y+2 &= -x+1 \\ y &= -x-1 \end{aligned}$$

41. The y-intercept is 3, and the slope is: $m=\frac{3-0}{0-(-1)}=\frac{3}{1}=3$

The slope-intercept form is $y=3x+3$.

43. The y-intercept is –1, and the slope is: $m=\frac{0-(-1)}{4-0}=\frac{0+1}{4}=\frac{1}{4}$

The slope-intercept form is $y=\frac{1}{4}x-1$.

45. The slope is given by: $m=\frac{0-2}{3-0}=-\frac{2}{3}$

Since $b=2$, the equation is $y=-\frac{2}{3}x+2$.

47. The slope is given by: $m=\frac{0-(-5)}{-2-0}=-\frac{5}{2}$

Since $b=-5$, the equation is $y=-\frac{5}{2}x-5$.

49. Its equation is $x=3$.

51. Its equation is $y=3$.

53. Finding the amount:

$$\begin{aligned} (0.25)(300) &= x \\ x &= 75 \end{aligned}$$

75 is 25% of 300.

55. Finding the percent:

$$\begin{aligned} p \cdot 125 &= 25 \\ p &= \frac{25}{125} \\ p &= 0.2 = 20\% \end{aligned}$$

25 is 20% of 125.

57. Finding the base:

$$\begin{aligned} 0.15x &= 60 \\ x &= \frac{60}{0.15} \\ x &= 400 \end{aligned}$$

60 is 15% of 400.

59. Graphing the line:

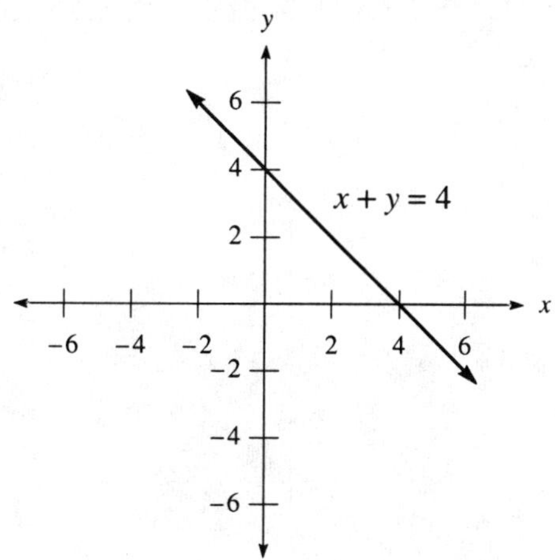

61. Graphing the line:

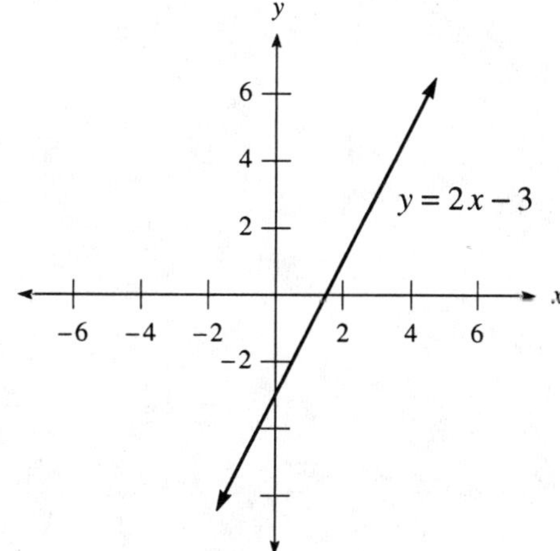

63. Graphing the line:

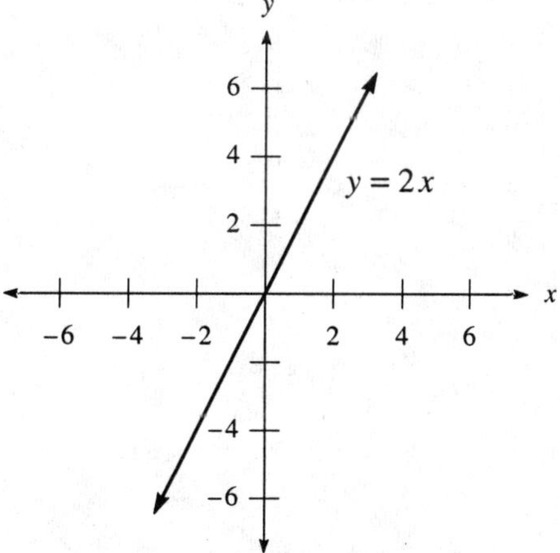

3.7 Linear Inequalities in Two Variables

1. Checking the point $(0,0)$:

$$2(0)-3(0)=0-0<6 \quad \text{(true)}$$

Graphing the linear inequality:

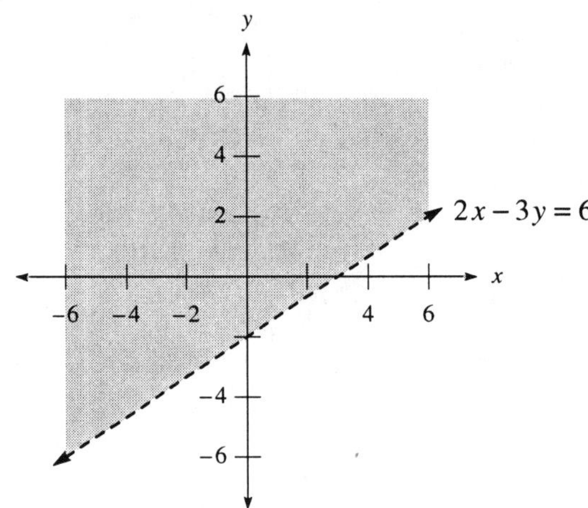

3. Checking the point $(0,0)$:

$$0-2(0)=0-0\le 4 \quad \text{(true)}$$

Graphing the linear inequality:

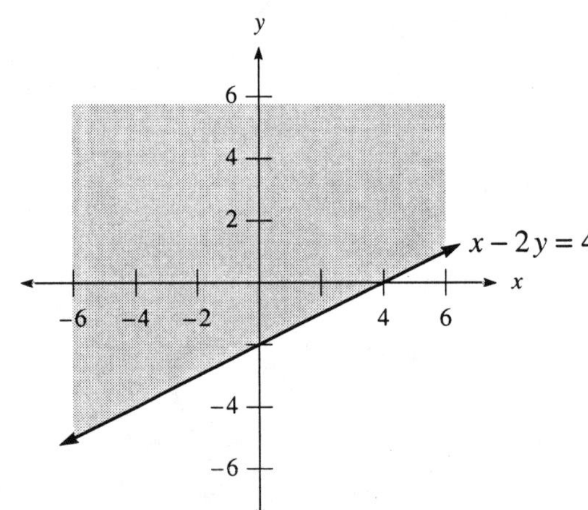

5. Checking the point $(0,0)$:

$$0-0\le 2 \quad \text{(true)}$$

Graphing the linear inequality:

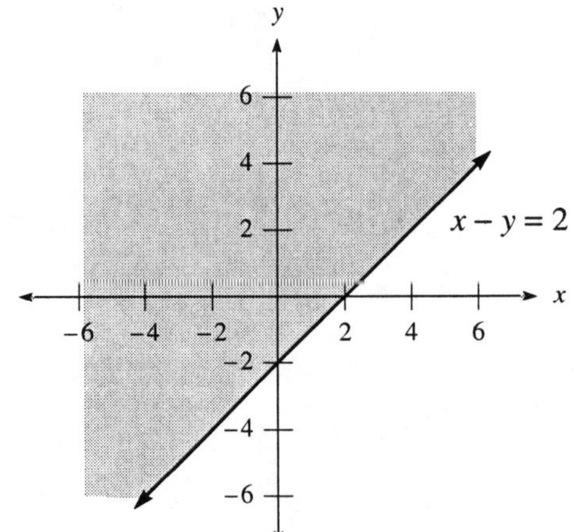

7. Checking the point $(0,0)$:

$$3(0)-4(0)=0-0\geq 12 \quad \text{(false)}$$

Graphing the linear inequality:

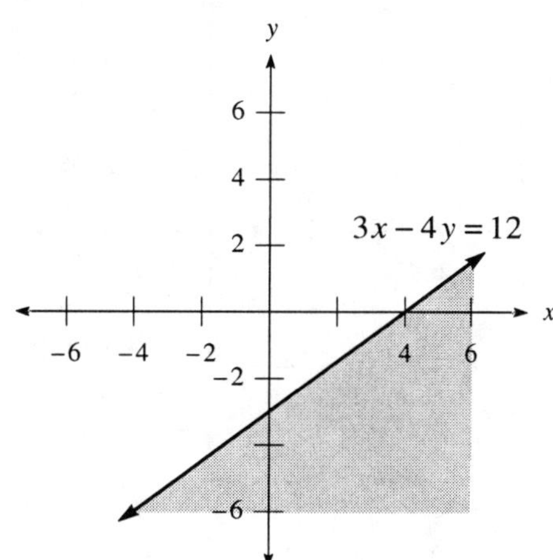

9. Checking the point $(0,0)$:

$$5(0)-0=0-0\leq 5 \quad \text{(true)}$$

Graphing the linear inequality:

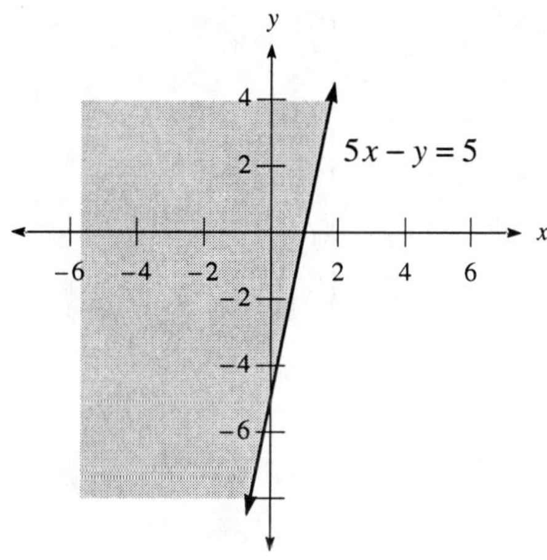

11. Checking the point $(0,0)$:

$$2(0)+6(0)=0+0\leq 12 \quad \text{(true)}$$

Graphing the linear inequality:

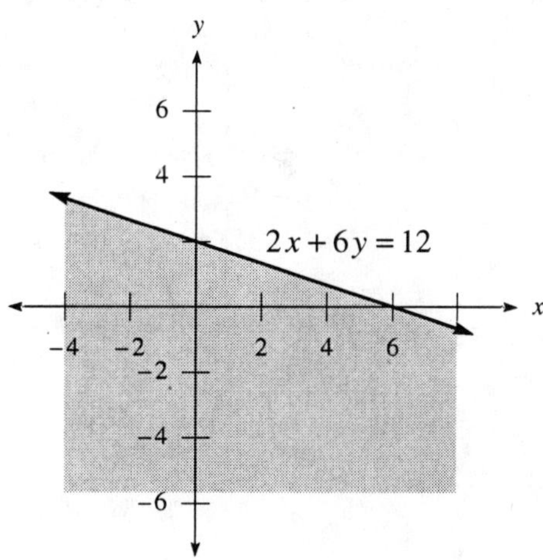

13. Graphing the linear inequality:

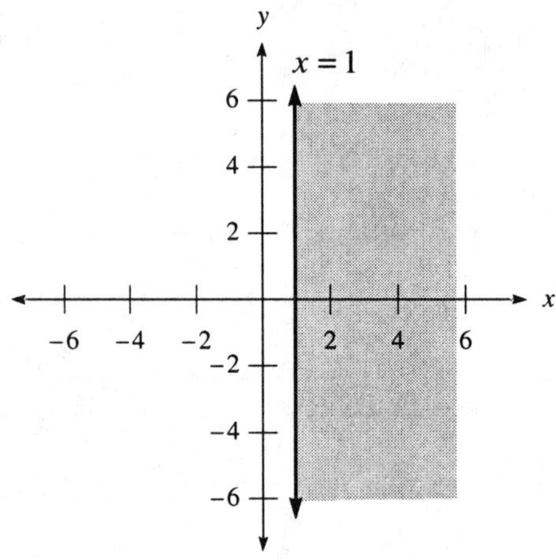

15. Graphing the linear inequality:

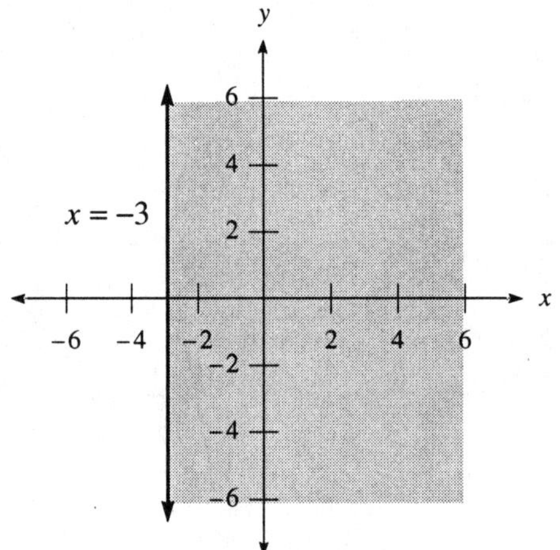

17. Graphing the linear inequality:

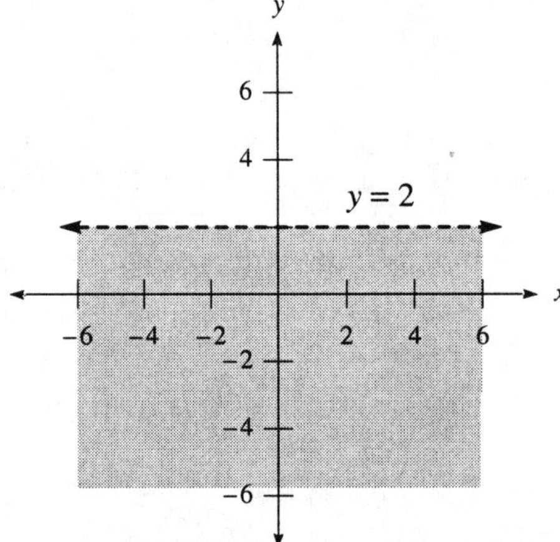

19. Checking the point $(0,0)$:

$$2(0)+0=0+0>3 \quad \text{(false)}$$

Graphing the linear inequality:

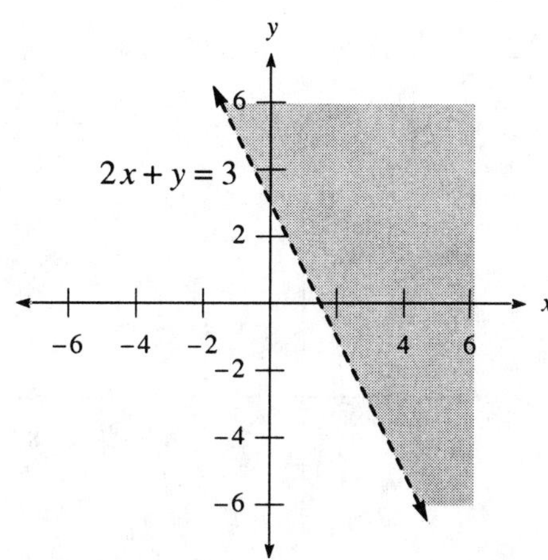

21. Checking the point $(0,0)$:

$$0 \le 3(0)-1$$
$$0 \le -1 \quad \text{(false)}$$

Graphing the linear inequality:

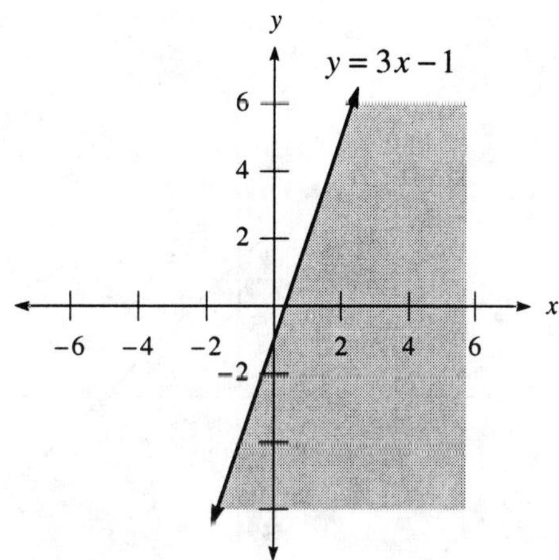

23. Checking the point $(0,0)$:

$$0 \ge 0-5$$
$$0 \ge -5 \quad \text{(true)}$$

Graphing the linear inequality:

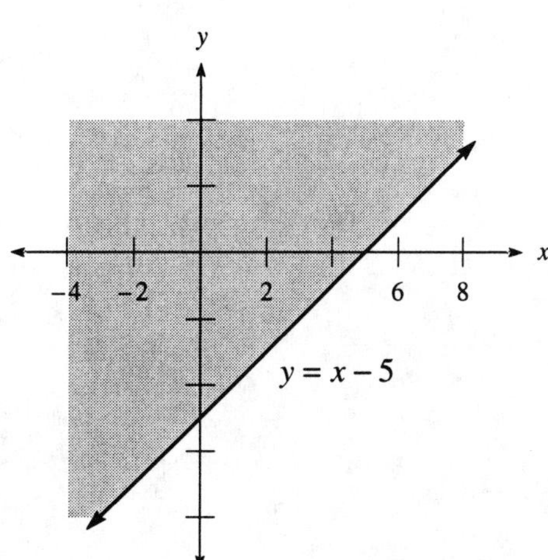

25. Checking the point $(0,0)$:

$$0 \le -\frac{1}{2}(0)+2$$
$$0 \le 2 \quad \text{(true)}$$

Graphing the linear inequality:

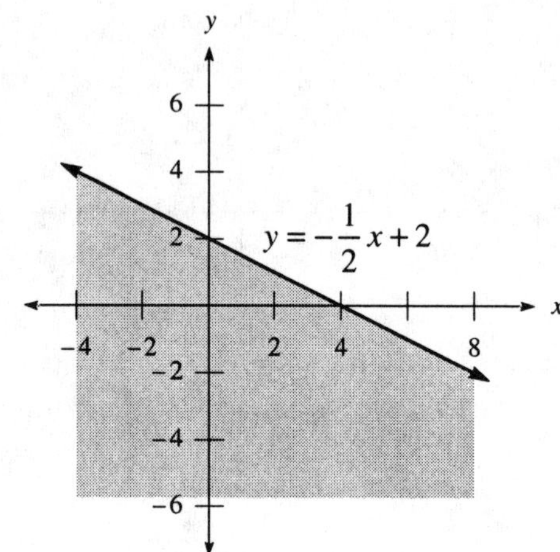

27. Checking the point $(0,0)$:

$$0 < -0+4$$
$$0 < 4 \quad \text{(true)}$$

Graphing the linear inequality:

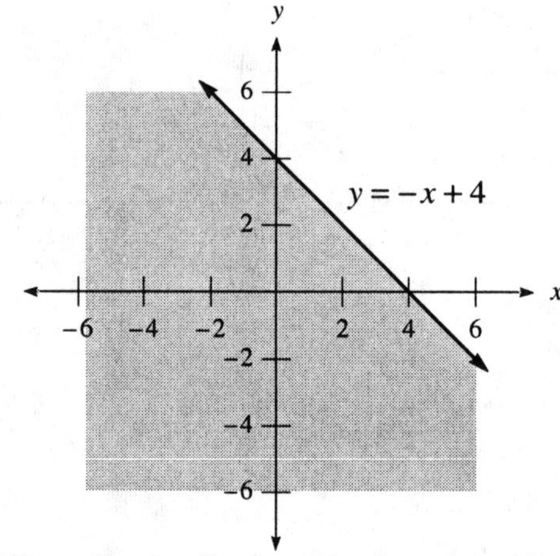

29. Simplifying the expression: $7-3(2x-4)-8=7-6x+12-8=-6x+11$

31. Solving the equation:

$$-\frac{3}{2}x=12$$
$$-\frac{2}{3}\left(-\frac{3}{2}x\right)=-\frac{2}{3}(12)$$
$$x=-8$$

33. Solving the equation:

$$8-2(x+7)=2$$
$$8-2x-14=2$$
$$-2x-6=2$$
$$-2x=8$$
$$x=-4$$

35. Solving for w:

$$\begin{aligned} 2l+2w &= P \\ 2w &= P-2l \\ w &= \frac{P-2l}{2} \end{aligned}$$

37. Solving the inequality:

$$\begin{aligned} 3-2x &> 5 \\ 3-3-2x &> 5-3 \\ -2x &> 2 \\ -\frac{1}{2}(-2x) &< -\frac{1}{2}(2) \\ x &< -1 \end{aligned}$$

Graphing the solution set:

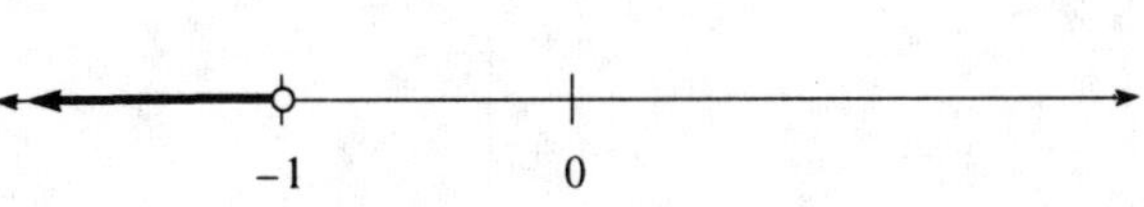

39. Solving for y:

$$\begin{aligned} 3x-2y &\le 12 \\ 3x-3x-2y &\le -3x+12 \\ -2y &\le -3x+12 \\ -\frac{1}{2}(-2y) &\ge -\frac{1}{2}(-3x+12) \\ y &\ge \frac{3}{2}x-6 \end{aligned}$$

41. Let w represent the width and $3w + 5$ represent the length. Using the perimeter formula:

$$\begin{aligned} 2(w)+2(3w+5) &= 26 \\ 2w+6w+10 &= 26 \\ 8w+10 &= 26 \\ 8w &= 16 \\ w &= 2 \\ 3w+5 = 3(2)+5 &= 11 \end{aligned}$$

The width is 2 inches and the length is 11 inches.

Chapter 3 Review

1. Substituting $x = 4$, $x = 0$, $y = 3$, and $y = 0$:

$$\begin{aligned} 3(4)+y &= 6 \\ 12+y &= 6 \\ y &= -6 \end{aligned} \qquad \begin{aligned} 3(0)+y &= 6 \\ 0+y &= 6 \\ y &= 6 \end{aligned} \qquad \begin{aligned} 3x+3 &= 6 \\ 3x &= 3 \\ x &= 1 \end{aligned} \qquad \begin{aligned} 3x+0 &= 6 \\ 3x &= 6 \\ x &= 2 \end{aligned}$$

The ordered pairs are $(4,-6),(0,6),(1,3)$, and $(2,0)$.

3. Substituting $x = 4$, $y = -2$, and $y = 3$:

$$\begin{aligned} y &= 2(4)-6 \\ y &= 8-6 \\ y &= 2 \end{aligned} \qquad \begin{aligned} -2 &= 2x-6 \\ 4 &= 2x \\ x &= 2 \end{aligned} \qquad \begin{aligned} 3 &= 2x-6 \\ 9 &= 2x \\ x &= \frac{9}{2} \end{aligned}$$

The ordered pairs are $(4,2),(2,-2)$, and $\left(\frac{9}{2},3\right)$.

5. Substituting $x = 2$, $x = -1$, and $x = -3$ results (in each case) in $y = -3$. The ordered pairs are $(2,-3),(-1,-3)$, and $(-3,-3)$.

7. Substituting each ordered pair into the equation:

$$\left(-2,\frac{9}{2}\right):\quad 3(-2)-4\left(\frac{9}{2}\right)=-6-18=-24\neq 12$$

$$(0,3):\quad 3(0)-4(3)=0-12=-12\neq 12$$

$$\left(2,-\frac{3}{2}\right):\quad 3(2)-4\left(-\frac{3}{2}\right)=6+6=12$$

Only the ordered pair $\left(2,-\frac{3}{2}\right)$ is a solution.

9. Graphing the ordered pair:

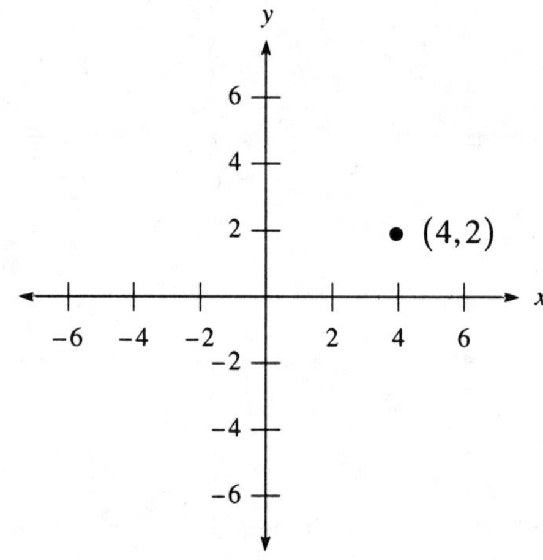

11. Graphing the ordered pair:

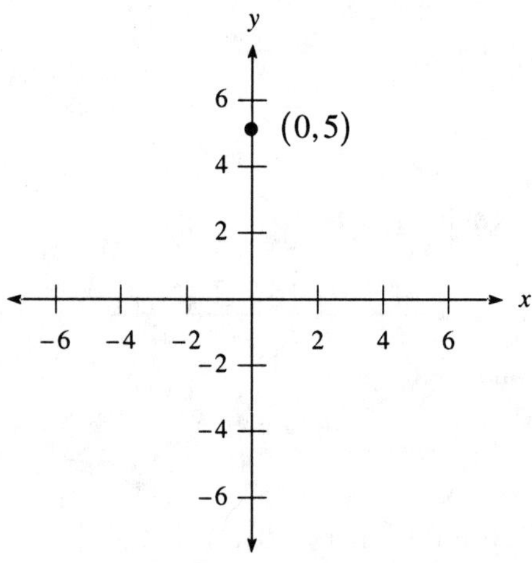

13. Graphing the ordered pair:

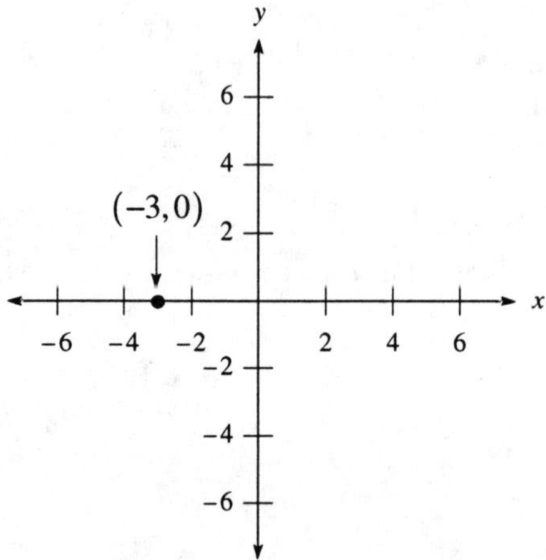

15. The ordered pairs are $(-2,0),(0,-2)$, and $(1,-3)$. Graphing the equation:

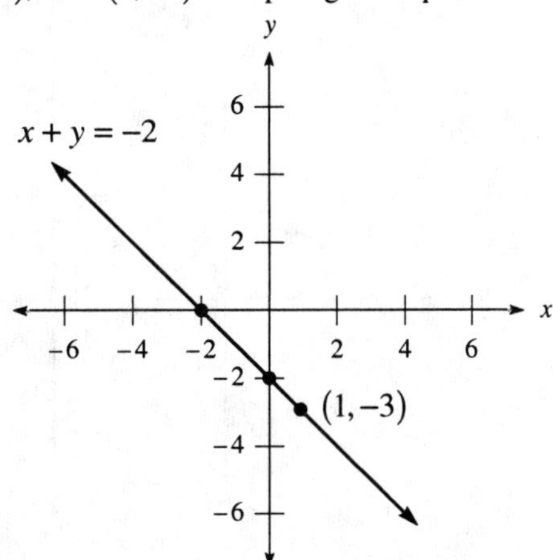

17. The ordered pairs are $(1,1),(0,-1)$, and $(-1,-3)$. Graphing the equation:

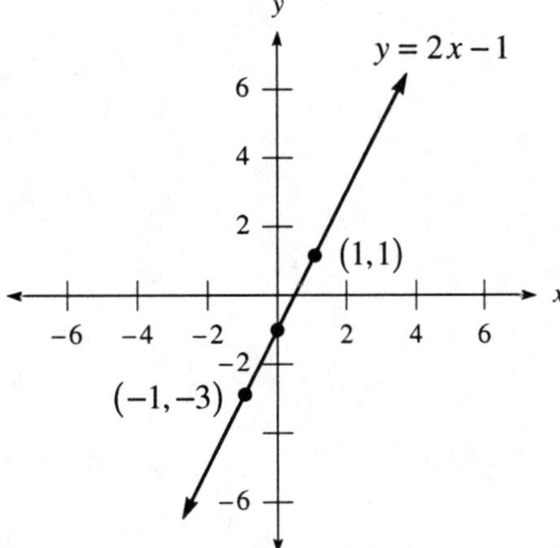

19. Graphing the equation:

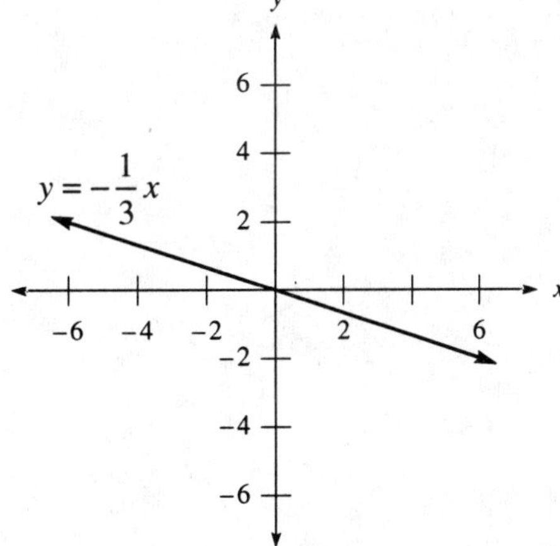

21. Graphing the equation:

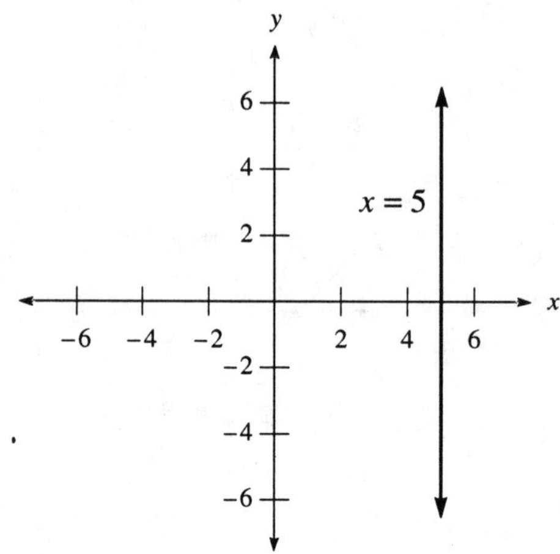

23. To find the x-intercept, let $y = 0$:

$$\begin{aligned} 3x - 0 &= 6 \\ 3x &= 6 \\ x &= 2 \end{aligned}$$

To find the y-intercept, let $x = 0$:

$$\begin{aligned} 3(0) - y &= 6 \\ -y &= 6 \\ y &= -6 \end{aligned}$$

25. To find the x-intercept, let $y = 0$:

$$\begin{aligned} 0 &= x - 3 \\ x &= 3 \end{aligned}$$

To find the y-intercept, let $x = 0$:

$$\begin{aligned} y &= 0 - 3 \\ y &= -3 \end{aligned}$$

27. The slope is given by: $m = \dfrac{5-3}{3-2} = \dfrac{2}{1} = 2$

29. The slope is given by: $m = \dfrac{-8-(-4)}{-3-(-1)} = \dfrac{-8+4}{-3+1} = \dfrac{-4}{-2} = 2$

31. Using the point-slope formula:

$$\begin{aligned} y - 4 &= -2(x - (-1)) \\ y - 4 &= -2(x + 1) \\ y - 4 &= -2x - 2 \\ y &= -2x + 2 \end{aligned}$$

33. Using the point-slope formula:

$$\begin{aligned} y - (-2) &= -\frac{3}{4}(x - 3) \\ y + 2 &= -\frac{3}{4}x + \frac{9}{4} \\ y &= -\frac{3}{4}x + \frac{1}{4} \end{aligned}$$

35. The slope-intercept form is $y = -x + 6$.

37. The equation is in slope-intercept form, so $m = 4$ and $b = -1$.

39. Solving for y:

$$\begin{aligned} 6x + 3y &= 9 \\ 3y &= -6x + 9 \\ y &= -2x + 3 \end{aligned}$$

The equation is in slope-intercept form, so $m = -2$ and $b = 3$.

41. Checking the point $(0,0)$:

$$0-0<3 \quad \text{(true)}$$

Graphing the linear inequality:

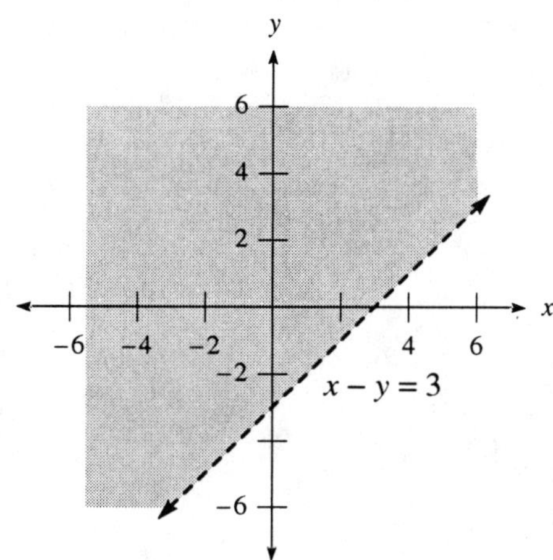

43. Checking the point $(0,0)$:

$$0 \le -2(0)+3$$
$$0 \le 3 \quad \text{(true)}$$

Graphing the linear inequality:

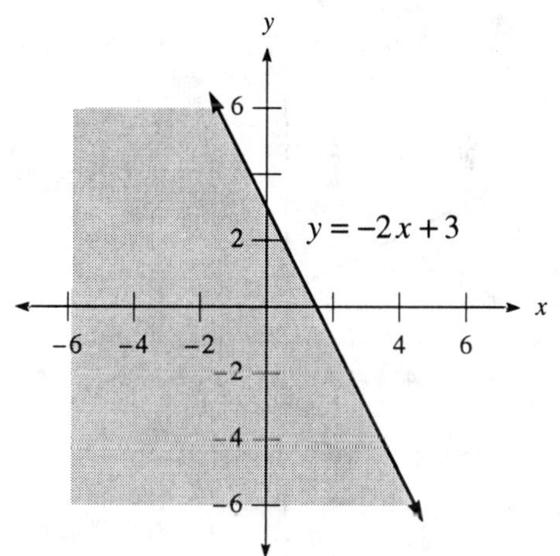

Cumulative Review: Chapters 1-3

1. Simplifying using order of operations: $7-2\cdot 6=7-12=-5$
3. Simplifying using order of operations: $4\cdot 6+12\div 4-3^2=24+3-9=18$
5. Simplifying using order of operations: $(4-9)(-3-8)=(-5)(-11)=55$
7. First factor the denominators to find the LCM:

$$60=2\cdot 2\cdot 3\cdot 5$$
$$84=2\cdot 2\cdot 3\cdot 7$$
$$\text{LCM}=2\cdot 2\cdot 3\cdot 5\cdot 7=420$$

Subtracting the fractions: $\frac{11}{60}-\frac{13}{84}=\frac{11\cdot 7}{60\cdot 7}-\frac{13\cdot 5}{84\cdot 5}=\frac{77}{420}-\frac{65}{420}=\frac{12}{420}=\frac{2\cdot 2\cdot 3}{2\cdot 2\cdot 3\cdot 5\cdot 7}=\frac{1}{5\cdot 7}=\frac{1}{35}$

9. Simplifying: $5a+3-4a-6=5a-4a+3-6=a-3$
11. Solving the equation:

$$4x-5=3$$
$$4x=8$$
$$x=2$$

13. Solving the equation:

$$\begin{aligned} 6(t+5)-4&=2 \\ 6t+30-4&=2 \\ 6t+26&=2 \\ 6t&=-24 \\ t&=-4 \end{aligned}$$

15. Solving the equation:

$$\begin{aligned} 0.05x+0.07(200-x)&=11 \\ 0.05x+14-0.07x&=11 \\ -0.02x+14&=11 \\ -0.02x&=-3 \\ x&=150 \end{aligned}$$

17. Solving the inequality:

$$\begin{aligned} 5-7x&\ge 19 \\ -5+5-7x&\ge -5+19 \\ -7x&\ge 14 \\ -\frac{1}{7}(-7x)&\le -\frac{1}{7}(14) \\ x&\le -2 \end{aligned}$$

Graphing the solution set:

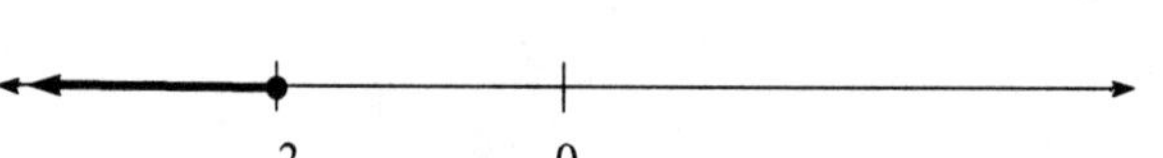

19. Graphing the equation:

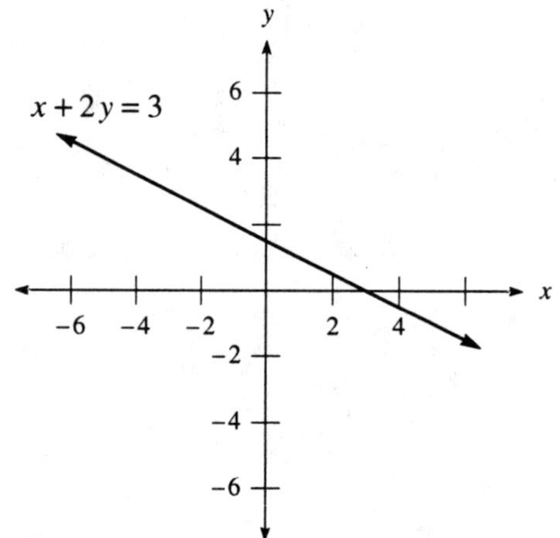

21. Checking the point $(0,0)$:

$$\begin{aligned} 0-2(0)&\le 4 \\ 0&\le 4 \qquad \text{(true)} \end{aligned}$$

Graphing the linear inequality:

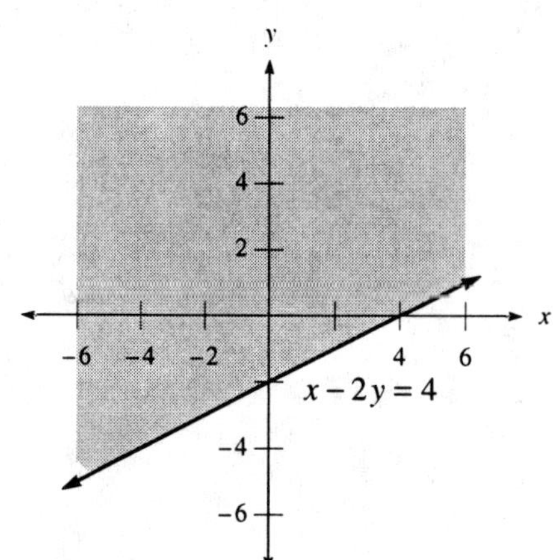

23. Graphing the line which passes through $(-3,-2)$ and $(2,3)$:

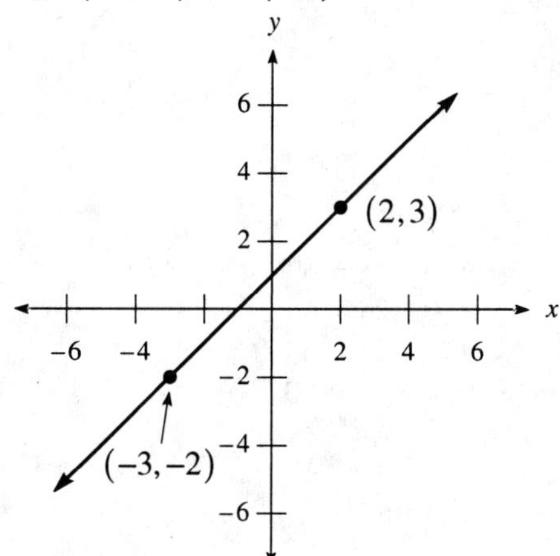

The point $(4,7)$ does not lie on the line, while $(1,2)$ does lie on the line.

25. To find the x-intercept, let $y = 0$:

$$\begin{aligned} 2x+5(0) &= 10 \\ 2x &= 10 \\ x &= 5 \end{aligned}$$

To find the y-intercept, let $x = 0$:

$$\begin{aligned} 2(0)+5y &= 10 \\ 5y &= 10 \\ y &= 2 \end{aligned}$$

Graphing the line:

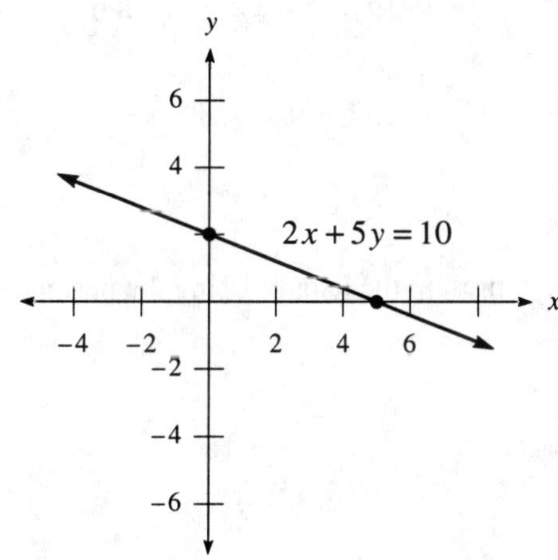

27. The slope is given by: $m = \dfrac{7-3}{5-2} = \dfrac{4}{3}$

29. Solving for y:

$$\begin{aligned} 2x+3y &= 6 \\ 3y &= -2x+6 \\ y &= -\frac{2}{3}x+2 \end{aligned}$$

The slope of the line is $-\dfrac{2}{3}$.

31. First find the slope: $m=\frac{7-3}{4-2}=\frac{4}{2}=2$

Using the point-slope formula:

$$\begin{aligned} y-3&=2(x-2)\\ y-3&=2x-4\\ y&=2x-1 \end{aligned}$$

33. Substituting $y=3$ and $x=0$:

$$\begin{aligned} 3x-2(3)&=6\\ 3x-6&=6\\ 3x&=12\\ x&=4 \end{aligned}$$

$$\begin{aligned} 3(0)-2y&=6\\ 0-2y&=6\\ -2y&=6\\ y&=-3 \end{aligned}$$

The completed table is:

x	y
4	3
0	–3

35. Translating into symbols: $x+7=4$

37. Each number in the sequence is the square of a fraction, since $\left(\frac{1}{2}\right)^2=\frac{1}{4}$, $\left(\frac{1}{3}\right)^2=\frac{1}{9}$, $\left(\frac{1}{4}\right)^2=\frac{1}{16}$, and $\left(\frac{1}{5}\right)^2=\frac{1}{25}$.

So the next number in the sequence is $\left(\frac{1}{6}\right)^2=\frac{1}{36}$.

39. Computing the expression: $6(-2)-5=-12-5=-17$

41. Reducing the fraction: $\frac{75}{135}=\frac{3\cdot5\cdot5}{3\cdot3\cdot3\cdot5}=\frac{5}{3\cdot3}=\frac{5}{9}$

43. Evaluating when $x=2$: $x^2+6x-7=(2)^2+6(2)-7=4+12-7=9$

45. Finding the percent:

$$\begin{aligned} p\cdot36&=27\\ p&=\frac{27}{36}\\ p&=0.75=75\% \end{aligned}$$

75% of 36 is 27.

47. Let w represent the width and $2w+5$ represent the length. Using the perimeter formula:

$$\begin{aligned} 2(w)+2(2w+5)&=44\\ 2w+4w+10&=44\\ 6w+10&=44\\ 6w&=34\\ w&=\frac{17}{3} \end{aligned}$$

$$2w+5=2\left(\frac{17}{3}\right)+5=\frac{34}{3}+5=\frac{49}{3}$$

The width is $\frac{17}{3}$ cm and the length is $\frac{49}{3}$ cm.

49. Completing the table:

	Dollars Invested at 8%	Dollars Invested at 6%
Number of	$x+900$	x
Interest on	$0.08(x+900)$	$0.06(x)$

The equation is:

$$\begin{aligned} 0.08(x+900)+0.06(x) &= 240 \\ 0.08x+72+0.06x &= 240 \\ 0.14x+72 &= 240 \\ 0.14x &= 168 \\ x &= 1200 \\ x+900 &= 2100 \end{aligned}$$

Barbara invested \$1,200 at 6% and \$2,100 at 8%.

Chapter 3 Test

1. Substituting $x=0$, $y=0$, $x=10$, and $y=-3$:

$$\begin{aligned} 2(0)-5y &= 10 \\ 0-5y &= 10 \\ -5y &= 10 \\ y &= -2 \end{aligned} \qquad \begin{aligned} 2x-5(0) &= 10 \\ 2x-0 &= 10 \\ 2x &= 10 \\ x &= 5 \end{aligned} \qquad \begin{aligned} 2(10)-5y &= 10 \\ 20-5y &= 10 \\ -5y &= -10 \\ y &= 2 \end{aligned} \qquad \begin{aligned} 2x-5(-3) &= 10 \\ 2x+15 &= 10 \\ 2x &= -5 \\ x &= -\frac{5}{2} \end{aligned}$$

The ordered pairs are $(0,-2),(5,0),(10,2)$, and $\left(-\frac{5}{2},-3\right)$.

2. Substituting each ordered pair into the equation:

$(2,5)$: $4(2)-3=8-3=5$

$(0,-3)$: $4(0)-3=0-3=-3$

$(3,0)$: $4(3)-3=12-3=9\neq 0$

$(-2,11)$: $4(-2)-3=-8-3=-11\neq 11$

The ordered pairs $(2,5)$ and $(0,-3)$ are solutions.

3. Graphing the line:

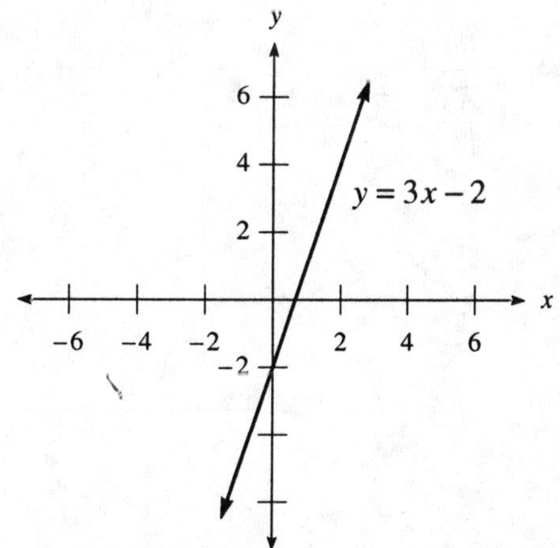

4. Graphing the line:

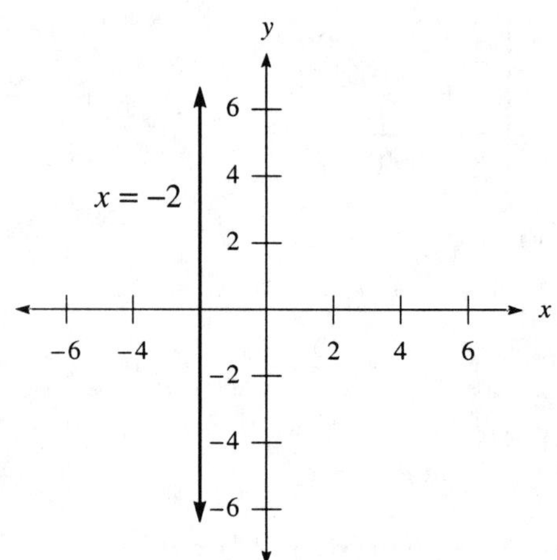

5. To find the x-intercept, let $y = 0$:

$$3x - 5(0) = 15$$
$$3x = 15$$
$$x = 5$$

To find the y-intercept, let $x = 0$:

$$3(0) - 5y = 15$$
$$-5y = 15$$
$$y = -3$$

6. To find the x-intercept, let $y = 0$:

$$0 = \frac{3}{2}x + 1$$
$$-1 = \frac{3}{2}x$$
$$x = -\frac{2}{3}$$

To find the y-intercept, let $x = 0$:

$$y = \frac{3}{2}(0) + 1 = 1$$

7. The slope is given by: $m = \frac{-7-(-3)}{4-2} = \frac{-4}{2} = -2$

8. The slope is given by: $m = \frac{-8-5}{2-(-3)} = -\frac{13}{5}$

9. The slope is given by: $m = \frac{d-b}{c-a}$

10. The slope is given by: $m = \frac{4-3}{2x-5x} = -\frac{1}{3x}$

11. Using the point-slope formula:

$$y - 5 = 3(x - (-2))$$
$$y - 5 = 3(x + 2)$$
$$y - 5 = 3x + 6$$
$$y = 3x + 11$$

12. The slope-intercept form is $y = 4x + 8$.

13. First find the slope: $m=\frac{4-1}{-2-3}=-\frac{3}{5}$

Using the point-slope formula:

$$y-1=-\frac{3}{5}(x-3)$$
$$y-1=-\frac{3}{5}x+\frac{9}{5}$$
$$y=-\frac{3}{5}x+\frac{14}{5}$$

14. First find the slope: $m=\frac{4-0}{3-1}=\frac{4}{2}=2$

Using the point-slope formula:

$$y-4=2(x-3)$$
$$y-4=2x-6$$
$$y=2x-2$$

15. Checking the point $(0,0)$:

$$0<0+4$$
$$0<4 \qquad \text{(true)}$$

Graphing the linear inequality:

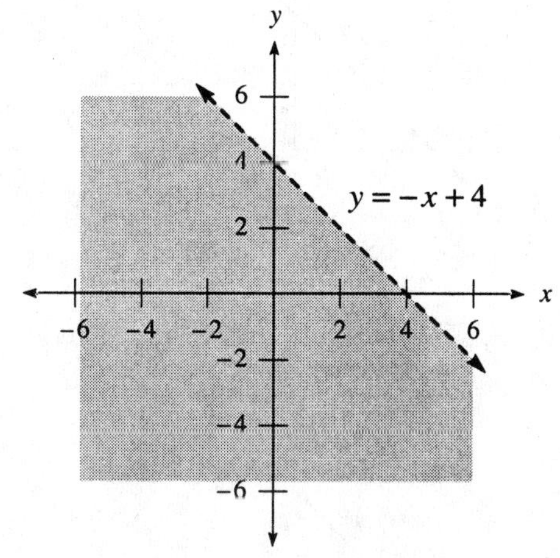

16. Checking the point $(0,0)$:

$$3(0)-4(0)\geq 12$$
$$0\geq 12 \qquad \text{(false)}$$

Graphing the linear inequality:

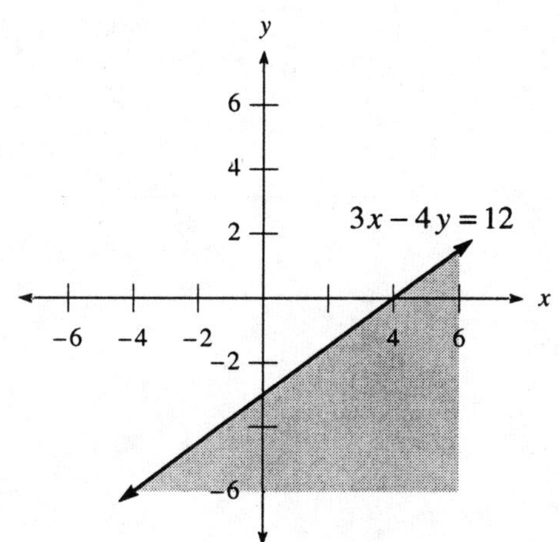

Chapter 4
Systems of Linear Equations

4.1 Solving Linear Systems by Graphing

1. Graphing both lines:

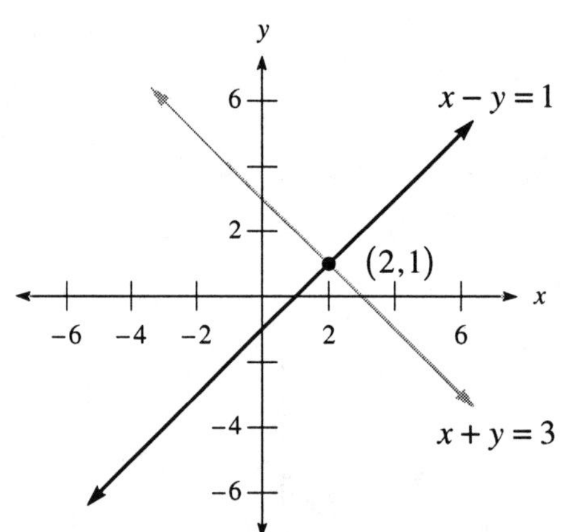

The intersection point is $(2,1)$.

3. Graphing both lines:

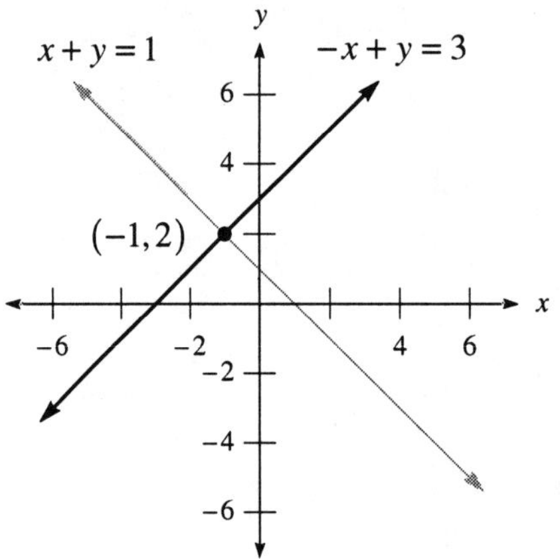

The intersection point is $(-1,2)$.

5. Graphing both lines:

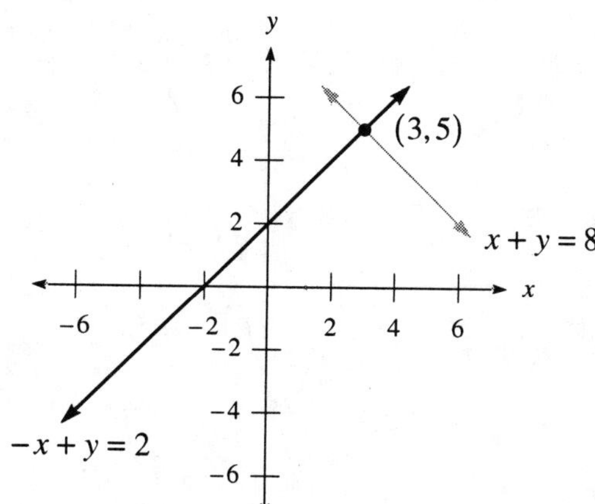

The intersection point is $(3,5)$.

7. Graphing both lines:

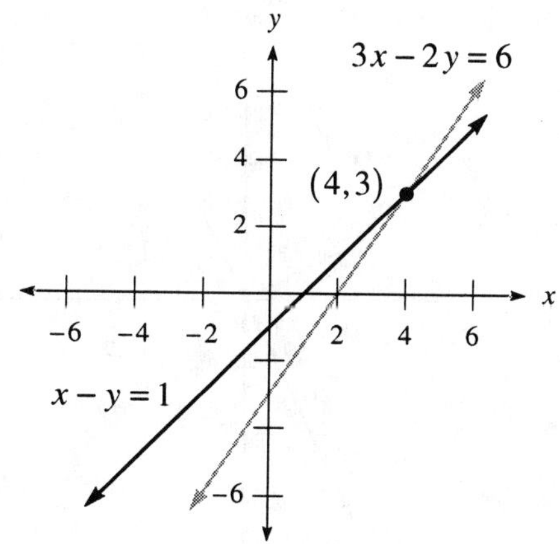

The intersection point is $(4,3)$.

9. Graphing both lines:

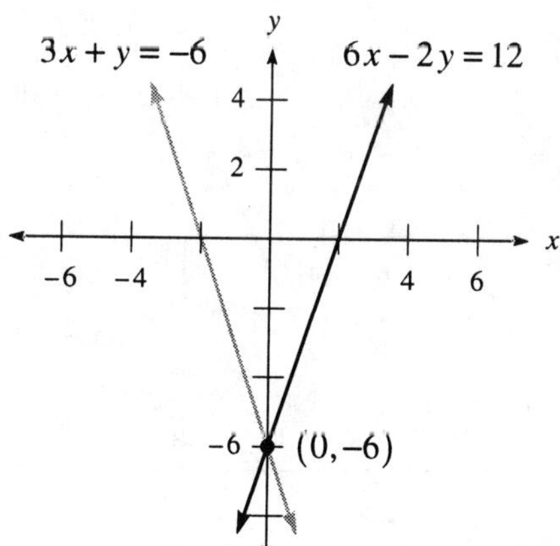

The intersection point is $(0,-6)$.

11. Graphing both lines:

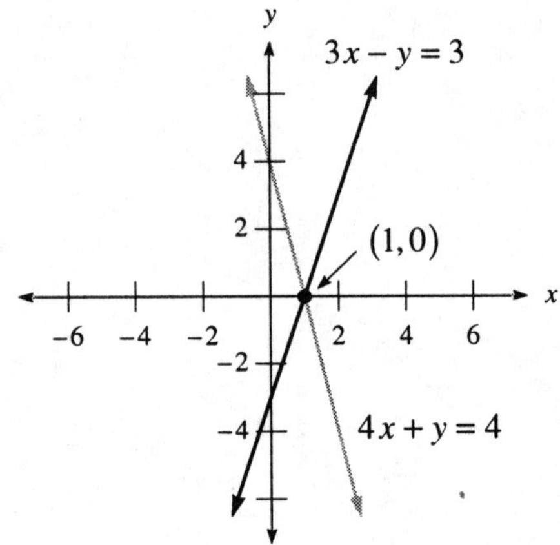

The intersection point is $(1,0)$.

13. Graphing both lines:

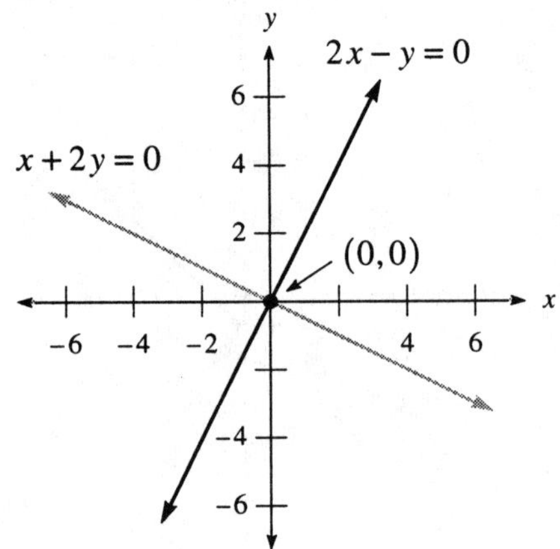

The intersection point is $(0,0)$.

15. Graphing both lines:

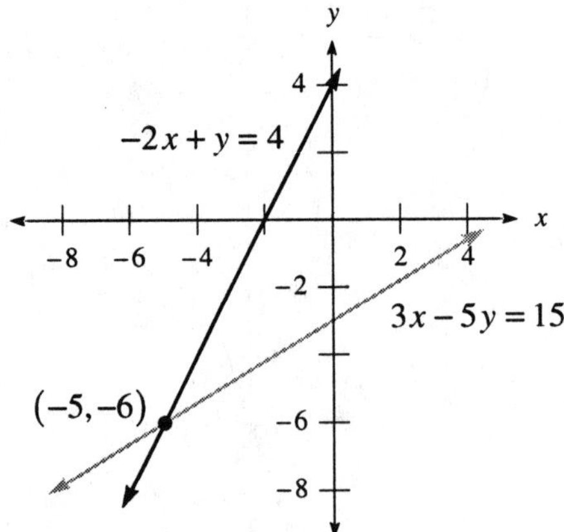

The intersection point is $(-5,-6)$.

17. Graphing both lines:

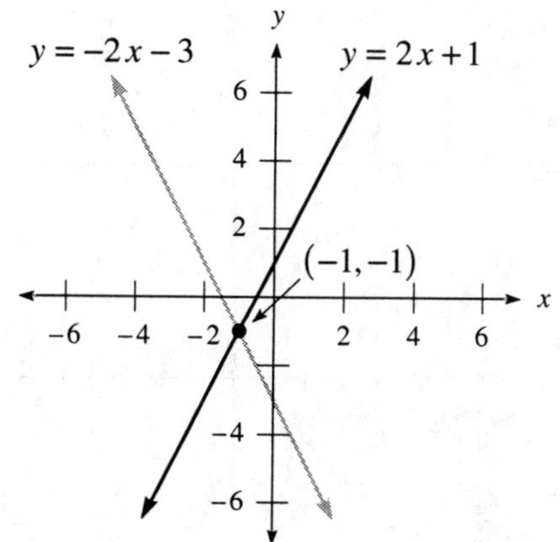

The intersection point is $(-1,-1)$.

19. Graphing both lines:

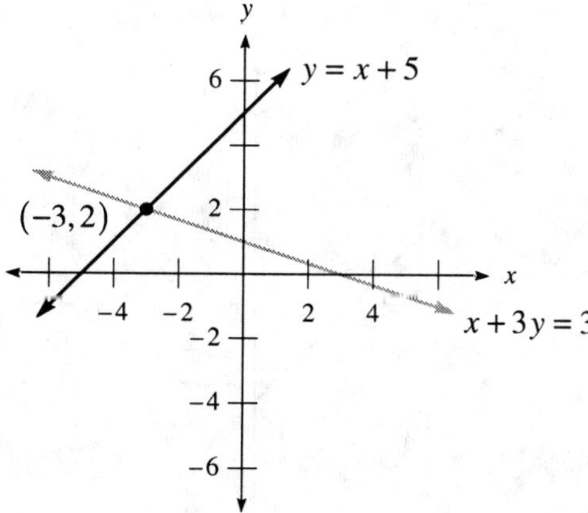

The intersection point is $(-3,2)$.

21. Graphing both lines:

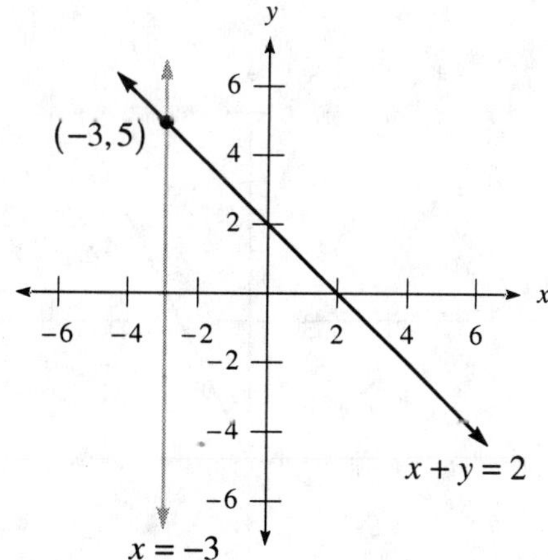

The intersection point is $(-3,5)$.

23. Graphing both lines:

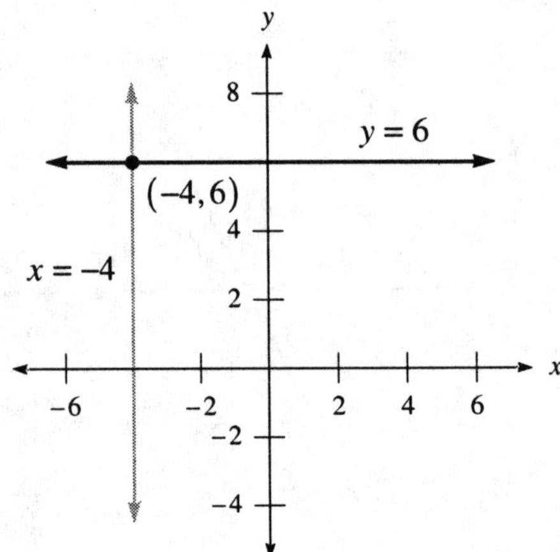

The intersection point is $(-4,6)$.

25. Graphing both lines:

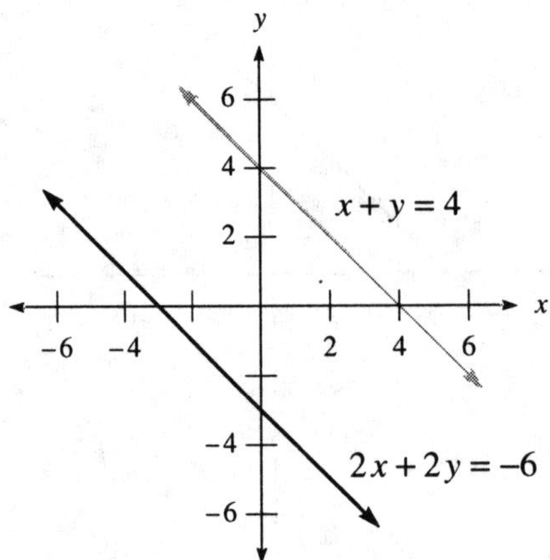

There is no intersection (the lines are parallel).

27. Graphing both lines:

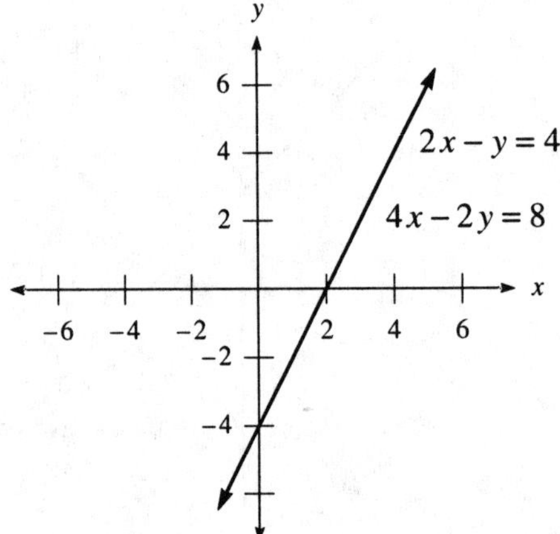

The system is dependent (both lines are the same, they coincide).

29. Graphing both lines:

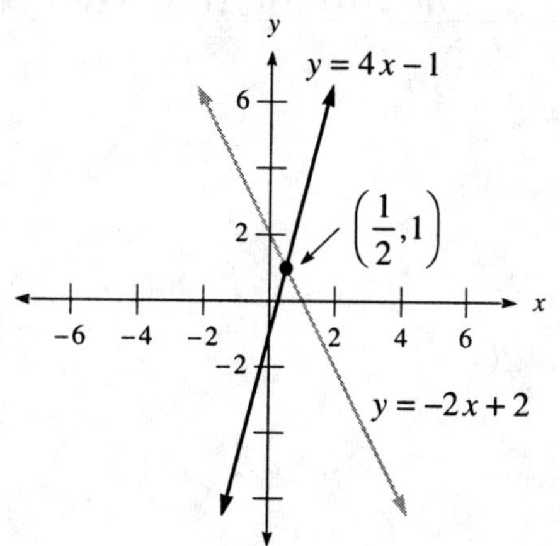

The intersection point is $\left(\frac{1}{2},1\right)$.

31. Graphing both lines:

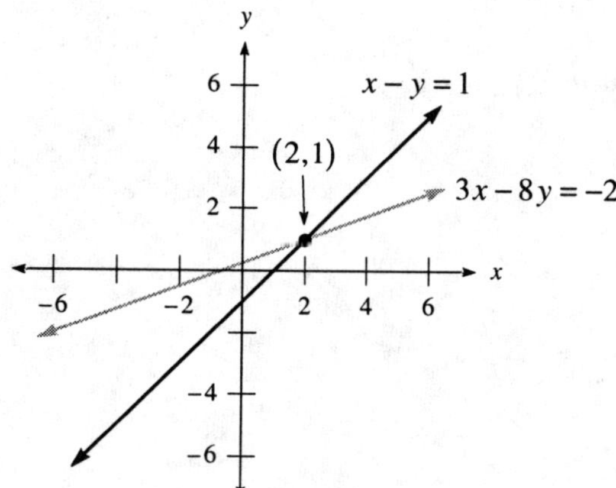

The intersection point is $(2,1)$.

33. Finding the slope: $m = \frac{1-(-5)}{3-(-3)} = \frac{1+5}{3+3} = \frac{6}{6} = 1$

35. Finding the slope: $m = \frac{0-3}{5-0} = -\frac{3}{5}$

37. Graphing the line:

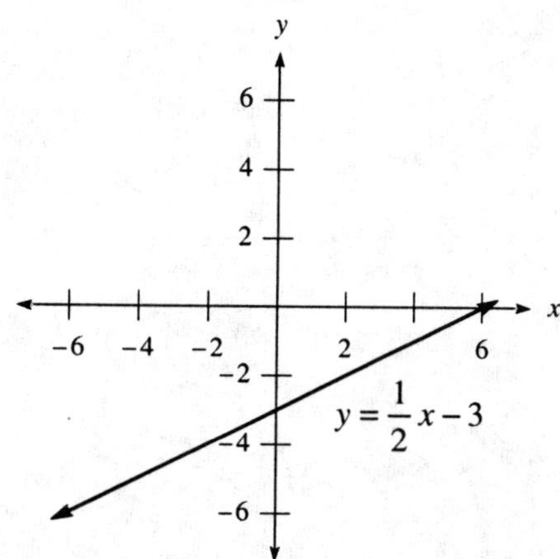

4.2 The Elimination Method

1. Adding the two equations:
$$2x = 4$$
$$x = 2$$
Substituting into the first equation:
$$2 + y = 3$$
$$y = 1$$
The solution is $(2,1)$.

3. Adding the two equations:
$$2y = 14$$
$$y = 7$$
Substituting into the first equation:
$$x + 7 = 10$$
$$x = 3$$
The solution is $(3,7)$.

5. Adding the two equations:
$$-2y = 10$$
$$y = -5$$
Substituting into the first equation:
$$x - (-5) = 7$$
$$x + 5 = 7$$
$$x = 2$$
The solution is $(2,-5)$.

7. Adding the two equations:
$$4x = -4$$
$$x = -1$$
Substituting into the first equation:
$$-1 + y = -1$$
$$y = 0$$
The solution is $(-1,0)$.

9. Adding the two equations:
$$0 = 0$$
The lines coincide (the system is dependent).

11. Multiplying the first equation by 2:
$$6x - 2y = 8$$
$$2x + 2y = 24$$
Adding the two equations:
$$8x = 32$$
$$x = 4$$
Substituting into the first equation:
$$3(4) - y = 4$$
$$12 - y = 4$$
$$-y = -8$$
$$y = 8$$
The solution is $(4,8)$.

13. Multiplying the second equation by –3:
$$5x - 3y = -2$$
$$-30x + 3y = -3$$
Adding the two equations:
$$-25x = -5$$
$$x = \frac{1}{5}$$

Substituting into the first equation:

$$5\left(\frac{1}{5}\right)-3y=-2$$
$$1-3y=-2$$
$$-3y=-3$$
$$y=1$$

The solution is $\left(\frac{1}{5},1\right)$.

15. Multiplying the second equation by 4:

$$11x-4y=11$$
$$20x+4y=20$$

Adding the two equations:

$$31x=31$$
$$x=1$$

Substituting into the second equation:

$$5(1)+y=5$$
$$5+y=5$$
$$y=0$$

The solution is $(1,0)$.

17. Multiplying the second equation by 3:

$$3x-5y=7$$
$$-3x+3y=-3$$

Adding the two equations:

$$-2y=4$$
$$y=-2$$

Substituting into the second equation:

$$-x-2=-1$$
$$-x=1$$
$$x=-1$$

The solution is $(-1,-2)$.

19. Multiplying the first equation by –2:

$$2x+16y=2$$
$$-2x+4y=13$$

Adding the two equations:

$$20y=15$$
$$y=\frac{3}{4}$$

Substituting into the first equation:

$$-x-8\left(\frac{3}{4}\right)=-1$$
$$-x-6=-1$$
$$-x=5$$
$$x=-5$$

The solution is $\left(-5,\frac{3}{4}\right)$.

21. Multiplying the first equation by 2:

$$-6x-2y=14$$
$$6x+7y=11$$

Adding the two equations:

$$5y=25$$
$$y=5$$

Substituting into the first equation:

$$\begin{aligned} -3x-5&=7 \\ -3x&=12 \\ x&=-4 \end{aligned}$$

The solution is $(-4,5)$.

23. Adding the two equations:

$$\begin{aligned} 8x&=-24 \\ x&=-3 \end{aligned}$$

Substituting into the second equation:

$$\begin{aligned} 2(-3)+y&=-16 \\ -6+y&=-16 \\ y&=-10 \end{aligned}$$

The solution is $(-3,-10)$.

25. Multiplying the second equation by 3:

$$\begin{aligned} x+3y&=9 \\ 6x-3y&=12 \end{aligned}$$

Adding the two equations:

$$\begin{aligned} 7x&=21 \\ x&=3 \end{aligned}$$

Substituting into the first equation:

$$\begin{aligned} 3+3y&=9 \\ 3y&=6 \\ y&=2 \end{aligned}$$

The solution is $(3,2)$.

27. Multiplying the second equation by 2:

$$\begin{aligned} x-6y&=3 \\ 8x+6y&=42 \end{aligned}$$

Adding the two equations:

$$\begin{aligned} 9x&=45 \\ x&=5 \end{aligned}$$

Substituting into the second equation:

$$\begin{aligned} 4(5)+3y&=21 \\ 20+3y&=21 \\ 3y&=1 \\ y&=\frac{1}{3} \end{aligned}$$

The solution is $\left(5,\frac{1}{3}\right)$.

29. Multiplying the second equation by –3:

$$\begin{aligned} 2x+9y&=2 \\ -15x-9y&=24 \end{aligned}$$

Adding the two equations:

$$\begin{aligned} -13x&=26 \\ x&=-2 \end{aligned}$$

Substituting into the first equation:

$$\begin{aligned} 2(-2)+9y&=2 \\ -4+9y&=2 \\ 9y&=6 \\ y&=\frac{2}{3} \end{aligned}$$

The solution is $\left(-2,\frac{2}{3}\right)$.

31. To clear each equation of fractions, multiply the first equation by 12 and the second equation by 6:

$$12\left(\frac{1}{3}x+\frac{1}{4}y\right)=12\left(\frac{7}{6}\right) \qquad 6\left(\frac{3}{2}x-\frac{1}{3}y\right)=6\left(\frac{7}{3}\right)$$
$$4x+3y=14 \qquad 9x-2y=14$$

The system of equations is:
$$4x+3y=14$$
$$9x-2y=14$$

Multiplying the first equation by 2 and the second equation by 3:
$$8x+6y=28$$
$$27x-6y=42$$

Adding the two equations:
$$35x=70$$
$$x=2$$

Substituting into $4x+3y=14$:
$$4(2)+3y=14$$
$$8+3y=14$$
$$3y=6$$
$$y=2$$

The solution is $(2,2)$.

33. Multiplying the first equation by –2:
$$-6x-4y=2$$
$$6x+4y=0$$

Adding the two equations:
$$0=2$$

Since this statement is false, the two lines are parallel, so the system has no solution.

35. Multiplying the first equation by 2 and the second equation by 3:
$$22x+12y=34$$
$$15x-12y=3$$

Adding the two equations:
$$37x=37$$
$$x=1$$

Substituting into the second equation:
$$5(1)-4y=1$$
$$5-4y=1$$
$$-4y=-4$$
$$y=1$$

The solution is $(1,1)$.

37. To clear each equation of fractions, multiply the first equation by 6 and the second equation by 6:

$$6\left(\frac{1}{2}x+\frac{1}{6}y\right)=6\left(\frac{1}{3}\right) \qquad 6\left(-x-\frac{1}{3}y\right)=6\left(-\frac{1}{6}\right)$$
$$3x+y=2 \qquad -6x-2y=-1$$

The system of equations is:
$$3x+y=2$$
$$-6x-2y=-1$$

Multiplying the first equation by 2:
$$6x+2y=4$$
$$-6x-2y=-1$$

Adding the two equations:
$$0=3$$

Since this statement is false, the two lines are parallel, so the system has no solution.

39. Adding $2x$ to each side of the first equation and $-3x$ to each side of the second equation:

$$6x-5y=17$$
$$-3x+5y=4$$

Adding the two equations:

$$3x=21$$
$$x=7$$

Substituting into $5y=3x+4$:

$$5y=3(7)+4$$
$$5y=21+4$$
$$5y=25$$
$$y=5$$

The solution is $(7,5)$.

41. Multiplying the second equation by 100 (to eliminate decimals):

$$x+y=22$$
$$5x+10y=170$$

Multiplying the first equation by –5:

$$-5x-5y=-110$$
$$5x+10y=170$$

Adding the two equations:

$$5y=60$$
$$y=12$$

Substituting into the first equation:

$$x+12=22$$
$$x=10$$

The solution is $(10,12)$.

43. Solving for y:

$$-2x+4y=8$$
$$4y=2x+8$$
$$y=\frac{1}{2}x+2$$

So $m=\frac{1}{2}$ and $b=2$.

45. Using the point-slope formula:

$$y-(-6)=3(x-(-2))$$
$$y+6=3(x+2)$$
$$y+6=3x+6$$
$$y=3x$$

47. First find the slope: $m=\frac{1-(-5)}{3-(-3)}=\frac{1+5}{3+3}=\frac{6}{6}=1$

Using the point-slope formula:

$$y-1=1(x-3)$$
$$y-1=x-3$$
$$y=x-2$$

4.3 The Substitution Method

1. Substituting into the first equation:

$$x+(2x-1)=11$$
$$3x-1=11$$
$$3x=12$$
$$x=4$$
$$y=2(4)-1=7$$

The solution is $(4,7)$.

3. Substituting into the first equation:

$$\begin{aligned} x+(5x+2)&=20\\ 6x+2&=20\\ 6x&=18\\ x&=3\\ y&=5(3)+2=17 \end{aligned}$$

The solution is $(3,17)$.

5. Substituting into the first equation:

$$\begin{aligned} -2x+(-4x+8)&=-1\\ -6x+8&=-1\\ -6x&=-9\\ x&=\frac{3}{2}\\ y&=-4\left(\frac{3}{2}\right)+8=-6+8=2 \end{aligned}$$

The solution is $\left(\frac{3}{2},2\right)$.

7. Substituting into the first equation:

$$\begin{aligned} 3(-y+6)-2y&=-2\\ -3y+18-2y&=-2\\ -5y+18&=-2\\ -5y&=-20\\ y&=4\\ x&=-4+6=2 \end{aligned}$$

The solution is $(2,4)$.

9. Substituting into the first equation:

$$\begin{aligned} 5x-4(4)&=-16\\ 5x-16&=-16\\ 5x&=0\\ x&=0 \end{aligned}$$

The solution is $(0,4)$.

11. Substituting into the first equation:

$$\begin{aligned} 5x+4(-3x)&=7\\ 5x-12x&=7\\ -7x&=7\\ x&=-1\\ y&=-3(-1)=3 \end{aligned}$$

The solution is $(-1,3)$.

13. Solving the second equation for x:

$$\begin{aligned} x-2y&=-1\\ x&=2y-1 \end{aligned}$$

Substituting into the first equation:

$$\begin{aligned} (2y-1)+3y&=4\\ 5y-1&=4\\ 5y&=5\\ y&=1\\ x&=2(1)-1=1 \end{aligned}$$

The solution is $(1,1)$.

15. Solving the second equation for x:

$$\begin{aligned} x-5y&=17 \\ x&=5y+17 \end{aligned}$$

Substituting into the first equation:

$$\begin{aligned} 2(5y+17)+y&=1 \\ 10y+34+y&=1 \\ 11y+34&=1 \\ 11y&=-33 \\ y&=-3 \\ x&=5(-3)+17=2 \end{aligned}$$

The solution is $(2,-3)$.

17. Solving the second equation for x:

$$\begin{aligned} x-5y&=-5 \\ x&=5y-5 \end{aligned}$$

Substituting into the first equation:

$$\begin{aligned} 3(5y-5)+5y&=-3 \\ 15y-15+5y&=-3 \\ 20y-15&=-3 \\ 20y&=12 \\ y&=\frac{3}{5} \\ x&=5\left(\frac{3}{5}\right)-5=3-5=-2 \end{aligned}$$

The solution is $\left(-2,\frac{3}{5}\right)$.

19. Solving the second equation for x:

$$\begin{aligned} x-3y&=-18 \\ x&=3y-18 \end{aligned}$$

Substituting into the first equation:

$$\begin{aligned} 5(3y-18)+3y&=0 \\ 15y-90+3y&=0 \\ 18y-90&=0 \\ 18y&=90 \\ y&=5 \\ x&=3(5)-18=-3 \end{aligned}$$

The solution is $(-3,5)$.

21. Solving the second equation for x:

$$\begin{aligned} x+3y&=12 \\ x&=-3y+12 \end{aligned}$$

Substituting into the first equation:

$$\begin{aligned} -3(-3y+12)-9y&=7 \\ 9y-36-9y&=7 \\ -36&=7 \end{aligned}$$

Since this statement is false, there is no solution to the system. The two lines are parallel.

23. Substituting into the first equation:

$$\begin{aligned} 5x-8(2x-5)&-7 \\ 5x-16x+40&=7 \\ -11x+40&=7 \\ -11x&=-33 \\ x&=3 \\ y&=2(3)-5=1 \end{aligned}$$

The solution is $(3,1)$.

25. Substituting into the first equation:

$$\begin{aligned} 7(2y-1)-6y&=-1 \\ 14y-7-6y&=-1 \\ 8y-7&=-1 \\ 8y&=6 \\ y&=\frac{3}{4} \\ x&=2\left(\frac{3}{4}\right)-1=\frac{3}{2}-1=\frac{1}{2} \end{aligned}$$

The solution is $\left(\frac{1}{2},\frac{3}{4}\right)$.

27. Substituting into the first equation:

$$\begin{aligned} -3x+2(3x)&=6 \\ -3x+6x&=6 \\ 3x&=6 \\ x&=2 \\ y&=3(2)=6 \end{aligned}$$

The solution is $(2,6)$.

29. Substituting into the first equation:

$$\begin{aligned} 5(y)-6y&=-4 \\ -y&=-4 \\ y&=4 \\ x&=4 \end{aligned}$$

The solution is $(4,4)$.

31. Substituting into the first equation:

$$\begin{aligned} 3x+3(2x-12)&=9 \\ 3x+6x-36&=9 \\ 9x-36&=9 \\ 9x&=45 \\ x&=5 \\ y&=2(5)-12=-2 \end{aligned}$$

The solution is $(5,-2)$.

33. Substituting into the first equation:

$$\begin{aligned} 7x-11(10)&=16 \\ 7x-110&=16 \\ 7x&=126 \\ x&=18 \\ y&=10 \end{aligned}$$

The solution is $(18,10)$.

35. Substituting into the first equation:

$$\begin{aligned} -4x+4(x-2)&=-8 \\ -4x+4x-8&=-8 \\ -8&=-8 \end{aligned}$$

Since this statement is true, the system is dependent. The two lines coincide.

37. Substituting into the first equation:

$$\begin{aligned} 0.05x+0.10(22-x)&=1.70 \\ 0.05x+2.2-0.10x&=1.70 \\ -0.05x+2.2&=1.7 \\ -0.05x&=-0.5 \\ x&=10 \\ y&=22-10=12 \end{aligned}$$

The solution is $(10,12)$.

39. Let x represent the number of gallons consumed. For the charges for each company to be the same, the equation is:

$$7.00+1.10x=5.00+1.15x$$
$$7.00-0.05x=5.00$$
$$-0.05x=-2$$
$$x=40$$

If 40 gallons are used in a month, the two companies charge the same amount.

41. Let w represent the width, and $3w$ represent the length. Using the perimeter formula:

$$2(w)+2(3w)=24$$
$$2w+6w=24$$
$$8w=24$$
$$w=3$$
$$3w=9$$

The length is 9 meters and the width is 3 meters.

43. Completing the table:

	Nickels	Dimes
Number	x	$x+3$
Value (cents)	$5(x)$	$10(x+3)$

The equation is:

$$5(x)+10(x+3)=210$$
$$5x+10x+30=210$$
$$15x+30=210$$
$$15x=180$$
$$x=12$$
$$x+3=15$$

The collection consists of 12 nickels and 15 dimes.

45. Finding the amount:

$$0.08(6000)=x$$
$$x=480$$

8% of 6,000 is 480.

47. Completing the table:

	Dollars Invested at 8%	Dollars Invested at 10%
Number of	x	$2x$
Interest on	$0.08(x)$	$0.10(2x)$

The equation is:

$$0.08(x)+0.10(2x)=224$$
$$0.08x+0.20x=224$$
$$0.28x=224$$
$$x=800$$
$$2x=1600$$

The man invested $800 at 8% interest and $1,600 at 10% interest.

4.4 Applications

1. Let x and y represent the two numbers. The system of equations is:

$$x+y=25$$
$$y=5+x$$

Substituting into the first equation:

$$x+(5+x)=25$$
$$2x+5=25$$
$$2x=20$$
$$x=10$$
$$y=5+10=15$$

The two numbers are 10 and 15.

3. Let x and y represent the two numbers. The system of equations is:

$$x+y=15$$
$$y=4x$$

Substituting into the first equation:

$$x+4x=15$$
$$5x=15$$
$$x=3$$
$$y=4(3)=12$$

The two numbers are 3 and 12.

5. Let x represent the larger number and y represent the smaller number. The system of equations is:

$$x-y=5$$
$$x=2y+1$$

Substituting into the first equation:

$$2y+1-y=5$$
$$y+1=5$$
$$y=4$$
$$x=2(4)+1=9$$

The two numbers are 4 and 9.

7. Let x and y represent the two numbers. The system of equations is:

$$y=4x+5$$
$$x+y=35$$

Substituting into the second equation:

$$x+4x+5=35$$
$$5x+5=35$$
$$5x=30$$
$$x=6$$
$$y=4(6)+5=29$$

The two numbers are 6 and 29.

9. Let x represent the amount invested at 6% and y represent the amount invested at 8%. The system of equations is:

$$x+y=20000$$
$$0.06x+0.08y=1380$$

Multiplying the first equation by –0.06:

$$-0.06x-0.06y=-1200$$
$$0.06x+0.08y=1380$$

Adding the two equations:

$$0.02y=180$$
$$y=9000$$

Substituting into the first equation:

$$x+9000=20000$$
$$x=11000$$

Mr. Wilson invested $9,000 at 8% and $11,000 at 6%.

11. Let x represent the amount invested at 5% and y represent the amount invested at 6%. The system of equations is:

$$x=4y$$
$$0.05x+0.06y=520$$

Substituting into the second equation:

$$0.05(4y)+0.06y=520$$
$$0.20y+0.06y=520$$
$$0.26y=520$$
$$y=2000$$
$$x=4(2000)=8000$$

She invested $8,000 at 5% and $2,000 at 6%.

13. Let x represent the number of nickels and y represent the number of quarters. The system of equations is:
$$x+y=14$$
$$0.05x+0.25y=2.30$$
Multiplying the first equation by –0.05:
$$-0.05x-0.05y=-0.7$$
$$0.05x+0.25y=2.30$$
Adding the two equations:
$$0.20y=1.6$$
$$y=8$$
Substituting into the first equation:
$$x+8=14$$
$$x=6$$
Ron has 6 nickels and 8 quarters.

15. Let x represent the number of dimes and y represent the number of quarters. The system of equations is:
$$x+y=21$$
$$0.10x+0.25y=3.45$$
Multiplying the first equation by –0.10:
$$-0.10x-0.10y=-2.10$$
$$0.10x+0.25y=3.45$$
Adding the two equations:
$$0.15y=1.35$$
$$y=9$$
Substituting into the first equation:
$$x+9=21$$
$$x=12$$
Tom has 12 dimes and 9 quarters.

17. Let x represent the liters of 50% alcohol solution and y represent the liters of 20% alcohol solution. The system of equations is:
$$x+y=18$$
$$0.50x+0.20y=0.30(18)$$
Multiplying the first equation by –0.20:
$$-0.20x-0.20y=-3.6$$
$$0.50x+0.20y=5.4$$
Adding the two equations:
$$0.30x=1.8$$
$$x=6$$
Substituting into the first equation:
$$6+y=18$$
$$y=12$$
The mixture contains 6 liters of 50% alcohol solution and 12 liters of 20% alcohol solution.

19. Let x represent the gallons of 10% disinfectant and y represent the gallons of 7% disinfectant. The system of equations is:
$$x+y=30$$
$$0.10x+0.07y=0.08(30)$$
Multiplying the first equation by –0.07:
$$-0.07x-0.07y=-2.1$$
$$0.10x+0.07y=2.4$$
Adding the two equations:
$$0.03x=0.3$$
$$x=10$$
Substituting into the first equation:
$$10+y=30$$
$$y=20$$
The mixture contains 10 gallons of 10% disinfectant and 20 gallons of 7% disinfectant.

21 . Let x represent the number of adult tickets and y represent the number of kids tickets. The system of equations is:

$$x+y=70$$
$$5.50x+4.00y=310$$

Multiplying the first equation by –4:

$$-4.00x-4.00y=-280$$
$$5.50x+4.00y=310$$

Adding the two equations:

$$1.5x=30$$
$$x=20$$

Substituting into the first equation:

$$20+y=70$$
$$y=50$$

The matinee had 20 adult tickets sold and 50 kids tickets sold.

23 . Let x represent the width and y represent the length. The system of equations is:

$$2x+2y=96$$
$$y=2x$$

Substituting into the first equation:

$$2x+2(2x)=96$$
$$2x+4x=96$$
$$6x=96$$
$$x=16$$
$$y=2(16)=32$$

The width is 16 feet and the length is 32 feet.

25 . Let x represent the number of \$5 chips and y represent the number of \$25 chips. The system of equations is:

$$x+y=45$$
$$5x+25y=465$$

Multiplying the first equation by –5:

$$-5x-5y=-225$$
$$5x+25y=465$$

Adding the two equations:

$$20y=240$$
$$y=12$$

Substituting into the first equation:

$$x+12=45$$
$$x=33$$

The gambler has 33 \$5 chips and 12 \$25 chips.

27 . Let x represent the number of shares of \$11 stock and y represent the number of shares of \$20 stock. The system of equations is:

$$x+y=150$$
$$11x+20y=2550$$

Multiplying the first equation by –11:

$$-11x-11y=-1650$$
$$11x+20y=2550$$

Adding the two equations:

$$9y=900$$
$$y=100$$

Substituting into the first equation:

$$x+100=150$$
$$x=50$$

She bought 50 shares at \$11 and 100 shares at \$20.

29. Substituting $x = -2$, $x = 0$, and $x = 2$:

$$y = \frac{1}{2}(-2) + 3 = -1 + 3 = 2$$
$$y = \frac{1}{2}(0) + 3 = 0 + 3 = 3$$
$$y = \frac{1}{2}(2) + 3 = 1 + 3 = 4$$

The ordered pairs are $(-2,2)$, $(0,3)$, and $(2,4)$.

31. Graphing the line:

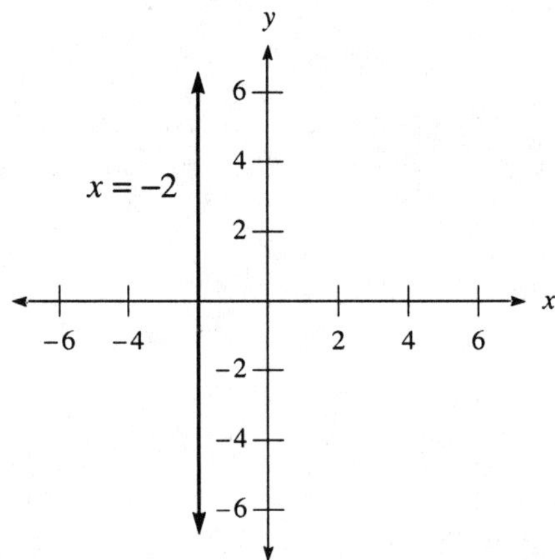

33. Computing the slope: $m = \frac{1-5}{0-2} = \frac{-4}{-2} = 2$

35. Using the point-slope formula:

$$y - 1 = \frac{1}{2}(x - (-2))$$
$$y - 1 = \frac{1}{2}(x + 2)$$
$$y - 1 = \frac{1}{2}x + 1$$
$$y = \frac{1}{2}x + 2$$

37. Computing the slope: $m = \frac{1-5}{0-2} = \frac{-4}{-2} = 2$

Using the point-slope formula:

$$y - 1 = 2(x - 0)$$
$$y - 1 = 2x$$
$$y = 2x + 1$$

Chapter 4 Review

1. Graphing both lines:

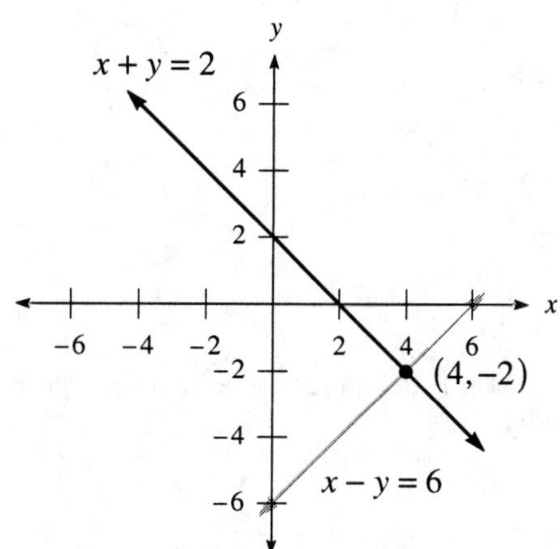

The intersection point is $(4,-2)$.

3. Graphing both lines:

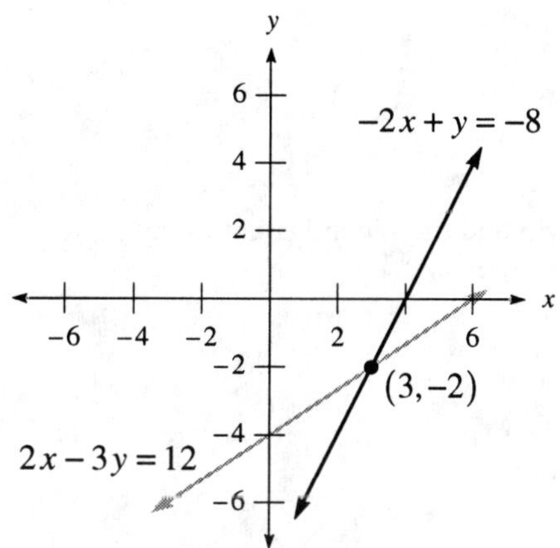

The intersection point is $(3,-2)$.

5. Graphing both lines:

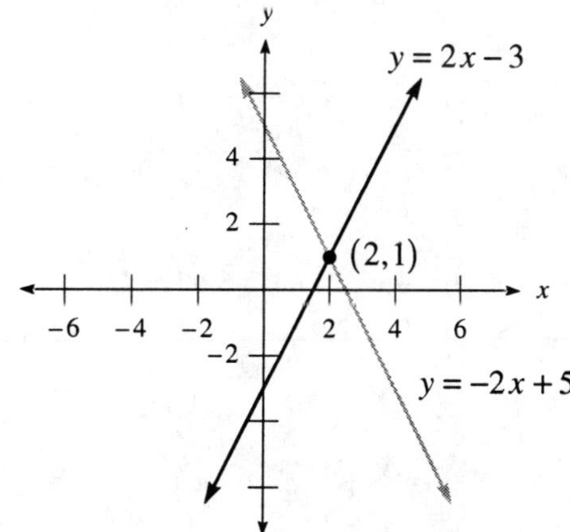

The intersection point is $(2,1)$.

7. Adding the two equations:

$$2x = 2$$
$$x = 1$$

Substituting into the second equation:

$$1 + y = -2$$
$$y = -3$$

The solution is $(1,-3)$.

9. Multiplying the first equation by 2:

$$10x - 6y = 4$$
$$-10x + 6y = -4$$

Adding the two equations:

$$0 = 0$$

Since this statement is true, the system is dependent. The two lines coincide.

11. Multiplying the second equation by –4:

$$-3x + 4y = 1$$
$$16x - 4y = 12$$

Adding the two equations:

$$13x = 13$$
$$x = 1$$

Substituting into the second equation:

$$-4(1) + y = -3$$
$$-4 + y = -3$$
$$y = 1$$

The solution is $(1,1)$.

13. Multiplying the first equation by 3 and the second equation by 5:

$$-6x + 15y = -33$$
$$35x - 15y = -25$$

Adding the two equations:

$$29x = -58$$
$$x = -2$$

Substituting into the first equation:

$$-2(-2) + 5y = -11$$
$$4 + 5y = -11$$
$$5y = -15$$
$$y = -3$$

The solution is $(-2,-3)$.

15. Substituting into the first equation:

$$x + (-3x + 1) = 5$$
$$-2x + 1 = 5$$
$$-2x = 4$$
$$x = -2$$

Substituting into the second equation: $y = -3(-2) + 1 = 6 + 1 = 7$

The solution is $(-2,7)$.

17. Substituting into the first equation:

$$4x - 3(3x + 7) = -16$$
$$4x - 9x - 21 = -16$$
$$-5x - 21 = -16$$
$$-5x = 5$$
$$x = -1$$

Substituting into the second equation: $y = 3(-1) + 7 = -3 + 7 = 4$

The solution is $(-1,4)$.

19. Solving the first equation for x:

$$x-4y=2$$
$$x=4y+2$$

Substituting into the second equation:

$$-3(4y+2)+12y=-8$$
$$-12y-6+12y=-8$$
$$-6=-8$$

Since this statement is false, there is no solution to the system. The two lines are parallel.

21. Solving the second equation for x:

$$x+6y=-11$$
$$x=-6y-11$$

Substituting into the first equation:

$$10(-6y-11)-5y=20$$
$$-60y-110-5y=20$$
$$-65y-110=20$$
$$-65y=130$$
$$y=-2$$

Substituting into $x=-6y-11$:

$$x=-6(-2)-11=12-11=1$$

The solution is $(1,-2)$.

23. Let x represent the smaller number and y represent the larger number. The system of equations is:

$$x+y=18$$
$$2x=6+y$$

Solving the first equation for y:

$$x+y=18$$
$$y=-x+18$$

Substituting into the second equation:

$$2x=6+(-x+18)$$
$$2x=-x+24$$
$$3x=24$$
$$x=8$$

Substituting into the first equation:

$$8+y=18$$
$$y=10$$

The two numbers are 8 and 10.

25. Let x represent the amount invested at 4% and y represent the amount invested at 5%. The system of equations is:

$$x+y=12000$$
$$0.04x+0.05y=560$$

Multiplying the first equation by –0.04:

$$-0.04x-0.04y=-480$$
$$0.04x+0.05y=560$$

Adding the two equations:

$$0.01y=80$$
$$y=8000$$

Substituting into the first equation:

$$x+8000=12000$$
$$x=4000$$

So \$4,000 was invested at 4% and \$8,000 was invested at 5%.

27. Let x represent the number of dimes and y represent the number of nickels. The system of equations is:

$$x+y=17$$
$$0.10x+0.05y=1.35$$

Multiplying the first equation by –0.05:

$$-0.05x-0.05y=-0.85$$
$$0.10x+0.05y=1.35$$

Adding the two equations:

$$0.05x=0.50$$
$$x=10$$

Substituting into the first equation:

$$10+y=17$$
$$y=7$$

Barbara has 10 dimes and 7 nickels.

29. Let x represent the liters of 20% alcohol solution and y represent the liters of 10% alcohol solution. The system of equations is:

$$x+y=50$$
$$0.20x+0.10y=0.12(50)$$

Multiplying the first equation by –0.10:

$$-0.10x-0.10y=-5$$
$$0.20x+0.10y=6$$

Adding the two equations:

$$0.10x=1$$
$$x=10$$

Substituting into the first equation:

$$10+y=50$$
$$y=40$$

The solution contains 40 liters of 10% alcohol solution and 10 liters of 20% alcohol solution.

Cumulative Review: Chapters 1-4

1. Simplifying using order of operations: $3\bullet 4+5=12+5=17$
3. Simplifying using order of operations: $7[8+(-5)]+3(-7+12)=7(3)+3(5)=21+15=36$
5. Simplifying using order of operations: $8-6(5-9)=8-6(-4)=8+24=32$
7. Simplifying: $\frac{2}{3}+\frac{3}{4}-\frac{1}{6}=\frac{2\bullet 4}{3\bullet 4}+\frac{3\bullet 3}{4\bullet 3}-\frac{1\bullet 2}{6\bullet 2}=\frac{8}{12}+\frac{9}{12}-\frac{2}{12}=\frac{15}{12}=\frac{5}{4}$
9. Solving the equation:

$$-5-6=-y-3+2y$$
$$-11=y-3$$
$$-11+3=y-3+3$$
$$y=-8$$

11. Solving the equation:

$$3(x-4)=9$$
$$3x-12=9$$
$$3x-12+12=9+12$$
$$3x=21$$
$$\frac{1}{3}(3x)=\frac{1}{3}(21)$$
$$x=7$$

13. Solving the inequality:

$$\begin{aligned} 0.3x + 0.7 &\le -2 \\ 0.3x + 0.7 - 0.7 &\le -2 - 0.7 \\ 0.3x &\le -2.7 \\ \frac{0.3x}{0.3} &\le \frac{-2.7}{0.3} \\ x &\le -9 \end{aligned}$$

Graphing the solution set:

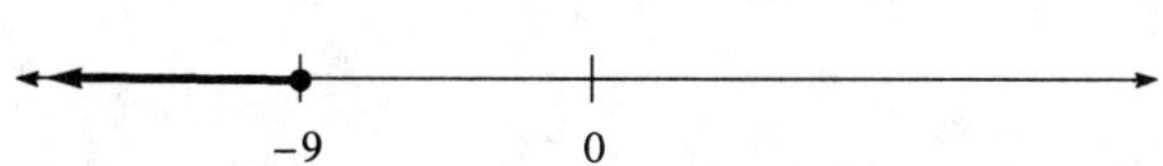

15. Graphing the line:

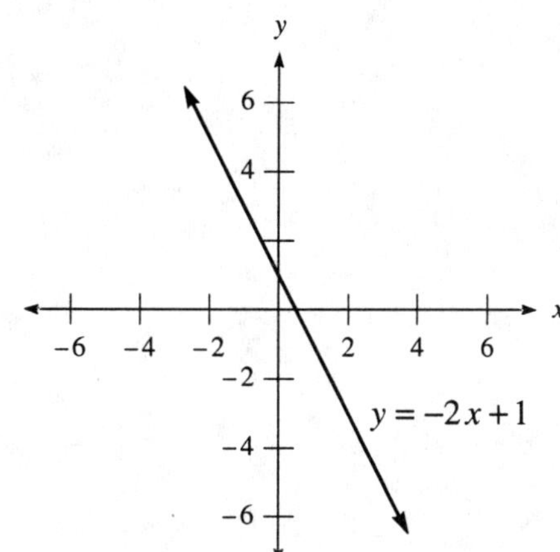

17. Graphing the line:

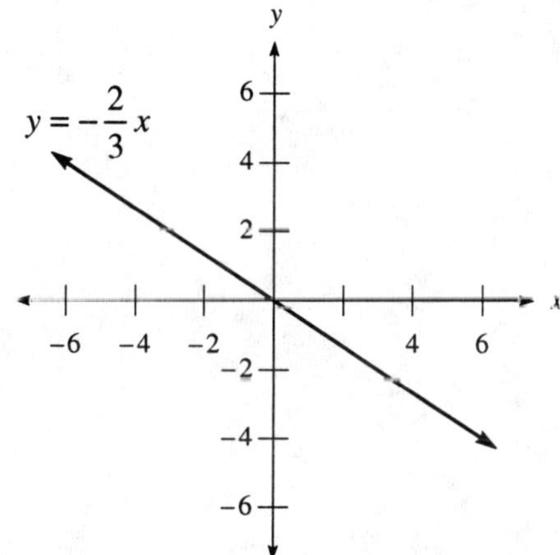

19. Graphing the two equations:

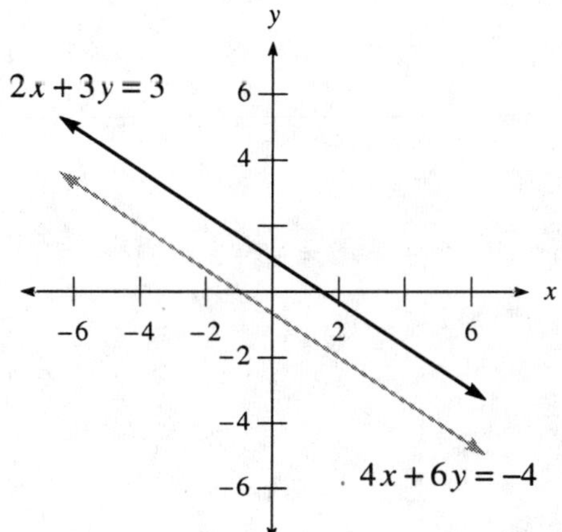

The two lines are parallel.

21. Multiplying the first equation by –2:

$$-2x-2y=-14$$
$$2x+2y=14$$

Adding the two equations:

$$0=0$$

Since this statement is true, the system is dependent. The two lines coincide.

23. Multiplying the second equation by 3:

$$2x+3y=13$$
$$3x-3y=-3$$

Adding the two equations:

$$5x=10$$
$$x=2$$

Substituting into the first equation:

$$2(2)+3y=13$$
$$4+3y=13$$
$$3y=9$$
$$y=3$$

The solution is $(2,3)$.

25. Multiplying the second equation by –2:

$$2x+5y=33$$
$$-2x+6y=0$$

Adding the two equations:

$$11y=33$$
$$y=3$$

Substituting into the second equation:

$$x-3(3)=0$$
$$x-9=0$$
$$x=9$$

The solution is $(9,3)$.

27. Multiplying the second equation by 7:

$$3x-7y=12$$
$$14x+7y=56$$

Adding the two equations:

$$17x=68$$
$$x=4$$

Substituting into the second equation:

$$2(4)+y=8$$
$$8+y=8$$
$$y=0$$

The solution is $(4,0)$.

29. Substituting into the first equation:

$$2x-3(5x+2)=7$$
$$2x-15x-6=7$$
$$-13x-6=7$$
$$-13x=13$$
$$x=-1$$

Substituting into the second equation: $y=5(-1)+2=-5+2=-3$

The solution is $(-1,-3)$.

31. commutative property of addition

33. The quotient is: $\frac{-30}{6}=-5$

35. Finding the percent:

$$p \bullet 82 = 20.5$$
$$p = \frac{20.5}{82}$$
$$p = 0.25 = 25\%$$

25% of 82 is 20.5.

37. Simplifying, then evaluating when $x = 3$: $-3x + 7 + 5x = 2x + 7 = 2(3) + 7 = 6 + 7 = 13$

39. Evaluating when $x = -2$: $4x - 5 = 4(-2) - 5 = -8 - 5 = -13$

41. The expression is: $5 - (-8) = 5 + 8 = 13$

43. To find the x-intercept, let $y = 0$:

$$3x - 4(0) = 12$$
$$3x = 12$$
$$x = 4$$

To find the y-intercept, let $x = 0$:

$$3(0) - 4y = 12$$
$$-4y = 12$$
$$y = -3$$

45. Computing the slope: $m = \frac{-4 - 1}{-5 - (-1)} = \frac{-5}{-4} = \frac{5}{4}$

47. The slope-intercept form is $y = \frac{2}{3}x + 3$.

49. First compute the slope: $m = \frac{6 - 3}{6 - 4} = \frac{3}{2}$

Using the point-slope formula:

$$y - 3 = \frac{3}{2}(x - 4)$$
$$y - 3 = \frac{3}{2}x - 6$$
$$y = \frac{3}{2}x - 3$$

Chapter 4 Test

1. Graphing the two equations:

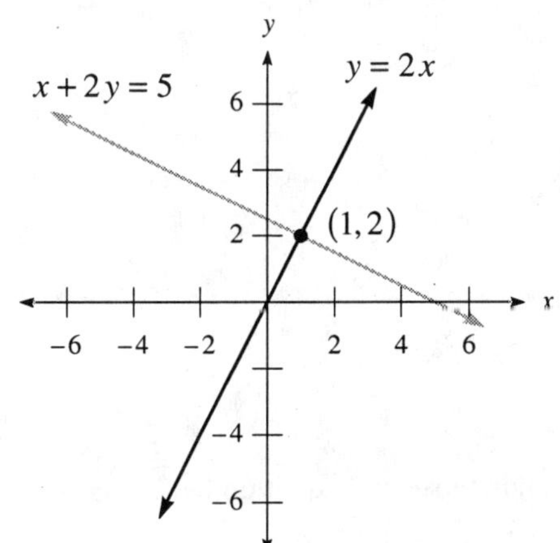

The intersection point is $(1, 2)$.

2. Graphing the two equations:

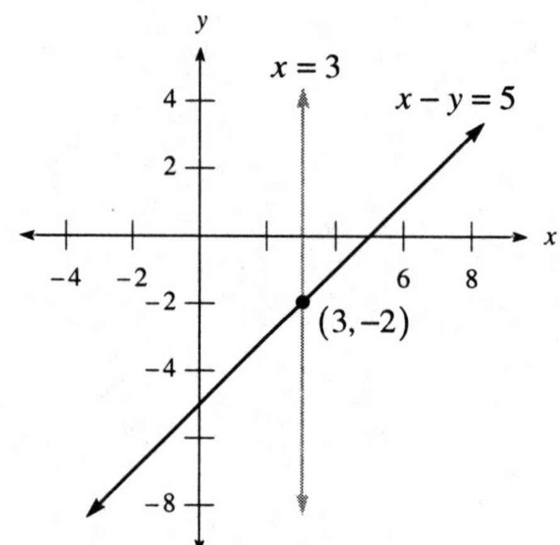

The intersection point is $(3,-2)$.

3. Adding the two equations:

$$3x=-9$$
$$x=-3$$

Substituting into the first equation:

$$-3-y=1$$
$$-y=4$$
$$y=-4$$

The solution is $(-3,-4)$.

4. Multiplying the first equation by –1:

$$-2x-y=-7$$
$$3x+y=12$$

Adding the two equations:

$$x=5$$

Substituting into the first equation:

$$2(5)+y=7$$
$$10+y=7$$
$$y=-3$$

The solution is $(5,-3)$.

5. Multiplying the second equation by 4:

$$7x+8y=-2$$
$$12x-8y=40$$

Adding the two equations:

$$19x=38$$
$$x=2$$

Substituting into the first equation:

$$7(2)+8y=-2$$
$$14+8y=-2$$
$$8y=-16$$
$$y=-2$$

The solution is $(2,-2)$.

6. Multiplying the first equation by –3 and the second equation by 2:

$$-18x+30y=-18$$
$$18x-30y=18$$

Adding the two equations:

$$0=0$$

Since this equation is true, the system is dependent. The two lines coincide.

7. Substituting into the first equation:

$$\begin{aligned} 3x+2(2x+3)&=20\\ 3x+4x+6&=20\\ 7x+6&=20\\ 7x&=14\\ x&=2 \end{aligned}$$

Substituting into the second equation: $y=2(2)+3=4+3=7$

The solution is $(2,7)$.

8. Substituting into the first equation:

$$\begin{aligned} 3(y+1)-6y&=-6\\ 3y+3-6y&=-6\\ -3y+3&=-6\\ -3y&=-9\\ y&=3 \end{aligned}$$

Substituting into the second equation: $x=3+1=4$

The solution is $(4,3)$.

9. Solving the second equation for y:

$$\begin{aligned} -3x+y&=3\\ y&=3x+3 \end{aligned}$$

Substituting into the first equation:

$$\begin{aligned} 7x-2(3x+3)&=-4\\ 7x-6x-6&=-4\\ x-6&=-4\\ x&=2 \end{aligned}$$

Substituting into $y=3x+3$: $y=3(2)+3=6+3=9$

The solution is $(2,9)$.

10. Solving the second equation for x:

$$\begin{aligned} x+3y&=-8\\ x&=-3y-8 \end{aligned}$$

Substituting into the first equation:

$$\begin{aligned} 2(-3y-8)-3y&=-7\\ -6y-16-3y&=-7\\ -9y-16&=-7\\ -9y&=9\\ y&=-1 \end{aligned}$$

Substituting into $x=-3y-8$: $x=-3(-1)-8=3-8=-5$

The solution is $(-5,-1)$.

11. Let x and y represent the two numbers. The system of equations is:

$$\begin{aligned} x+y&=12\\ x-y&=2 \end{aligned}$$

Adding the two equations:

$$\begin{aligned} 2x&=14\\ x&=7 \end{aligned}$$

Substituting into the first equation:

$$\begin{aligned} 7+y&=12\\ y&=5 \end{aligned}$$

The two numbers are 7 and 5.

12. Let x and y represent the two numbers. The system of equations is:
$$x+y=15$$
$$y=6+2x$$
Substituting into the first equation:
$$x+6+2x=15$$
$$3x+6=15$$
$$3x=9$$
$$x=3$$
Substituting into the second equation: $y=6+2(3)=6+6=12$
The two numbers are 3 and 12.

13. Let x represent the amount invested at 9% and y represent the amount invested at 11%. The system of equations is:
$$x+y=10000$$
$$0.09x+0.11y=980$$
Multiplying the first equation by –0.09:
$$-0.09x-0.09y=-900$$
$$0.09x+0.11y=980$$
Adding the two equations:
$$0.02y=80$$
$$y=4000$$
Substituting into the first equation:
$$x+4000=10000$$
$$x=6000$$
Dr. Stork should invest $6,000 at 9%.

14. Let x represent the number of nickels and y represent the number of quarters. The system of equations is:
$$x+y=12$$
$$0.05x+0.25y=1.60$$
Multiplying the first equation by –0.05:
$$-0.05x-0.05y=-0.60$$
$$0.05x+0.25y=1.60$$
Adding the two equations:
$$0.20y=1.00$$
$$y=5$$
Substituting into the first equation:
$$x+5=12$$
$$x=7$$
Diane has 7 nickels and 5 quarters.

Chapter 5
Exponents and Polynomials

5.1 Multiplication with Exponents

1. The base is 4 and the exponent is 2. Evaluating the expression: $4^2 = 4 \cdot 4 = 16$

3. The base is 0.3 and the exponent is 2. Evaluating the expression: $0.3^2 = 0.3 \cdot 0.3 = 0.09$

5. The base is 4 and the exponent is 3. Evaluating the expression: $4^3 = 4 \cdot 4 \cdot 4 = 64$

7. The base is –5 and the exponent is 2. Evaluating the expression: $(-5)^2 = (-5) \cdot (-5) = 25$

9. The base is 2 and the exponent is 3. Evaluating the expression: $-2^3 = -2 \cdot 2 \cdot 2 = -8$

11. The base is 3 and the exponent is 4. Evaluating the expression: $3^4 = 3 \cdot 3 \cdot 3 \cdot 3 = 81$

13. The base is $\frac{2}{3}$ and the exponent is 2. Evaluating the expression: $\left(\frac{2}{3}\right)^2 = \left(\frac{2}{3}\right) \cdot \left(\frac{2}{3}\right) = \frac{4}{9}$

15. The base is $\frac{1}{2}$ and the exponent is 4. Evaluating the expression: $\left(\frac{1}{2}\right)^4 = \left(\frac{1}{2}\right) \cdot \left(\frac{1}{2}\right) \cdot \left(\frac{1}{2}\right) \cdot \left(\frac{1}{2}\right) = \frac{1}{16}$

17. Simplifying the expression: $x^4 \cdot x^5 = x^{4+5} = x^9$

19. Simplifying the expression: $7^6 \cdot 7^1 = 7^{6+1} = 7^7$

21. Simplifying the expression: $y^{10} \cdot y^{20} = y^{10+20} = y^{30}$

23. Simplifying the expression: $2^5 \cdot 2^4 \cdot 2^3 = 2^{5+4+3} = 2^{12}$

25. Simplifying the expression: $x^4 \cdot x^6 \cdot x^8 \cdot x^{10} = x^{4+6+8+10} = x^{28}$

27. Simplifying the expression: $\left(x^2\right)^5 = x^{2 \cdot 5} = x^{10}$

29. Simplifying the expression: $\left(5^4\right)^3 = 5^{4 \cdot 3} = 5^{12}$

31. Simplifying the expression: $\left(y^3\right)^3 = y^{3 \cdot 3} = y^9$

33. Simplifying the expression: $\left(2^5\right)^{10} = 2^{5 \cdot 10} = 2^{50}$

35. Simplifying the expression: $\left(a^3\right)^x = a^{3x}$

37. Simplifying the expression: $\left(b^x\right)^y = b^{xy}$

39. Simplifying the expression: $(4x)^2 = 4^2 \cdot x^2 = 16x^2$

41. Simplifying the expression: $(2y)^5 = 2^5 \cdot y^5 = 32y^5$

43. Simplifying the expression: $(-3x)^4 = (-3)^4 \cdot x^4 = 81x^4$

45. Simplifying the expression: $(0.5ab)^2 = (0.5)^2 \cdot a^2b^2 = 0.25a^2b^2$

47. Simplifying the expression: $(4xyz)^3 = 4^3 \cdot x^3y^3z^3 = 64x^3y^3z^3$

49. Simplifying using properties of exponents: $\left(2x^4\right)^3 = 2^3\left(x^4\right)^3 = 8x^{12}$

51. Simplifying using properties of exponents: $\left(4a^3\right)^2 = 4^2\left(a^3\right)^2 = 16a^6$

53. Simplifying using properties of exponents: $\left(x^2\right)^3\left(x^4\right)^2 = x^6 \bullet x^8 = x^{14}$

55. Simplifying using properties of exponents: $\left(a^3\right)^1\left(a^2\right)^4 = a^3 \bullet a^8 = a^{11}$

57. Simplifying using properties of exponents: $(2x)^3(2x)^4 = (2x)^7 = 2^7 x^7 = 128x^7$

59. Simplifying using properties of exponents: $\left(3x^2\right)^3(2x)^4 = 3^3 x^6 \bullet 2^4 x^4 = 27x^6 \bullet 16x^4 = 432x^{10}$

61. Simplifying using properties of exponents: $\left(4x^2y^3\right)^2 = 4^2 x^4 y^6 = 16x^4y^6$

63. Simplifying using properties of exponents: $\left(\frac{2}{3}a^4b^5\right)^3 = \left(\frac{2}{3}\right)^3 a^{12}b^{15} = \frac{8}{27}a^{12}b^{15}$

65. Simplifying using properties of exponents: $\left(3x^2\right)\left(2x^3\right)\left(5x^4\right) = 30x^9$

67. Simplifying using properties of exponents: $\left(4x^2y\right)^3\left(\frac{1}{8}xy\right)^2 = 64x^6y^3 \bullet \frac{1}{64}x^2y^2 = x^8y^5$

69. Writing in scientific notation: $43,200 = 4.32 \times 10^4$
71. Writing in scientific notation: $570 = 5.7 \times 10^2$
73. Writing in scientific notation: $238,000 = 2.38 \times 10^5$
75. Writing in expanded form: $2.49 \times 10^3 = 2,490$
77. Writing in expanded form: $3.52 \times 10^2 = 352$
79. Writing in expanded form: $2.8 \times 10^4 = 28,000$
81. The volume is given by: $V = (3 \text{ in.})^3 = 27 \text{ inches}^3$
83. The volume is given by: $V = (2.5 \text{ in.})^3 \approx 15.6 \text{ inches}^3$
85. The volume is given by: $V = (8 \text{ in.})(4.5 \text{ in.})(1 \text{ in.}) = 36 \text{ inches}^3$
87. Yes. If the box was 7 ft x 3 ft x 2 ft, you could fit inside of it.
89. Writing in scientific notation: $650,000,000$ seconds $= 6.5 \times 10^8$ seconds
91. Writing in expanded form: 7.4×10^5 dollars $= \$740,000$
93. Writing in expanded form: 1.8×10^5 dollars $= \$180,000$
95. $\left(2^3\right)^2 = 2^6 = 64$, while $2^{3^2} = 2^9 = 512$.
97. (b) 4
99. Subtracting: $4 - 7 = 4 + (-7) = -3$
101. Subtracting: $4 - (-7) = 4 + 7 = 11$
103. Subtracting: $15 - 20 = 15 + (-20) = -5$
105. Subtracting: $-15 - (-20) = -15 + 20 = 5$

5.2 Division with Exponents

1. Writing with positive exponents: $3^{-2} = \frac{1}{3^2} = \frac{1}{9}$

3. Writing with positive exponents: $6^{-2} = \frac{1}{6^2} = \frac{1}{36}$

5. Writing with positive exponents: $8^{-2} = \frac{1}{8^2} = \frac{1}{64}$

7. Writing with positive exponents: $5^{-3} = \frac{1}{5^3} = \frac{1}{125}$

9. Writing with positive exponents: $2x^{-3} = 2 \bullet \frac{1}{x^3} = \frac{2}{x^3}$

11. Writing with positive exponents: $(2x)^{-3} = \frac{1}{(2x)^3} = \frac{1}{8x^3}$

13. Writing with positive exponents: $(5y)^{-2} = \frac{1}{(5y)^2} = \frac{1}{25y^2}$

15. Writing with positive exponents: $10^{-2} = \frac{1}{10^2} = \frac{1}{100}$

17. Simplifying the expression: $\frac{5^3}{5^1} = 5^{3-1} = 5^2 = 25$

19. Simplifying the expression: $\frac{5^1}{5^3} = 5^{1-3} = 5^{-2} = \frac{1}{5^2} = \frac{1}{25}$

21. Simplifying the expression: $\frac{x^{10}}{x^4} = x^{10-4} = x^6$

23. Simplifying the expression: $\frac{4^3}{4^0} = 4^{3-0} = 4^3 = 64$

25. Simplifying the expression: $\frac{(2x)^7}{(2x)^4} = (2x)^{7-4} = (2x)^3 = 2^3x^3 = 8x^3$

27. Simplifying the expression: $\frac{6^{11}}{6} = \frac{6^{11}}{6^1} = 6^{11-1} = 6^{10} \quad (= 60,466,176)$

29. Simplifying the expression: $\frac{6}{6^{11}} = \frac{6^1}{6^{11}} = 6^{1-11} = 6^{-10} = \frac{1}{6^{10}} \quad \left(= \frac{1}{60,466,176}\right)$

31. Simplifying the expression: $\frac{2^{-5}}{2^3} = 2^{-5-3} = 2^{-8} = \frac{1}{2^8} = \frac{1}{256}$

33. Simplifying the expression: $\frac{2^5}{2^{-3}} = 2^{5-(-3)} = 2^{5+3} = 2^8 = 256$

35. Simplifying the expression: $\frac{(3x)^{-5}}{(3x)^{-8}} = (3x)^{-5-(-8)} = (3x)^{-5+8} = (3x)^3 = 3^3x^3 = 27x^3$

37. Simplifying the expression: $(3xy)^4 = 3^4x^4y^4 = 81x^4y^4$

39. Simplifying the expression: $10^0 = 1$

41. Simplifying the expression: $\left(2a^2b\right)^1 = 2a^2b$

43. Simplifying the expression: $\left(7y^3\right)^{-2} = \frac{1}{\left(7y^3\right)^2} = \frac{1}{49y^6}$

45. Simplifying the expression: $x^{-3} \bullet x^{-5} = x^{-3-5} = x^{-8} = \frac{1}{x^8}$

47. Simplifying the expression: $y^7 \bullet y^{-10} = y^{7-10} = y^{-3} = \frac{1}{y^3}$

49. Simplifying the expression: $\frac{\left(x^2\right)^3}{x^4} = \frac{x^6}{x^4} = x^{6-4} = x^2$

51. Simplifying the expression: $\frac{\left(a^4\right)^3}{\left(a^3\right)^2} = \frac{a^{12}}{a^6} = a^{12-6} = a^6$

53. Simplifying the expression: $\frac{y^7}{\left(y^2\right)^8}=\frac{y^7}{y^{16}}=y^{7-16}=y^{-9}=\frac{1}{y^9}$

55. Simplifying the expression: $\left(\frac{y^7}{y^2}\right)^8=\left(y^{7-2}\right)^8=\left(y^5\right)^8=y^{40}$

57. Simplifying the expression: $\frac{\left(x^{-2}\right)^3}{x^{-5}}=\frac{x^{-6}}{x^{-5}}=x^{-6-(-5)}=x^{-6+5}=x^{-1}=\frac{1}{x}$

59. Simplifying the expression: $\left(\frac{x^{-2}}{x^{-5}}\right)^3=\left(x^{-2+5}\right)^3=\left(x^3\right)^3=x^9$

61. Simplifying the expression: $\frac{\left(a^3\right)^2\left(a^4\right)^5}{\left(a^5\right)^2}=\frac{a^6\bullet a^{20}}{a^{10}}=\frac{a^{26}}{a^{10}}=a^{26-10}=a^{16}$

63. Simplifying the expression: $\frac{\left(a^{-2}\right)^3\left(a^4\right)^2}{\left(a^{-3}\right)^{-2}}=\frac{a^{-6}\bullet a^8}{a^6}=\frac{a^2}{a^6}=a^{2-6}=a^{-4}=\frac{1}{a^4}$

65. Simplifying the expression: $\frac{\left(x^{-7}\right)^3\left(x^4\right)^5}{\left(x^3\right)^2\left(x^{-1}\right)^8}=\frac{x^{-21}\bullet x^{20}}{x^6\bullet x^{-8}}=\frac{x^{-1}}{x^{-2}}=x^{-1+2}=x^1=x$

67. Writing in scientific notation: $0.000357=3.57\times10^{-4}$
69. Writing in scientific notation: $35,700=3.57\times10^4$
71. Writing in scientific notation: $0.0048=4.8\times10^{-3}$
73. Writing in scientific notation: $25=2.5\times10^1$
75. Writing in scientific notation: $0.000009=9\times10^{-6}$
77. Writing in expanded form: $4.23\times10^{-3}=0.00423$
79. Writing in expanded form: $5.6\times10^4=56,000$
81. Writing in expanded form: $8\times10^{-5}=0.00008$
83. Writing in expanded form: $7.89\times10^1=78.9$
85. Writing in expanded form: $4.2\times10^0=4.2$
87. Writing in expanded form: 2×10^{-3} seconds $=0.002$ seconds
89. Writing in scientific notation: 0.006 inches $=6\times10^{-3}$ inches
91. Writing in scientific notation: $25\times10^3=2.5\times10^4$
93. Writing in scientific notation: $23.5\times10^4=2.35\times10^5$
95. Writing in scientific notation: $0.82\times10^{-3}=8.2\times10^{-4}$

97. The area of the smaller square is $(10\text{ in.})^2=100$ inches^2, while the area of the larger square is $(20\text{ in.})^2=400$ inches^2. It would take 4 smaller squares to cover the larger square.

99. The area of the smaller square is x^2, while the area of the larger square is $(2x)^2=4x^2$. It would take 4 smaller squares to cover the larger square.

101. The volume of the smaller box is $(6\text{ in.})^3=216$ inches^3, while the volume of the larger box is $(12\text{ in.})^3=1,728$ inches^3. Thus 8 smaller boxes will fit inside the larger box ($8\bullet216=1,728$).

103. The volume of the smaller box is x^3, while the volume of the larger box is $(2x)^3=8x^3$. Thus 8 smaller boxes will fit inside the larger box.

105. Simplifying by combining like terms: $4x+3x=(4+3)x=7x$
107. Simplifying by combining like terms: $5a-3a=(5-3)a=2a$
109. Simplifying by combining like terms: $4y+5y+y=(4+5+1)y=10y$

5.3 Operations with Monomials

1. Multiplying the monomials: $(3x^4)(4x^3) = 12x^{4+3} = 12x^7$
3. Multiplying the monomials: $(-2y^4)(8y^7) = -16y^{4+7} = -16y^{11}$
5. Multiplying the monomials: $(8x)(4x) = 32x^{1+1} = 32x^2$
7. Multiplying the monomials: $(10a^3)(10a)(2a^2) = 200a^{3+1+2} = 200a^6$
9. Multiplying the monomials: $(6ab^2)(-4a^2b) = -24a^{1+2}b^{2+1} = -24a^3b^3$
11. Multiplying the monomials: $(4x^2y)(3x^3y^3)(2xy^4) = 24x^{2+3+1}y^{1+3+4} = 24x^6y^8$
13. Dividing the monomials: $\dfrac{15x^3}{5x^2} = \dfrac{15}{5} \bullet \dfrac{x^3}{x^2} = 3x$
15. Dividing the monomials: $\dfrac{18y^9}{3y^{12}} = \dfrac{18}{3} \bullet \dfrac{y^9}{y^{12}} = 6 \bullet \dfrac{1}{y^3} = \dfrac{6}{y^3}$
17. Dividing the monomials: $\dfrac{32a^3}{64a^4} = \dfrac{32}{64} \bullet \dfrac{a^3}{a^4} = \dfrac{1}{2} \bullet \dfrac{1}{a} = \dfrac{1}{2a}$
19. Dividing the monomials: $\dfrac{21a^2b^3}{-7ab^5} = \dfrac{21}{-7} \bullet \dfrac{a^2}{a} \bullet \dfrac{b^3}{b^5} = -3 \bullet a \bullet \dfrac{1}{b^2} = -\dfrac{3a}{b^2}$
21. Dividing the monomials: $\dfrac{3x^3y^2z}{27xy^2z^3} = \dfrac{3}{27} \bullet \dfrac{x^3}{x} \bullet \dfrac{y^2}{y^2} \bullet \dfrac{z}{z^3} = \dfrac{1}{9} \bullet x^2 \bullet \dfrac{1}{z^2} = \dfrac{x^2}{9z^2}$
23. Dividing the monomials: $\dfrac{144x^9y^2}{-12x^{10}y^8} = \dfrac{144}{-12} \bullet \dfrac{x^9}{x^{10}} \bullet \dfrac{y^2}{y^8} = -12 \bullet \dfrac{1}{x} \bullet \dfrac{1}{y^6} = -\dfrac{12}{xy^6}$
25. Finding the product: $(3 \times 10^3)(2 \times 10^5) = 6 \times 10^8$
27. Finding the product: $(3.5 \times 10^4)(5 \times 10^{-6}) = 17.5 \times 10^{-2} = 1.75 \times 10^{-1}$
29. Finding the product: $(5.5 \times 10^{-3})(2.2 \times 10^{-4}) = 12.1 \times 10^{-7} = 1.21 \times 10^{-6}$
31. Finding the quotient: $\dfrac{8.4 \times 10^5}{2 \times 10^2} = 4.2 \times 10^3$
33. Finding the quotient: $\dfrac{6 \times 10^8}{2 \times 10^{-2}} = 3 \times 10^{10}$
35. Finding the quotient: $\dfrac{2.5 \times 10^{-6}}{5 \times 10^{-4}} = 0.5 \times 10^{-2} = 5.0 \times 10^{-3}$
37. Combining the monomials: $3x^2 + 5x^2 = (3+5)x^2 = 8x^2$
39. Combining the monomials: $8x^5 - 19x^5 = (8-19)x^5 = -11x^5$
41. Combining the monomials: $2a + a - 3a = (2+1-3)a = 0a = 0$
43. Combining the monomials: $10x^3 - 8x^3 + 2x^3 = (10-8+2)x^3 = 4x^3$
45. Combining the monomials: $20ab^2 - 19ab^2 + 30ab^2 = (20-19+30)ab^2 = 31ab^2$
47. Combining the monomials: $-4abc - 9abc - abc = (-4-9-1)abc = -14abc$
49. Simplifying the expression: $\dfrac{(3x^2)(8x^5)}{6x^4} = \dfrac{24x^7}{6x^4} = \dfrac{24}{6} \bullet \dfrac{x^7}{x^4} = 4x^3$
51. Simplifying the expression: $\dfrac{(9a^2b)(2a^3b^4)}{18a^5b^7} = \dfrac{18a^5b^5}{18a^5b^7} = \dfrac{18}{18} \bullet \dfrac{a^5}{a^5} \bullet \dfrac{b^5}{b^7} = 1 \bullet \dfrac{1}{b^2} = \dfrac{1}{b^2}$
53. Simplifying the expression: $\dfrac{(4x^3y^2)(9x^4y^{10})}{(3x^5y)(2x^6y)} = \dfrac{36x^7y^{12}}{6x^{11}y^2} = \dfrac{36}{6} \bullet \dfrac{x^7}{x^{11}} \bullet \dfrac{y^{12}}{y^2} = 6 \bullet \dfrac{1}{x^4} \bullet y^{10} = \dfrac{6y^{10}}{x^4}$

55. Simplifying the expression: $\dfrac{(6\times10^8)(3\times10^5)}{9\times10^7}=\dfrac{18\times10^{13}}{9\times10^7}=2\times10^6$

57. Simplifying the expression: $\dfrac{(5\times10^3)(4\times10^{-5})}{2\times10^{-2}}=\dfrac{20\times10^{-2}}{2\times10^{-2}}=10=1\times10^1$

59. Simplifying the expression: $\dfrac{(2.8\times10^{-7})(3.6\times10^4)}{2.4\times10^3}=\dfrac{10.08\times10^{-3}}{2.4\times10^3}=4.2\times10^{-6}$

61. Simplifying the expression: $\dfrac{18x^4}{3x}+\dfrac{21x^7}{7x^4}=6x^3+3x^3=9x^3$

63. Simplifying the expression: $\dfrac{45a^6}{9a^4}-\dfrac{50a^8}{2a^6}=5a^2-25a^2=-20a^2$

65. Simplifying the expression: $\dfrac{6x^7y^4}{3x^2y^2}+\dfrac{8x^5y^8}{2y^6}=2x^5y^2+4x^5y^2=6x^5y^2$

67. Solving the equation:

$$4^x\cdot4^5=4^7$$
$$4^{x+5}=4^7$$
$$x+5=7$$
$$x=2$$

69. Solving the equation:

$$(7^3)^x=7^{12}$$
$$7^{3x}=7^{12}$$
$$3x=12$$
$$x=4$$

71. Simplifying each value:

$$(a+b)^2=(4+5)^2=9^2=81$$
$$a^2+b^2=4^2+5^2=16+25=41$$

Note that the values are not equal.

73. Simplifying each value:

$$(a+b)^2=(3+4)^2=7^2=49$$
$$a^2+2ab+b^2=3^2+2(3)(4)+4^2=9+24+16=49$$

Note that the values are equal.

75. Let x represent the width and $2x$ represent the length. The perimeter and area are given by:

$$P=2(x)+2(2x)=2x+4x=6x$$
$$A=x\cdot2x=2x^2$$

77. Let x represent the width and $2x$ represent the length. The volume is given by:

$$V=(x)(2x)(4)=8x^2\text{ inches}^3$$

79. Let x respresent the width, $4x$ represent the length, and $\frac{1}{2}x$ represent the height. The volume is given by:

$$V=(x)(4x)\left(\frac{1}{2}x\right)=2x^3$$

81. Evaluating when $x=-2$: $4x=4(-2)=-8$

83. Evaluating when $x=-2$: $-2x+5=-2(-2)+5=4+5=9$

85. Evaluating when $x=-2$: $x^2+5x+6=(-2)^2+5(-2)+6=4-10+6=0$

87. The ordered pairs are $(-2,-2)$, $(0,2)$, and $(2,6)$. Graphing the equation:

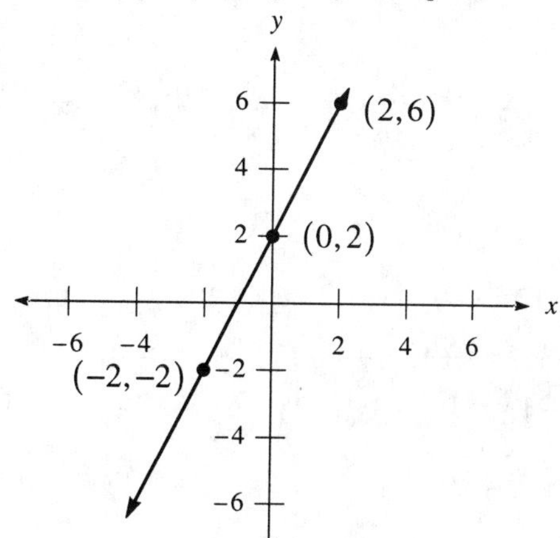

89. The ordered pairs are $(-3,0)$, $(0,1)$, and $(3,2)$. Graphing the equation:

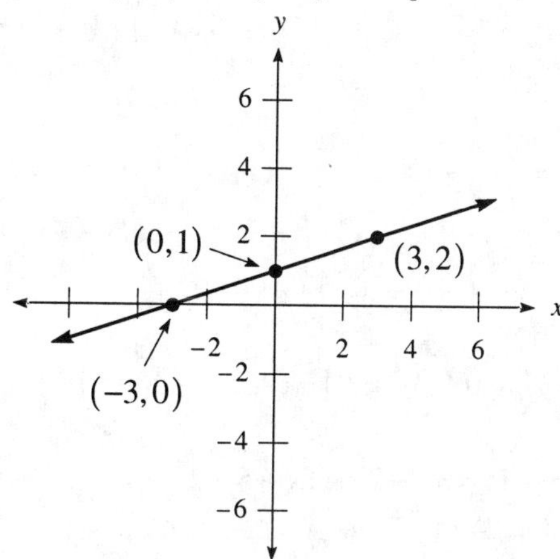

5.4 Addition and Subtraction of Polynomials

1. This is a trinomial of degree 3.
3. This is a trinomial of degree 3.
5. This is a binomial of degree 1.
7. This is a binomial of degree 2.
9. This is a monomial of degree 2.
11. This is a monomial of degree 0.
13. Combining the polynomials: $(2x^2+3x+4)+(3x^2+2x+5)=(2x^2+3x^2)+(3x+2x)+(4+5)=5x^2+5x+9$
15. Combining the polynomials: $(3a^2-4a+1)+(2a^2-5a+6)=(3a^2+2a^2)+(-4a-5a)+(1+6)=5a^2-9a+7$
17. Combining the polynomials: $x^2+4x+2x+8=x^2+(4x+2x)+8=x^2+6x+8$
19. Combining the polynomials: $6x^2-3x-10x+5=6x^2+(-3x-10x)+5=6x^2-13x+5$
21. Combining the polynomials: $x^2-3x+3x-9=x^2+(-3x+3x)-9=x^2-9$
23. Combining the polynomials: $3y^2-5y-6y+10=3y^2+(-5y-6y)+10=3y^2-11y+10$
25. Combining the polynomials:
$$(6x^3-4x^2+2x)+(9x^2-6x+3)=6x^3+(-4x^2+9x^2)+(2x-6x)+3=6x^3+5x^2-4x+3$$
27. Combining the polynomials:
$$\left(\frac{2}{3}x^2-\frac{1}{5}x-\frac{3}{4}\right)+\left(\frac{4}{3}x^2-\frac{4}{5}x+\frac{7}{4}\right)=\left(\frac{2}{3}x^2+\frac{4}{3}x^2\right)+\left(-\frac{1}{5}x-\frac{4}{5}x\right)+\left(-\frac{3}{4}+\frac{7}{4}\right)=2x^2-x+1$$

29. Combining the polynomials: $\left(a^2-a-1\right)-\left(-a^2+a+1\right)=a^2-a-1+a^2-a-1=2a^2-2a-2$

31. Combining the polynomials:
$$\left(\frac{5}{9}x^3+\frac{1}{3}x^2-2x+1\right)-\left(\frac{2}{3}x^3+x^2+\frac{1}{2}x-\frac{3}{4}\right)=\frac{5}{9}x^3+\frac{1}{3}x^2-2x+1-\frac{2}{3}x^3-x^2-\frac{1}{2}x+\frac{3}{4}$$
$$=-\frac{1}{9}x^3-\frac{2}{3}x^2-\frac{5}{2}x+\frac{7}{4}$$

33. Combining the polynomials:
$$\begin{aligned}&\left(4y^2-3y+2\right)+\left(5y^2+12y-4\right)-\left(13y^2-6y+20\right)\\&=4y^2-3y+2+5y^2+12y-4-13y^2+6y-20\\&=\left(4y^2+5y^2-13y^2\right)+(-3y+12y+6y)+(2-4-20)\\&=-4y^2+15y-22\end{aligned}$$

35. Performing the subtraction:
$$\begin{aligned}\left(11x^2-10x+13\right)-\left(10x^2+23x-50\right)&=11x^2-10x+13-10x^2-23x+50\\&=\left(11x^2-10x^2\right)+(-10x-23x)+(13+50)\\&=x^2-33x+63\end{aligned}$$

37. Performing the subtraction:
$$\begin{aligned}\left(11y^2+11y+11\right)-\left(3y^2+7y-15\right)&=11y^2+11y+11-3y^2-7y+15\\&=\left(11y^2-3y^2\right)+(11y-7y)+(11+15)\\&=8y^2+4y+26\end{aligned}$$

39. Performing the addition:
$$\left(25x^2-50x+75\right)+\left(50x^2-100x-150\right)=\left(25x^2+50x^2\right)+(-50x-100x)+(75-150)=75x^2-150x-75$$

41. Performing the operations:
$$(3x-2)+(11x+5)-(2x+1)=3x-2+11x+5-2x-1=(3x+11x-2x)+(-2+5-1)=12x+2$$

43. Evaluating when $x=3$: $x^2-2x+1=(3)^2-2(3)+1=9-6+1=4$

45. Evaluating when $y=10$: $(y-5)^2=(10-5)^2=(5)^2=25$

47. Evaluating when $a=2$: $a^2+4a+4=(2)^2+4(2)+4=4+8+4=16$

49. Multiplying the monomials: $2x(5x)=10x^2$

51. Multiplying the monomials: $3x(-5x)=-15x^2$

53. Multiplying the monomials: $2x\left(3x^2\right)=6x^{1+2}=6x^3$

55. Multiplying the monomials: $3x^2\left(2x^2\right)=6x^{2+2}=6x^4$

5.5 Multiplication with Polynomials

1. Using the distributive property: $2x(3x+1)=2x(3x)+2x(1)=6x^2+2x$

3. Using the distributive property: $2x^2\left(3x^2-2x+1\right)=2x^2\left(3x^2\right)-2x^2(2x)+2x^2(1)=6x^4-4x^3+2x^2$

5. Using the distributive property: $2ab\left(a^2-ab+1\right)=2ab\left(a^2\right)-2ab(ab)+2ab(1)=2a^3b-2a^2b^2+2ab$

7. Using the distributive property: $y^2\left(3y^2+9y+12\right)=y^2\left(3y^2\right)+y^2(9y)+y^2(12)=3y^4+9y^3+12y^2$

9. Using the distributive property:
$$4x^2y\left(2x^3y+3x^2y^2+8y^3\right)=4x^2y\left(2x^3y\right)+4x^2y\left(3x^2y^2\right)+4x^2y\left(8y^3\right)=8x^5y^2+12x^4y^3+32x^2y^4$$

11. Multiplying using the FOIL method: $(x+3)(x+4)=x^2+3x+4x+12=x^2+7x+12$

13. Multiplying using the FOIL method: $(x+6)(x+1)=x^2+6x+1x+6=x^2+7x+6$

15. Multiplying using the FOIL method: $\left(x+\frac{1}{2}\right)\left(x+\frac{3}{2}\right)=x^2+\frac{1}{2}x+\frac{3}{2}x+\frac{3}{4}=x^2+2x+\frac{3}{4}$

17. Multiplying using the FOIL method: $(a+5)(a-3)=a^2+5a-3a-15=a^2+2a-15$

19. Multiplying using the FOIL method: $(x-a)(y+b)=xy-ay+bx-ab$

21. Multiplying using the FOIL method: $(x+6)(x-6)=x^2+6x-6x-36=x^2-36$

23. Multiplying using the FOIL method: $\left(y+\frac{5}{6}\right)\left(y-\frac{5}{6}\right)=y^2+\frac{5}{6}y-\frac{5}{6}y-\frac{25}{36}=y^2-\frac{25}{36}$

25. Multiplying using the FOIL method: $(2x-3)(x-4)=2x^2-3x-8x+12=2x^2-11x+12$

27. Multiplying using the FOIL method: $(a+2)(2a-1)=2a^2+4a-a-2=2a^2+3a-2$

29. Multiplying using the FOIL method: $(2x-5)(3x-2)=6x^2-15x-4x+10=6x^2-19x+10$

31. Multiplying using the FOIL method: $(2x+3)(a+4)=2ax+3a+8x+12$

33. Multiplying using the FOIL method: $(5x-4)(5x+4)=25x^2-20x+20x-16=25x^2-16$

35. Multiplying using the FOIL method: $\left(2x-\frac{1}{2}\right)\left(x+\frac{3}{2}\right)=2x^2-\frac{1}{2}x+3x-\frac{3}{4}=2x^2+\frac{5}{2}x-\frac{3}{4}$

37. Multiplying using the FOIL method: $(1-2a)(3-4a)=3-6a-4a+8a^2=3-10a+8a^2$

39. Multiplying using the FOIL method: $(7-6x)(8-5x)=56-48x-35x+30x^2=56-83x+30x^2$

41. Multiplying using the column method:

$$\begin{array}{r} x^2+3x-4 \\ x+1 \\ \hline x^3+3x^2-4x \\ x^2+3x-4 \\ \hline x^3+4x^2-x-4 \end{array}$$

43. Multiplying using the column method:

$$\begin{array}{r} a^2-3a+2 \\ a-3 \\ \hline a^3-3a^2+2a \\ -3a^2+9a-6 \\ \hline a^3-6a^2+11a-6 \end{array}$$

45. Multiplying using the column method:

$$\begin{array}{r} x^2-2x+4 \\ x+2 \\ \hline x^3-2x^2+4x \\ 2x^2-4x+8 \\ \hline x^3+8 \end{array}$$

47. Multiplying using the column method:

$$\begin{array}{r} x^2+8x+9 \\ 2x+1 \\ \hline 2x^3+16x^2+18x \\ x^2+8x+9 \\ \hline 2x^3+17x^2+26x+9 \end{array}$$

49. Multiplying using the column method:

$$\begin{array}{r} 5x^2+2x+1 \\ \underline{x^2-3x+5} \\ 5x^4+2x^3+x^2 \\ -15x^3-6x^2-3x \\ 25x^2+10x+5 \\ \hline 5x^4-13x^3+20x^2+7x+5 \end{array}$$

51. Multiplying using the FOIL method: $(x^2+3)(2x^2-5)=2x^4-5x^2+6x^2-15=2x^4+x^2-15$

53. Multiplying using the FOIL method: $(3a^4+2)(2a^2+5)=6a^6+15a^4+4a^2+10$

55. First multiply two polynomials using the FOIL method: $(x+3)(x+4)=x^2+3x+4x+12=x^2+7x+12$
Now using the column method:

$$\begin{array}{r} x^2+7x+12 \\ \underline{x+5} \\ x^3+7x^2+12x \\ 5x^2+35x+60 \\ \hline x^3+12x^2+47x+60 \end{array}$$

57. Let x represent the width and $2x + 5$ represent the length. The area is given by: $A=x(2x+5)=2x^2+5x$

59. Let x and $x + 1$ represent the width and length, respectively. The area is given by: $A=x(x+1)=x^2+x$

61. **a.** The completed table is:

Side	Length	Width	Height	Volume
0	17	11	0	0 in.^{3}
1	15	9	1	135 in.^{3}
1.5	14	8	1.5	168 in.^{3}
2	13	7	2	182 in.^{3}
2.5	12	6	2.5	180 in.^{3}
3	11	5	3	165 in.^{3}
3.5	10	4	3.5	140 in.^{3}
4	9	3	4	108 in.^{3}
4.5	8	2	4.5	72 in.^{3}
5	7	1	5	35 in.^{3}
5.5	6	0	5.5	0 in.^{3}

b. Constructing a line graph:

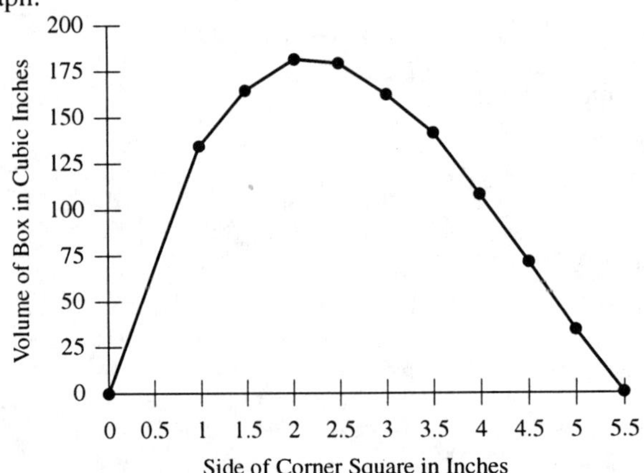

63. Graphing each line:

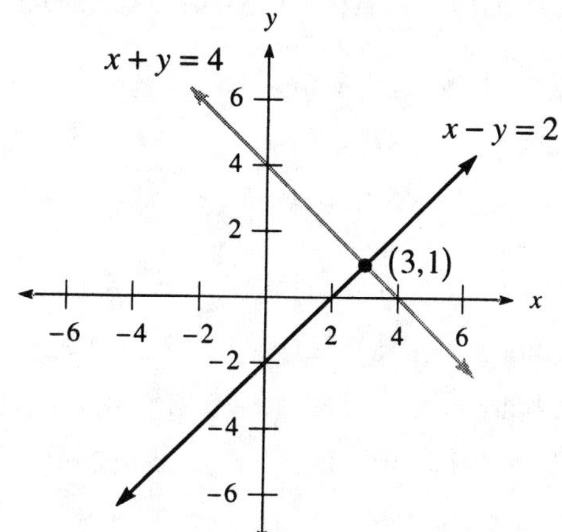

The intersection point is $(3,1)$.

65. Multiplying the first equation by 3 and the second equation by –2:

$$6x+9y=-3$$
$$-6x-10y=4$$

Adding the two equations:

$$-y=1$$
$$y=-1$$

Substituting into the first equation:

$$2x+3(-1)=-1$$
$$2x-3=-1$$
$$2x=2$$
$$x=1$$

The solution is $(1,-1)$.

67. Substituting into the first equation:

$$2x-6(3x+1)=2$$
$$2x-18x-6=2$$
$$-16x-6=2$$
$$-16x=8$$
$$x=-\frac{1}{2}$$

Substituting into the second equation: $y=3\left(-\frac{1}{2}\right)+1=-\frac{3}{2}+1=-\frac{1}{2}$

The solution is $\left(-\frac{1}{2},-\frac{1}{2}\right)$.

69. Let x represent the number of dimes and y represent the number of quarters. The system of equations is:

$$x+y=11$$
$$0.10x+0.25y=1.85$$

Multiplying the first equation by –0.10:

$$-0.10x-0.10y=-1.10$$
$$0.10x+0.25y=1.85$$

Adding the two equations:

$$0.15y=0.75$$
$$y=5$$

Substituting into the first equation:

$$x+5=11$$
$$x=6$$

Amy has 6 dimes and 5 quarters.

5.6 Binomial Squares and Other Special Products

1. Multiplying using the FOIL method: $(x-2)^2=(x-2)(x-2)=x^2-2x-2x+4=x^2-4x+4$
3. Multiplying using the FOIL method: $(a+3)^2=(a+3)(a+3)=a^2+3a+3a+9=a^2+6a+9$
5. Multiplying using the FOIL method: $(x-5)^2=(x-5)(x-5)=x^2-5x-5x+25=x^2-10x+25$
7. Multiplying using the FOIL method: $\left(a-\frac{1}{2}\right)^2=\left(a-\frac{1}{2}\right)\left(a-\frac{1}{2}\right)=a^2-\frac{1}{2}a-\frac{1}{2}a+\frac{1}{4}=a^2-a+\frac{1}{4}$
9. Multiplying using the FOIL method: $(x+10)^2=(x+10)(x+10)=x^2+10x+10x+100=x^2+20x+100$
11. Multiplying using the square of binomial formula: $(a+0.8)^2=a^2+2(a)(0.8)+(0.8)^2=a^2+1.6a+0.64$
13. Multiplying using the square of binomial formula: $(2x-1)^2=(2x)^2-2(2x)(1)+(1)^2=4x^2-4x+1$
15. Multiplying using the square of binomial formula: $(4a+5)^2=(4a)^2+2(4a)(5)+(5)^2=16a^2+40a+25$
17. Multiplying using the square of binomial formula: $(3x-2)^2=(3x)^2-2(3x)(2)+(2)^2=9x^2-12x+4$
19. Multiplying using the square of binomial formula: $(3a+5b)^2=(3a)^2+2(3a)(5b)+(5b)^2=9a^2+30ab+25b^2$
21. Multiplying using the square of binomial formula: $(4x-5y)^2=(4x)^2-2(4x)(5y)+(5y)^2=16x^2-40xy+25y^2$
23. Multiplying using the square of binomial formula:
$(7m+2n)^2=(7m)^2+2(7m)(2n)+(2n)^2=49m^2+28mn+4n^2$
25. Multiplying using the square of binomial formula:
$(6x-10y)^2=(6x)^2-2(6x)(10y)+(10y)^2=36x^2-120xy+100y^2$
27. Multiplying using the square of binomial formula: $\left(x^2+5\right)^2=\left(x^2\right)^2+2\left(x^2\right)(5)+(5)^2=x^4+10x^2+25$
29. Multiplying using the square of binomial formula: $\left(a^2+1\right)^2=\left(a^2\right)^2+2\left(a^2\right)(1)+(1)^2=a^4+2a^2+1$
31. Multiplying using the square of binomial formula: $\left(x^3-7\right)^2=\left(x^3\right)^2-2\left(x^3\right)(7)+(7)^2=x^6-14x^3+49$
33. Multiplying using the FOIL method: $(x-3)(x+3)=x^2-3x+3x-9=x^2-9$
35. Multiplying using the FOIL method: $(a+5)(a-5)=a^2+5a-5a-25=a^2-25$
37. Multiplying using the FOIL method: $(y-1)(y+1)=y^2-y+y-1=y^2-1$
39. Multiplying using the difference of squares formula: $(9+x)(9-x)=(9)^2-(x)^2=81-x^2$
41. Multiplying using the difference of squares formula: $(2x+5)(2x-5)=(2x)^2-(5)^2=4x^2-25$
43. Multiplying using the difference of squares formula: $\left(4x+\frac{1}{3}\right)\left(4x-\frac{1}{3}\right)=(4x)^2-\left(\frac{1}{3}\right)^2=16x^2-\frac{1}{9}$
45. Multiplying using the difference of squares formula: $(2a+7)(2a-7)=(2a)^2-(7)^2=4a^2-49$
47. Multiplying using the difference of squares formula: $(6-7x)(6+7x)=(6)^2-(7x)^2=36-49x^2$
49. Multiplying using the difference of squares formula: $\left(x^2+3\right)\left(x^2-3\right)=\left(x^2\right)^2-(3)^2=x^4-9$
51. Multiplying using the difference of squares formula: $\left(a^2+4\right)\left(a^2-4\right)=\left(a^2\right)^2-(4)^2=a^4-16$
53. Multiplying using the difference of squares formula: $\left(5y^4-8\right)\left(5y^4+8\right)=\left(5y^4\right)^2-(8)^2=25y^8-64$
55. Multiplying and simplifying: $(x+3)(x-3)+(x+5)(x-5)=\left(x^2-9\right)+\left(x^2-25\right)=2x^2-34$
57. Multiplying and simplifying:
$(2x+3)^2-(4x-1)^2=\left(4x^2+12x+9\right)-\left(16x^2-8x+1\right)=4x^2+12x+9-16x^2+8x-1=-12x^2+20x+8$

59. Multiplying and simplifying:

$$\begin{aligned}(a+1)^2-(a+2)^2+(a+3)^2&=\left(a^2+2a+1\right)-\left(a^2+4a+4\right)+\left(a^2+6a+9\right)\\&=a^2+2a+1-a^2-4a-4+a^2+6a+9\\&=a^2+4a+6\end{aligned}$$

61. Multiplying and simplifying:

$$\begin{aligned}(2x+3)^3&=(2x+3)(2x+3)^2\\&=(2x+3)\left(4x^2+12x+9\right)\\&=8x^3+24x^2+18x+12x^2+36x+27\\&=8x^3+36x^2+54x+27\end{aligned}$$

63. Finding the product: $49(51)=(50-1)(50+1)=(50)^2-(1)^2=2{,}500-1=2{,}499$

65. Evaluating when $x=2$:

$$(x+3)^2=(2+3)^2=(5)^2=25$$
$$x^2+6x+9=(2)^2+6(2)+9=4+12+9=25$$

67. Let x and $x+1$ represent the two integers. The expression can be written as:

$$(x)^2+(x+1)^2=x^2+\left(x^2+2x+1\right)=2x^2+2x+1$$

69. Let x, $x+1$, and $x+2$ represent the three integers. The expression can be written as:

$$(x)^2+(x+1)^2+(x+2)^2=x^2+\left(x^2+2x+1\right)+\left(x^2+4x+4\right)=3x^2+6x+5$$

71. Verifying the areas: $(a+b)^2=a^2+ab+ab+b^2=a^2+2ab+b^2$

73. Simplifying: $\dfrac{10x^3}{5x}=\dfrac{10}{5}\bullet\dfrac{x^3}{x}=2x^2$

75. Simplifying: $\dfrac{15x^2y}{3xy}=\dfrac{15}{3}\bullet\dfrac{x^2}{x}\bullet\dfrac{y}{y}=5x$

77. Simplifying: $\dfrac{35a^6b^8}{70a^2b^{10}}=\dfrac{35}{70}\bullet\dfrac{a^6}{a^2}\bullet\dfrac{b^8}{b^{10}}=\dfrac{1}{2}\bullet a^4\bullet\dfrac{1}{b^2}=\dfrac{a^4}{2b^2}$

79. Graphing both lines:

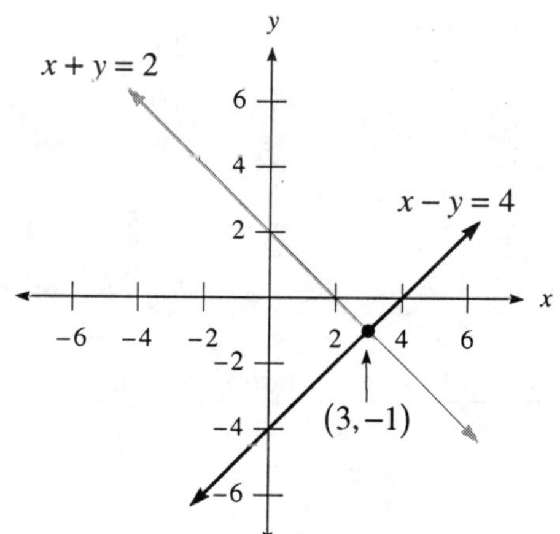

The intersection point is $(3,-1)$.

81. Graphing both lines:

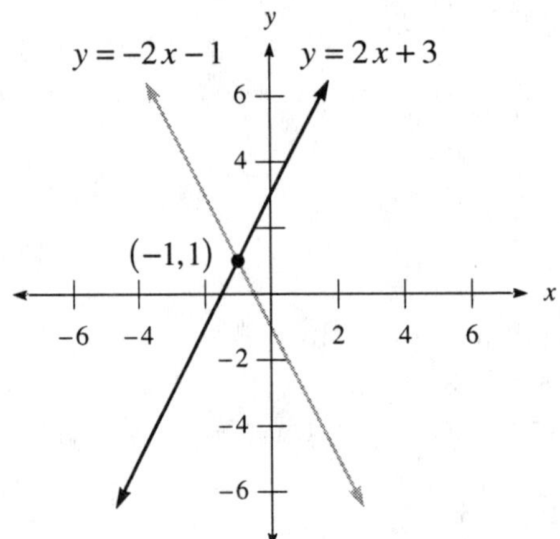

The intersection point is $(-1,1)$.

5.7 Dividing a Polynomial by a Monomial

1. Performing the division: $\dfrac{5x^2-10x}{5x}=\dfrac{5x^2}{5x}-\dfrac{10x}{5x}=x-2$

3. Performing the division: $\dfrac{15x-10x^3}{5x}=\dfrac{15x}{5x}-\dfrac{10x^3}{5x}=3-2x^2$

5. Performing the division: $\dfrac{25x^2y-10xy}{5x}=\dfrac{25x^2y}{5x}-\dfrac{10xy}{5x}=5xy-2y$

7. Performing the division: $\dfrac{35x^5-30x^4+25x^3}{5x}=\dfrac{35x^5}{5x}-\dfrac{30x^4}{5x}+\dfrac{25x^3}{5x}=7x^4-6x^3+5x^2$

9. Performing the division: $\dfrac{50x^5-25x^3+5x}{5x}=\dfrac{50x^5}{5x}-\dfrac{25x^3}{5x}+\dfrac{5x}{5x}=10x^4-5x^2+1$

11. Performing the division: $\dfrac{8a^2-4a}{-2a}=\dfrac{8a^2}{-2a}+\dfrac{-4a}{-2a}=-4a+2$

13. Performing the division: $\dfrac{16a^5+24a^4}{-2a}=\dfrac{16a^5}{-2a}+\dfrac{24a^4}{-2a}=-8a^4-12a^3$

15. Performing the division: $\dfrac{8ab+10a^2}{-2a}=\dfrac{8ab}{-2a}+\dfrac{10a^2}{-2a}=-4b-5a$

17. Performing the division: $\dfrac{12a^3b-6a^2b^2+14ab^3}{-2a}=\dfrac{12a^3b}{-2a}+\dfrac{-6a^2b^2}{-2a}+\dfrac{14ab^3}{-2a}=-6a^2b+3ab^2-7b^3$

19. Performing the division: $\dfrac{a^2+2ab+b^2}{-2a}=\dfrac{a^2}{-2a}+\dfrac{2ab}{-2a}+\dfrac{b^2}{-2a}=-\dfrac{a}{2}-b-\dfrac{b^2}{2a}$

21. Performing the division: $\dfrac{6x+8y}{2}=\dfrac{6x}{2}+\dfrac{8y}{2}=3x+4y$

23. Performing the division: $\dfrac{7y-21}{-7}=\dfrac{7y}{-7}+\dfrac{-21}{-7}=-y+3$

25. Performing the division: $\dfrac{10xy-8x}{2x}=\dfrac{10xy}{2x}-\dfrac{8x}{2x}=5y-4$

27. Performing the division: $\dfrac{x^2y-x^3y^2}{x}=\dfrac{x^2y}{x}-\dfrac{x^3y^2}{x}=xy-x^2y^2$

29. Performing the division: $\dfrac{x^2y-x^3y^2}{-x^2y}=\dfrac{x^2y}{-x^2y}+\dfrac{-x^3y^2}{-x^2y}=-1+xy$

31. Performing the division: $\frac{a^2b^2 - ab^2}{-ab^2} = \frac{a^2b^2}{-ab^2} + \frac{-ab^2}{-ab^2} = -a+1$

33. Performing the division: $\frac{x^3 - 3x^2y + xy^2}{x} = \frac{x^3}{x} - \frac{3x^2y}{x} + \frac{xy^2}{x} = x^2 - 3xy + y^2$

35. Performing the division: $\frac{10a^2 - 15a^2b + 25a^2b^2}{5a^2} = \frac{10a^2}{5a^2} - \frac{15a^2b}{5a^2} + \frac{25a^2b^2}{5a^2} = 2 - 3b + 5b^2$

37. Performing the division: $\frac{26x^2y^2 - 13xy}{-13xy} = \frac{26x^2y^2}{-13xy} + \frac{-13xy}{-13xy} = -2xy + 1$

39. Performing the division: $\frac{4x^2y^2 - 2xy}{4xy} = \frac{4x^2y^2}{4xy} - \frac{2xy}{4xy} = xy - \frac{1}{2}$

41. Performing the division: $\frac{5a^2x - 10ax^2 + 15a^2x^2}{20a^2x^2} = \frac{5a^2x}{20a^2x^2} - \frac{10ax^2}{20a^2x^2} + \frac{15a^2x^2}{20a^2x^2} = \frac{1}{4x} - \frac{1}{2a} + \frac{3}{4}$

43. Performing the division: $\frac{16x^5 + 8x^2 + 12x}{12x^3} = \frac{16x^5}{12x^3} + \frac{8x^2}{12x^3} + \frac{12x}{12x^3} = \frac{4x^2}{3} + \frac{2}{3x} + \frac{1}{x^2}$

45. Performing the division: $\frac{9a^{5m} - 27a^{3m}}{3a^{2m}} = \frac{9a^{5m}}{3a^{2m}} - \frac{27a^{3m}}{3a^{2m}} = 3a^{5m-2m} - 9a^{3m-2m} = 3a^{3m} - 9a^m$

47. Performing the division:

$$\frac{10x^{5m} - 25x^{3m} + 35x^m}{5x^m} = \frac{10x^{5m}}{5x^m} - \frac{25x^{3m}}{5x^m} + \frac{35x^m}{5x^m} = 2x^{5m-m} - 5x^{3m-m} + 7x^{m-m} = 2x^{4m} - 5x^{2m} + 7$$

49. Simplifying and then dividing:

$$\begin{aligned} \frac{2x^3(3x+2) - 3x^2(2x-4)}{2x^2} &= \frac{6x^4 + 4x^3 - 6x^3 + 12x^2}{2x^2} \\ &= \frac{6x^4 - 2x^3 + 12x^2}{2x^2} \\ &= \frac{6x^4}{2x^2} - \frac{2x^3}{2x^2} + \frac{12x^2}{2x^2} \\ &= 3x^2 - x + 6 \end{aligned}$$

51. Simplifying and then dividing:

$$\frac{(x+2)^2 - (x-2)^2}{2x} = \frac{\left(x^2 + 4x + 4\right) - \left(x^2 - 4x + 4\right)}{2x} = \frac{x^2 + 4x + 4 - x^2 + 4x - 4}{2x} = \frac{8x}{2x} = 4$$

53. Simplifying and then dividing:

$$\frac{(x+5)^2 + (x+5)(x-5)}{2x} = \frac{\left(x^2 + 10x + 25\right) + \left(x^2 - 25\right)}{2x} = \frac{2x^2 + 10x}{2x} = \frac{2x^2}{2x} + \frac{10x}{2x} = x + 5$$

55. Evaluating each expression when $x = 2$:

$$\frac{10x+15}{5} = \frac{10(2)+15}{5} = \frac{20+15}{5} = \frac{35}{5} = 7$$

$$2x + 3 = 2(2) + 3 = 4 + 3 = 7$$

57. Evaluating each expression when $x = 10$:

$$\frac{3x+8}{2} = \frac{3(10)+8}{2} = \frac{30+8}{2} = \frac{38}{2} = 19$$

$$3x + 4 = 3(10) + 4 = 30 + 4 = 34$$

Thus $\frac{3x+8}{2} \neq 3x + 4$.

59. Adding the two equations:

$$2x = 14$$
$$x = 7$$

Substituting into the first equation:

$$7 + y = 6$$
$$y = -1$$

The solution is $(7, -1)$.

61. Multiplying the second equation by 3:

$$2x - 3y = -5$$
$$3x + 3y = 15$$

Adding the two equations:

$$5x = 10$$
$$x = 2$$

Substituting into the second equation:

$$2 + y = 5$$
$$y = 3$$

The solution is $(2, 3)$.

63. Substituting into the first equation:

$$x + 2x - 1 = 2$$
$$3x - 1 = 2$$
$$3x = 3$$
$$x = 1$$

Substituting into the second equation: $y = 2(1) - 1 = 2 - 1 = 1$

The solution is $(1, 1)$.

65. Substituting into the first equation:

$$4x + 2(-2x + 4) = 8$$
$$4x - 4x + 8 = 8$$
$$8 = 8$$

Since this statement is true, the system is dependent. The two lines coincide.

5.8 Dividing a Polynomial by a Polynomial

1. Using long division:

$$\begin{array}{r} x - 2 \\ x - 3 \overline{) x^2 - 5x + 6} \\ \underline{x^2 - 3x} \\ -2x + 6 \\ \underline{-2x + 6} \\ 0 \end{array}$$

The quotient is $x - 2$.

3. Using long division:

$$\begin{array}{r} a + 4 \\ a + 5 \overline{) a^2 + 9a + 20} \\ \underline{a^2 + 5a} \\ 4a + 20 \\ \underline{4a + 20} \\ 0 \end{array}$$

The quotient is $a + 4$.

5. Using long division:

$$\begin{array}{r}
x-3 \\
x-3\overline{\big)\,x^2-6x+9} \\
\underline{x^2-3x} \\
-3x+9 \\
\underline{-3x+9} \\
0
\end{array}$$

The quotient is $x-3$.

7. Using long division:

$$\begin{array}{r}
x+3 \\
2x-1\overline{\big)\,2x^2+5x-3} \\
\underline{2x^2-x} \\
6x-3 \\
\underline{6x-3} \\
0
\end{array}$$

The quotient is $x+3$.

9. Using long division:

$$\begin{array}{r}
a-5 \\
2a+1\overline{\big)\,2a^2-9a-5} \\
\underline{2a^2+a} \\
-10a-5 \\
\underline{-10a-5} \\
0
\end{array}$$

The quotient is $a-5$.

11. Using long division:

$$\begin{array}{r}
x+2 \\
x+3\overline{\big)\,x^2+5x+8} \\
\underline{x^2+3x} \\
2x+8 \\
\underline{2x+6} \\
2
\end{array}$$

The quotient is $x+2+\dfrac{2}{x+3}$.

13. Using long division:

$$\begin{array}{r}
a-2 \\
a+5\overline{\big)\,a^2+3a+2} \\
\underline{a^2+5a} \\
-2u+2 \\
\underline{-2a-10} \\
12
\end{array}$$

The quotient is $a-2+\dfrac{12}{a+5}$.

15. Using long division:

$$\begin{array}{r} x+4 \\ x-2\overline{\big)x^2+2x+1} \\ \underline{x^2-2x} \\ 4x+1 \\ \underline{4x-8} \\ 9 \end{array}$$

The quotient is $x+4+\dfrac{9}{x-2}$.

17. Using long division:

$$\begin{array}{r} x+4 \\ x+1\overline{\big)x^2+5x-6} \\ \underline{x^2+\ x} \\ 4x-6 \\ \underline{4x+4} \\ -10 \end{array}$$

The quotient is $x+4+\dfrac{-10}{x+1}$.

19. Using long division:

$$\begin{array}{r} a+1 \\ a+2\overline{\big)a^2+3a+1} \\ \underline{a^2+2a} \\ a+1 \\ \underline{a+2} \\ -1 \end{array}$$

The quotient is $a+1+\dfrac{-1}{a+2}$.

21. Using long division:

$$\begin{array}{r} x-3 \\ 2x+4\overline{\big)2x^2-2x+5} \\ \underline{2x^2+4x} \\ -6x+5 \\ \underline{-6x-12} \\ 17 \end{array}$$

The quotient is $x-3+\dfrac{17}{2x+4}$.

23. Using long division:

$$\begin{array}{r} 3a-2 \\ 2a+3\overline{\big)6a^2+5a+1} \\ \underline{6a^2+9a} \\ -4a+1 \\ \underline{-4a-6} \\ 7 \end{array}$$

The quotient is $3a-2+\dfrac{7}{2a+3}$.

25. Using long division:

$$\begin{array}{r} 2a^2 - a - 3 \\ 3a-5\overline{)6a^3 - 13a^2 - 4a + 15} \\ \underline{6a^3 - 10a^2} \\ -3a^2 - 4a \\ \underline{-3a^2 + 5a} \\ -9a + 15 \\ \underline{-9a + 15} \\ 0 \end{array}$$

The quotient is $2a^2 - a - 3$.

27. Using long division:

$$\begin{array}{r} x^2 - x + 5 \\ x+1\overline{)x^3 + 0x^2 + 4x + 5} \\ \underline{x^3 + x^2} \\ -x^2 + 4x \\ \underline{-x^2 - x} \\ 5x + 5 \\ \underline{5x + 5} \\ 0 \end{array}$$

The quotient is $x^2 - x + 5$.

29. Using long division:

$$\begin{array}{r} x^2 + x + 1 \\ x-1\overline{)x^3 + 0x^2 + 0x - 1} \\ \underline{x^3 - x^2} \\ x^2 + 0x \\ \underline{x^2 - x} \\ x - 1 \\ \underline{x - 1} \\ 0 \end{array}$$

The quotient is $x^2 + x + 1$.

31. Using long division:

$$\begin{array}{r} x^2 + 2x + 4 \\ x-2\overline{)x^3 + 0x^2 + 0x - 8} \\ \underline{x^3 - 2x^2} \\ 2x^2 + 0x \\ \underline{2x^2 - 4x} \\ 4x - 8 \\ \underline{4x - 8} \\ 0 \end{array}$$

The quotient is $x^2 + 2x + 4$.

33. Let x and y represent the two numbers. The system of equations is:

$$x+y=25$$
$$y=4x$$

Substituting into the first equation:

$$x+4x=25$$
$$5x=25$$
$$x=5$$
$$y=4(5)=20$$

The two numbers are 5 and 20.

35. Let x represent the amount invested at 8% and y represent the amount invested at 9%. The system of equations is:

$$x+y=1200$$
$$0.08x+0.09y=100$$

Multiplying the first equation by –0.08:

$$-0.08x-0.08y=-96$$
$$0.08x+0.09y=100$$

Adding the two equations:

$$0.01y=4$$
$$y=400$$

Substituting into the first equation:

$$x+400=1200$$
$$x=800$$

You have \$800 invested at 8% and \$400 invested at 9%.

37. Let x represent the number of \$5 bills and $x + 4$ represent the number of \$10 bills. The equation is:

$$5(x)+10(x+4)=160$$
$$5x+10x+40=160$$
$$15x+40=160$$
$$15x=120$$
$$x=8$$
$$x+4=12$$

You have 8 \$5 bills and 12 \$10 bills.

39. Let x represent the gallons of 20% antifreeze and y represent the gallons of 60% antifreeze. The system of equations is:

$$x+y=16$$
$$0.20x+0.60y=0.35(16)$$

Multiplying the first equation by –0.20:

$$-0.20x-0.20y=-3.2$$
$$0.20x+0.60y=5.6$$

Adding the two equations:

$$0.40y=2.4$$
$$y=6$$

Substituting into the first equation:

$$x+6=16$$
$$x=10$$

The mixture contains 10 gallons of 20% antifreeze and 6 gallons of 60% antifreeze.

Chapter 5 Review

1. Simplifying the expression: $(-1)^3=(-1)(-1)(-1)=-1$

3. Simplifying the expression: $\left(\frac{3}{7}\right)^2=\left(\frac{3}{7}\right)\bullet\left(\frac{3}{7}\right)=\frac{9}{49}$

5. Simplifying the expression: $x^{15}\bullet x^7\bullet x^5\bullet x^3=x^{15+7+5+3}=x^{30}$

7. Simplifying the expression: $\left(2^6\right)^4=2^{6\bullet 4}=2^{24}$

9. Simplifying the expression: $(-2xyz)^3=(-2)^3\bullet x^3y^3z^3=-8x^3y^3z^3$

11. Writing with positive exponents: $4x^{-5} = 4 \bullet \dfrac{1}{x^5} = \dfrac{4}{x^5}$

13. Simplifying the expression: $\dfrac{a^9}{a^3} = a^{9-3} = a^6$

15. Simplifying the expression: $\dfrac{x^9}{x^{-6}} = x^{9-(-6)} = x^{9+6} = x^{15}$

17. Simplifying the expression: $(-3xy)^0 = 1$

19. Simplifying the expression: $\left(3x^3y^2\right)^2 = 3^2x^6y^4 = 9x^6y^4$

21. Simplifying the expression: $\left(-3xy^2\right)^{-3} = \dfrac{1}{\left(-3xy^2\right)^3} = \dfrac{1}{(-3)^3x^3y^6} = -\dfrac{1}{27x^3y^6}$

23. Simplifying the expression: $\dfrac{\left(x^{-3}\right)^3\left(x^6\right)^{-1}}{\left(x^{-5}\right)^{-4}} = \dfrac{x^{-9} \bullet x^{-6}}{x^{20}} = \dfrac{x^{-15}}{x^{20}} = x^{-15-20} = x^{-35} = \dfrac{1}{x^{35}}$

25. Simplifying the expression: $\dfrac{\left(10x^3y^5\right)\left(21x^2y^6\right)}{\left(7xy^3\right)\left(5x^9y\right)} = \dfrac{210x^5y^{11}}{35x^{10}y^4} = \dfrac{210}{35} \bullet \dfrac{x^5}{x^{10}} \bullet \dfrac{y^{11}}{y^4} = 6 \bullet \dfrac{1}{x^5} \bullet y^7 = \dfrac{6y^7}{x^5}$

27. Simplifying the expression: $\dfrac{8x^8y^3}{2x^3y} - \dfrac{10x^6y^9}{5xy^7} = 4x^5y^2 - 2x^5y^2 = 2x^5y^2$

29. Finding the quotient: $\dfrac{4.6 \times 10^5}{2 \times 10^{-3}} - 2.3 \times 10^8$

31. Performing the operations: $\left(3a^2 - 5a + 5\right) + \left(5a^2 - 7a - 8\right) = \left(3a^2 + 5a^2\right) + (-5a - 7a) + (5 - 8) = 8a^2 - 12a - 3$

33. Performing the operations:

$$\begin{aligned}\left(4x^2 - 3x - 2\right) - \left(8x^2 + 3x - 2\right) &= 4x^2 - 3x - 2 - 8x^2 - 3x + 2 \\ &= \left(4x^2 - 8x^2\right) + (-3x - 3x) + (-2 + 2) \\ &= -4x^2 - 6x\end{aligned}$$

35. Multiplying: $3x(4x - 7) = 3x(4x) - 3x(7) = 12x^2 - 21x$

37. Multiplying using the column method:

$$\begin{array}{r} a^2 + 5a - 4 \\ a + 1 \\ \hline a^3 + 5a^2 - 4a \\ a^2 + 5a - 4 \\ \hline a^3 + 6a^2 + a - 4 \end{array}$$

39. Multiplying using the FOIL method: $(3x - 7)(2x - 5) = 6x^2 - 14x - 15x + 35 = 6x^2 - 29x + 35$

41. Multiplying using the difference of squares formula: $\left(a^2 - 3\right)\left(a^2 + 3\right) = \left(a^2\right)^2 - (3)^2 = a^4 - 9$

43. Multiplying using the square of binomial formula: $(3x + 4)^2 = (3x)^2 + 2(3x)(4) + (4)^2 = 9x^2 + 24x + 16$

45. Performing the division: $\dfrac{10ab + 20a^2}{-5a} = \dfrac{10ab}{-5a} + \dfrac{20a^2}{-5a} = -2b - 4a$

47. Using long division:

$$\begin{array}{r} x+9 \\ x+6\overline{)x^2+15x+54} \\ \underline{x^2+6x} \\ 9x+54 \\ \underline{9x+54} \\ 0 \end{array}$$

The quotient is $x+9$.

49. Using long division:

$$\begin{array}{r} x^2-4x+16 \\ x+4\overline{)x^3+0x^2+0x+64} \\ \underline{x^3+4x^2} \qquad\qquad \\ -4x^2+0x \qquad \\ \underline{-4x^2-16x} \qquad \\ 16x+64 \\ \underline{16x+64} \\ 0 \end{array}$$

The quotient is $x^2-4x+16$.

51. Using long division:

$$\begin{array}{r} x^2-4x+5 \\ 2x+1\overline{)2x^3-7x^2+6x+10} \\ \underline{2x^3+x^2} \qquad\qquad \\ -8x^2+6x \qquad \\ \underline{-8x^2-4x} \qquad \\ 10x+10 \\ \underline{10x+5} \\ 5 \end{array}$$

The quotient is $x^2-4x+5+\dfrac{5}{2x+1}$.

53. If the width is 2 feet, the volume is $3\bullet 2^3=3\bullet 8=24$ cubic feet. Yes, this would hold as much food as your refrigerator.

Cumulative Review: Chapters 1-5

1. Simplifying the expression: $-\left(-\frac{3}{4}\right)=\frac{3}{4}$
3. Simplifying the expression: $6\bullet 7+7\bullet 9=42+63=105$
5. Simplifying the expression: $6(4a+2)-3(5a-1)=24a+12-15a+3=24a-15a+12+3=9a+15$
7. Simplifying the expression: $-15-(-3)=-15+3=-12$
9. Simplifying the expression: $(-9)(-5)=45$
11. Simplifying the expression: $(2y)^4=2^4y^4=16y^4$
13. Simplifying the expression: $\dfrac{\left(12xy^5\right)^3\left(16x^2y^2\right)}{\left(8x^3y^3\right)\left(3x^5y\right)}=\dfrac{1728x^3y^{15}\bullet 16x^2y^2}{24x^8y^4}=\dfrac{27648x^5y^{17}}{24x^8y^4}=\dfrac{1152y^{13}}{x^3}$
15. Multiplying using the square of binomial formula: $(5x-1)^2=(5x)^2-2(5x)(1)+(1)^2=25x^2-10x+1$

17. Multiplying using the column method:

$$\begin{array}{r} x^2+x+1 \\ x-1 \\ \hline x^3+x^2+x \\ -x^2-x-1 \\ \hline x^3-1 \end{array}$$

19. Solving the equation:

$$\begin{aligned} 6a-5&=4a \\ -5&=-2a \\ a&=\frac{5}{2} \end{aligned}$$

21. Solving the equation:

$$\begin{aligned} 2(3x+5)+8&=2x+10 \\ 6x+10+8&=2x+10 \\ 6x+18&=2x+10 \\ 4x+18&=10 \\ 4x&=-8 \\ x&=-2 \end{aligned}$$

23. Solving the inequality:

$$\begin{aligned} 3(2t-5)-7&\le 5(3t+1)+5 \\ 6t-15-7&\le 15t+5+5 \\ 6t-22&\le 15t+10 \\ -9t-22&\le 10 \\ -9t&\le 32 \\ t&\ge -32/9 \end{aligned}$$

Graphing the solution set:

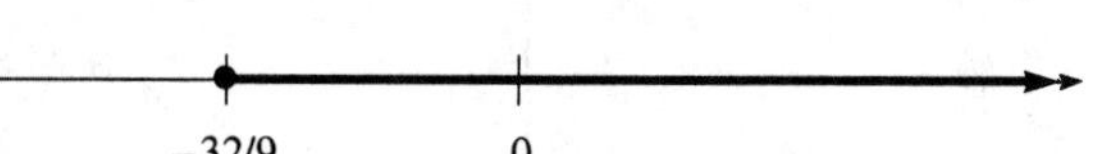

25. Solving the compound inequality:

$$\begin{aligned} -5\le 4x+3\le 11 \\ -8\le 4x\le 8 \\ -2\le x\le 2 \end{aligned}$$

Graphing the solution set:

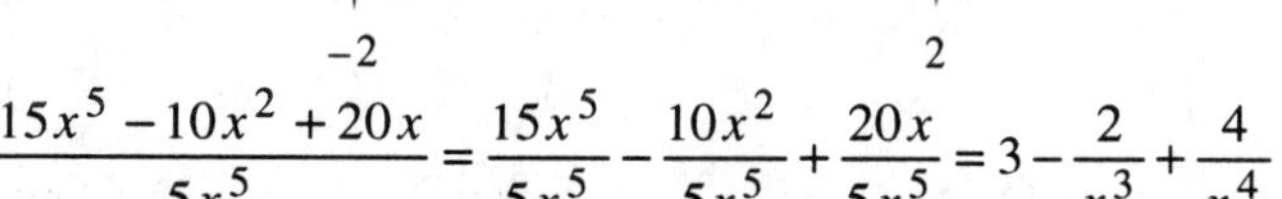

27. Performing the division: $\dfrac{15x^5-10x^2+20x}{5x^5}=\dfrac{15x^5}{5x^5}-\dfrac{10x^2}{5x^5}+\dfrac{20x}{5x^5}=3-\dfrac{2}{x^3}+\dfrac{4}{x^4}$

29. Graphing the equation:

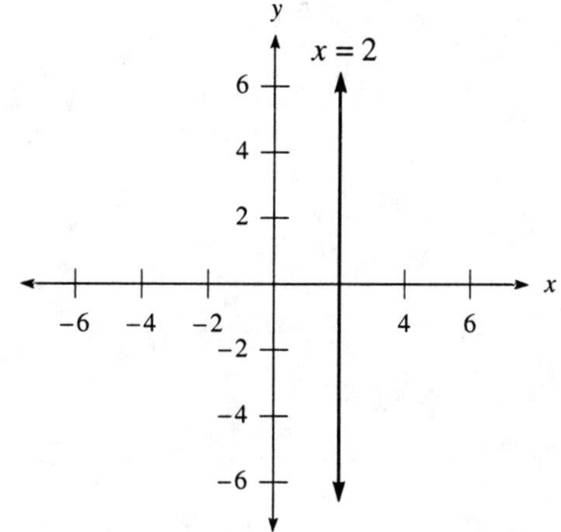

31. Graphing the two lines:

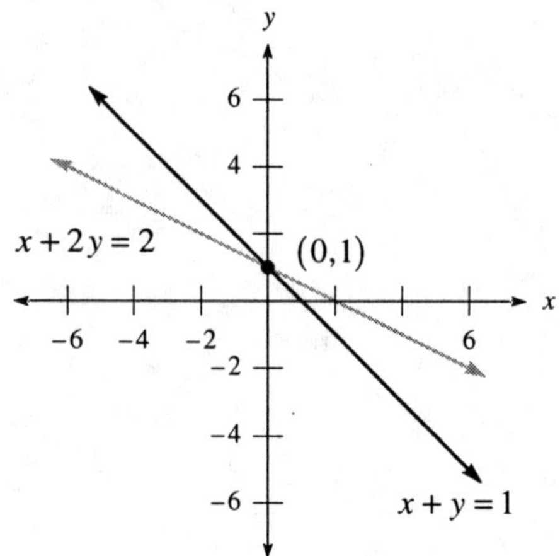

The intersection point is $(0,1)$.

33. Graphing the two lines:

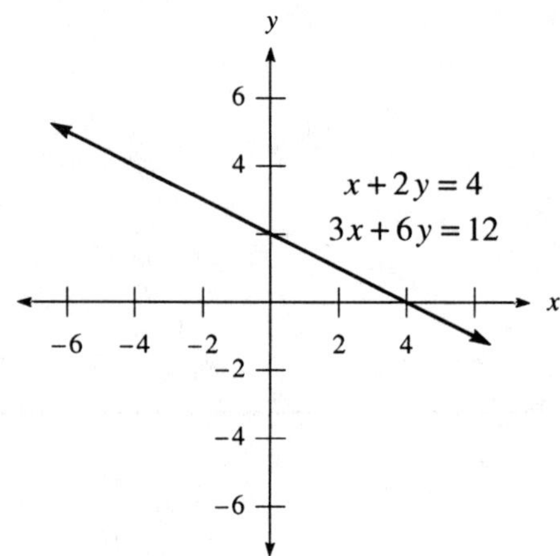

The system is dependent. The two lines coincide.

35. Multiplying the first equation by –3:

$$-3x - 6y = -15$$
$$3x + 6y = 14$$

Adding the two equations:

$$0 = -1$$

Since this statement is false, there is no solution to the system. The two lines are parallel.

37. To clear each equation of fractions, multiply the first equation by 12 and the second equation by 5:

$$12\left(\frac{1}{6}x + \frac{1}{4}y\right) = 12(1) \qquad 5\left(\frac{6}{5}x - y\right) = 5\left(\frac{8}{5}\right)$$
$$2x + 3y = 12 \qquad 6x - 5y = 8$$

The system of equations is:

$$2x + 3y = 12$$
$$6x - 5y = 8$$

Multiplying the first equation by –3:

$$-6x - 9y = -36$$
$$6x - 5y = 8$$

Adding the two equations:

$$-14y = -28$$
$$y = 2$$

Substituting into $2x+3y=12$:

$$\begin{aligned} 2x+3(2) &= 12 \\ 2x+6 &= 12 \\ 2x &= 6 \\ x &= 3 \end{aligned}$$

The solution is $(3,2)$.

39. Solving the second equation for x:

$$\begin{aligned} 2y &= x-3 \\ x &= 2y+3 \end{aligned}$$

Substituting into the first equation:

$$\begin{aligned} 4(2y+3)+5y &= 25 \\ 8y+12+5y &= 25 \\ 13y+12 &= 25 \\ 13y &= 13 \\ y &= 1 \\ x &= 2(1)+3=5 \end{aligned}$$

The solution is $(5,1)$.

41. Evaluating when $x = 3$: $8x-3=8(3)-3=24-3=21$

43. The irrational numbers are $-\sqrt{2}$ and π.

45. Substituting each ordered pair:

$(0,3)$: $3(0)-4(3)=0-12=-12 \neq 12$

$(4,0)$: $3(4)-4(0)=12-0=12$

$\left(\frac{16}{3},1\right)$: $3\left(\frac{16}{3}\right)-4(1)=16-4-12$

The ordered pairs $(4,0)$ and $\left(\frac{16}{3},1\right)$ are solutions to the equation.

47. First find the slope of the line: $m=\dfrac{-1-3}{6-(-2)}=\dfrac{-4}{8}=-\dfrac{1}{2}$

Using the point-slope formula:

$$\begin{aligned} y-3 &= -\frac{1}{2}(x-(-2)) \\ y-3 &= -\frac{1}{2}(x+2) \\ y-3 &= -\frac{1}{2}x-1 \\ y &= -\frac{1}{2}x+2 \end{aligned}$$

To find the x-intercept, let $y = 0$:

$$\begin{aligned} 0 &= -\frac{1}{2}x+2 \\ -2 &= -\frac{1}{2}x \\ x &= 4 \end{aligned}$$

To find the y-intercept, let $x = 0$: $y=-\frac{1}{2}(0)+2=0+2=2$

49. First find the slope: $m=\frac{8-4}{1-(-2)}=\frac{4}{3}$

Using the point-slope formula:

$$y-8=\frac{4}{3}(x-1)$$
$$y-8=\frac{4}{3}x-\frac{4}{3}$$
$$y=\frac{4}{3}x+\frac{20}{3}$$

Chapter 5 Test

1. Simplifying the expression: $(-3)^4=(-3)(-3)(-3)(-3)=81$
2. Simplifying the expression: $\left(\frac{3}{4}\right)^2=\left(\frac{3}{4}\right)\left(\frac{3}{4}\right)=\frac{9}{16}$
3. Simplifying the expression: $\left(3x^3\right)^2\left(2x^4\right)^3=9x^6\bullet 8x^{12}=72x^{18}$
4. Simplifying the expression: $3^{-2}=\frac{1}{3^2}=\frac{1}{9}$
5. Simplifying the expression: $\left(3a^4b^2\right)^0=1$
6. Simplifying the expression: $\frac{a^{-3}}{a^{-5}}=a^{-3-(-5)}=a^{-3+5}=a^2$
7. Simplifying the expression: $\frac{\left(x^{-2}\right)^3\left(x^{-3}\right)^{-5}}{\left(x^{-4}\right)^{-2}}=\frac{x^{-6}x^{15}}{x^8}=\frac{x^9}{x^8}=x^{9-8}=x$
8. Writing in scientific notation: $0.0278=2.78\times10^{-2}$
9. Writing in expanded form: $2.43\times10^5=243,000$
10. Simplifying the expression: $\frac{35x^2y^4z}{70x^6y^2z}=\frac{35}{70}\bullet\frac{x^2}{x^6}\bullet\frac{y^4}{y^2}\bullet\frac{z}{z}=\frac{1}{2}\bullet\frac{1}{x^4}\bullet y^2=\frac{y^2}{2x^4}$
11. Simplifying the expression: $\frac{\left(6a^2b\right)\left(9a^3b^2\right)}{18a^4b^3}=\frac{54a^5b^3}{18a^4b^3}=3a$
12. Simplifying the expression: $\frac{24x^7}{3x^2}+\frac{14x^9}{7x^4}=8x^5+2x^5=10x^5$
13. Simplifying the expression: $\frac{\left(2.4\times10^5\right)\left(4.5\times10^{-2}\right)}{1.2\times10^{-6}}=\frac{10.8\times10^3}{1.2\times10^{-6}}=9.0\times10^9$
14. Performing the operations: $8x^2-4x+6x+2=8x^2+2x+2$
15. Performing the operations: $\left(5x^2-3x+4\right)-\left(2x^2-7x-2\right)=5x^2-3x+4-2x^2+7x+2=3x^2+4x+6$
16. Performing the operations: $(6x-8)-(3x-4)=6x-8-3x+4=3x-4$
17. Evaluating when $y=-2$: $2y^2-3y-4=2(-2)^2-3(-2)-4=2(4)+6-4=8+6-4=10$
18. Multiplying using the distributive property:
$$2a^2\left(3a^2-5a+4\right)=2a^2\left(3a^2\right)-2a^2(5a)+2a^2(4)=6a^4-10a^3+8a^2$$
19. Multiplying using the FOIL method: $\left(x+\frac{1}{2}\right)\left(x+\frac{1}{3}\right)=x^2+\frac{1}{2}x+\frac{1}{3}x+\frac{1}{6}=x^2+\frac{5}{6}x+\frac{1}{6}$
20. Multiplying using the FOIL method: $(4x-5)(2x+3)=8x^2-10x+12x-15=8x^2+2x-15$

21. Multiplying using the column method:

$$\begin{array}{r l} & x^2+3x+9 \\ & \quad\;\; x-3 \\ \hline & x^3+3x^2+9x \\ & \quad -3x^2-9x-27 \\ \hline & x^3-27 \end{array}$$

22. Multiplying using the square of binomial formula: $(x+5)^2=x^2+2(x)(5)+(5)^2=x^2+10x+25$

23. Multiplying using the square of binomial formula: $(3a-2b)^2=(3a)^2-2(3a)(2b)+(2b)^2=9a^2-12ab+4b^2$

24. Multiplying using the difference of squares formula: $(3x-4y)(3x+4y)=(3x)^2-(4y)^2=9x^2-16y^2$

25. Multiplying using the difference of squares formula: $\left(a^2-3\right)\left(a^2+3\right)=\left(a^2\right)^2-(3)^2=a^4-9$

26. Performing the division: $\dfrac{10x^3+15x^2-5x}{5x}=\dfrac{10x^3}{5x}+\dfrac{15x^2}{5x}-\dfrac{5x}{5x}=2x^2+3x-1$

27. Using long division:

$$\begin{array}{r l} & \quad\quad 4x+3 \\ 2x-3 \big) & 8x^2-\;6x-5 \\ & \underline{8x^2-12x} \\ & \quad\quad\;\; 6x-5 \\ & \quad\quad\;\; \underline{6x-9} \\ & \quad\quad\quad\quad 4 \end{array}$$

The quotient is $4x+3+\dfrac{4}{2x-3}$.

28. Using long division:

$$\begin{array}{r l} & \quad 3x^2+9x+25 \\ x-3 \big) & 3x^3+0x^2-2x+1 \\ & \underline{3x^3-9x^2} \\ & \quad\;\; 9x^2-\;2x \\ & \quad\;\; \underline{9x^2-27x} \\ & \quad\quad\quad\;\; 25x+1 \\ & \quad\quad\quad\;\; \underline{25x-75} \\ & \quad\quad\quad\quad\;\; 76 \end{array}$$

The quotient is $3x^2+9x+25+\dfrac{76}{x-3}$.

29. Using the volume formula: $V=(2.5\text{ cm})^3=15.625\text{ cm}^3$

30. Let x represent the width, $5x$ represent the length, and $\frac{1}{5}x$ represent the height. The volume is given by:

$$V=(x)(5x)\left(\frac{1}{5}x\right)=x^3$$

Chapter 6
Factoring

6.1 The Greatest Common Factor and Factoring by Grouping

1. Factoring out the greatest common factor: $15x+25=5(3x+5)$
3. Factoring out the greatest common factor: $6a+9=3(2a+3)$
5. Factoring out the greatest common factor: $4x-8y=4(x-2y)$
7. Factoring out the greatest common factor: $3x^2-6x-9=3(x^2-2x-3)$
9. Factoring out the greatest common factor: $3a^2-3a-60=3(a^2-a-20)$
11. Factoring out the greatest common factor: $24y^2-52y+24=4(6y^2-13y+6)$
13. Factoring out the greatest common factor: $9x^2-8x^3=x^2(9-8x)$
15. Factoring out the greatest common factor: $13a^2-26a^3=13a^2(1-2a)$
17. Factoring out the greatest common factor: $21x^2y-28xy^2=7xy(3x-4y)$
19. Factoring out the greatest common factor: $22a^2b^2-11ab^2=11ab^2(2a-1)$
21. Factoring out the greatest common factor: $7x^3+21x^2-28x=7x(x^2+3x-4)$
23. Factoring out the greatest common factor: $121y^4-11x^4=11(11y^4-x^4)$
25. Factoring out the greatest common factor: $100x^4-50x^3+25x^2=25x^2(4x^2-2x+1)$
27. Factoring out the greatest common factor: $8a^2+16b^2+32c^2=8(a^2+2b^2+4c^2)$
29. Factoring out the greatest common factor: $4a^2b-16ab^2+32a^2b^2=4ab(a-4b+8ab)$
31. Factoring out the greatest common factor: $121a^3b^2-22a^2b^3+33a^3b^3=11a^2b^2(11a-2b+3ab)$
33. Factoring out the greatest common factor: $12x^2y^3-72x^5y^3-36x^4y^4=12x^2y^3(1-6x^3-3x^2y)$
35. Factoring by grouping: $xy+5x+3y+15=x(y+5)+3(y+5)=(y+5)(x+3)$
37. Factoring by grouping: $xy+6x+2y+12=x(y+6)+2(y+6)=(y+6)(x+2)$
39. Factoring by grouping: $ab+7a-3b-21=a(b+7)-3(b+7)=(b+7)(a-3)$
41. Factoring by grouping: $ax-bx+ay-by=x(a-b)+y(a-b)=(a-b)(x+y)$
43. Factoring by grouping: $2ax+6x-5a-15=2x(a+3)-5(a+3)=(a+3)(2x-5)$
45. Factoring by grouping: $3xb-4b-6x+8=b(3x-4)-2(3x-4)=(3x-4)(b-2)$
47. Factoring by grouping: $x^2+ax+2x+2a=x(x+a)+2(x+a)=(x+a)(x+2)$
49. Factoring by grouping: $x^2-ax-bx+ab=x(x-a)-b(x-a)=(x-a)(x-b)$
51. Factoring by grouping: $ax+ay+bx+by+cx+cy=a(x+y)+b(x+y)+c(x+y)=(x+y)(a+b+c)$
53. Factoring by grouping: $6x^2+9x+4x+6=3x(2x+3)+2(2x+3)=(2x+3)(3x+2)$

55. Factoring by grouping: $20x^2-2x+50x-5=2x(10x-1)+5(10x-1)=(10x-1)(2x+5)$
57. Factoring by grouping: $20x^2+4x+25x+5=4x(5x+1)+5(5x+1)=(5x+1)(4x+5)$
59. Factoring by grouping: $x^3+2x^2+3x+6=x^2(x+2)+3(x+2)=(x+2)(x^2+3)$
61. Factoring by grouping: $6x^3-4x^2+15x-10=2x^2(3x-2)+5(3x-2)=(3x-2)(2x^2+5)$
63. Its greatest common factor is $3 \bullet 2=6$.
65. The correct factoring is: $12x^2+6x+3=3(4x^2+2x+1)$
67. The factored form is: $A=1000+1000r=1000(1+r)$
    Substituting $r=0.12$: $A=1000(1+0.12)=1000(1.12)=\$1,120$
69. Multiplying using the FOIL method: $(x+3)(x+4)=x^2+3x+4x+12=x^2+7x+12$
71. Multiplying using the FOIL method: $(x+7)(x-2)=x^2+7x-2x-14=x^2+5x-14$
73. Multiplying using the FOIL method: $(x-7)(x+2)=x^2-7x+2x-14=x^2-5x-14$
75. Multiplying using the FOIL method: $(x-3)(x+2)=x^2-3x+2x-6=x^2-x-6$
77. Multiplying using the column method:

$$\begin{array}{r} x^2-3x+9 \\ x+3 \\ \hline x^3-3x^2+9x \quad\quad \\ 3x^2-9x+27 \\ \hline x^3 \quad\quad\quad\quad +27 \end{array}$$

79. Multiplying using the column method:

$$\begin{array}{r} x^2+4x-3 \\ 2x+1 \\ \hline 2x^3+8x^2-6x \quad\quad \\ x^2+4x-3 \\ \hline 2x^3+9x^2-2x-3 \end{array}$$

6.2 Factoring Trinomials

1. Factoring the trinomial: $x^2+7x+12=(x+3)(x+4)$
3. Factoring the trinomial: $x^2+3x+2=(x+2)(x+1)$
5. Factoring the trinomial: $a^2+10a+21=(a+7)(a+3)$
7. Factoring the trinomial: $x^2-7x+10=(x-5)(x-2)$
9. Factoring the trinomial: $y^2-10y+21=(y-7)(y-3)$
11. Factoring the trinomial: $x^2-x-12=(x-4)(x+3)$
13. Factoring the trinomial: $y^2+y-12=(y+4)(y-3)$
15. Factoring the trinomial: $x^2+5x-14=(x+7)(x-2)$
17. Factoring the trinomial: $r^2-8r-9=(r-9)(r+1)$
19. Factoring the trinomial: $x^2-x-30=(x-6)(x+5)$
21. Factoring the trinomial: $a^2+15a+56=(a+7)(a+8)$
23. Factoring the trinomial: $y^2-y-42=(y-7)(y+6)$
25. Factoring the trinomial: $x^2+13x+42=(x+7)(x+6)$
27. Factoring the trinomial: $2x^2+6x+4=2(x^2+3x+2)=2(x+2)(x+1)$

29. Factoring the trinomial: $3a^2-3a-60=3\left(a^2-a-20\right)=3(a-5)(a+4)$
31. Factoring the trinomial: $100x^2-500x+600=100\left(x^2-5x+6\right)=100(x-3)(x-2)$
33. Factoring the trinomial: $100p^2-1300p+4000=100\left(p^2-13p+40\right)=100(p-8)(p-5)$
35. Factoring the trinomial: $x^4-x^3-12x^2=x^2\left(x^2-x-12\right)=x^2(x-4)(x+3)$
37. Factoring the trinomial: $2r^3+4r^2-30r=2r\left(r^2+2r-15\right)=2r(r+5)(r-3)$
39. Factoring the trinomial: $2y^4-6y^3-8y^2=2y^2\left(y^2-3y-4\right)=2y^2(y-4)(y+1)$
41. Factoring the trinomial: $x^5+4x^4+4x^3=x^3\left(x^2+4x+4\right)=x^3(x+2)(x+2)=x^3(x+2)^2$
43. Factoring the trinomial: $3y^4-12y^3-15y^2=3y^2\left(y^2-4y-5\right)=3y^2(y-5)(y+1)$
45. Factoring the trinomial: $4x^4-52x^3+144x^2=4x^2\left(x^2-13x+36\right)=4x^2(x-9)(x-4)$
47. Factoring the trinomial: $x^2+5xy+6y^2=(x+2y)(x+3y)$
49. Factoring the trinomial: $x^2-9xy+20y^2=(x-4y)(x-5y)$
51. Factoring the trinomial: $a^2+2ab-8b^2=(a+4b)(a-2b)$
53. Factoring the trinomial: $a^2-10ab+25b^2=(a-5b)(a-5b)=(a-5b)^2$
55. Factoring the trinomial: $a^2+10ab+25b^2=(a+5b)(a+5b)=(a+5b)^2$
57. Factoring the trinomial: $x^2+2xa-48a^2=(x+8a)(x-6a)$
59. Factoring the trinomial: $x^2-5xb-36b^2=(x-9b)(x+4b)$
61. Factoring the trinomial: $x^4-5x^2+6=\left(x^2-2\right)\left(x^2-3\right)$
63. Factoring the trinomial: $x^2-80x-2000=(x-100)(x+20)$
65. Factoring the trinomial: $x^2-x-\frac{1}{4}=\left(x-\frac{1}{2}\right)\left(x-\frac{1}{2}\right)=\left(x-\frac{1}{2}\right)^2$
67. Factoring the trinomial: $x^2+0.6x+0.08=(x+0.4)(x+0.2)$
69. We can use long division to find the other factor:

$$
\begin{array}{r}
x+16 \\
x+8\overline{)x^2+24x+128} \\
\underline{x^2+\ 8x} \\
16x+128 \\
\underline{16x+128} \\
0
\end{array}
$$

The other factor is $x+16$.

71. Using FOIL to multiply out the factors: $(4x+3)(x-1)=4x^2+3x-4x-3=4x^2-x-3$
73. Multiplying using the FOIL method: $(6a+1)(a+2)=6a^2+a+12a+2=6a^2+13a+2$
75. Multiplying using the FOIL method: $(3a+2)(2a+1)=6a^2+4a+3a+2=6a^2+7a+2$
77. Multiplying using the FOIL method: $(6a+2)(a+1)=6a^2+2a+6a+2=6a^2+8a+2$
79. Subtracting the polynomials: $\left(5x^2+5x-4\right)-\left(3x^2-2x+7\right)=5x^2+5x-4-3x^2+2x-7=2x^2+7x-11$
81. Subtracting the polynomials: $(7x+3)-(4x-5)=7x+3-4x+5=3x+8$
83. Subtracting the polynomials: $\left(5x^2-5\right)-\left(2x^2-4x\right)=5x^2-5-2x^2+4x=3x^2+4x-5$

6.3 More Trinomials to Factor

1. Factoring the trinomial: $2x^2+7x+3=(2x+1)(x+3)$
3. Factoring the trinomial: $2a^2-a-3=(2a-3)(a+1)$
5. Factoring the trinomial: $3x^2+2x-5=(3x+5)(x-1)$
7. Factoring the trinomial: $3y^2-14y-5=(3y+1)(y-5)$
9. Factoring the trinomial: $6x^2+13x+6=(3x+2)(2x+3)$
11. Factoring the trinomial: $4x^2-12xy+9y^2=(2x-3y)(2x-3y)=(2x-3y)^2$
13. Factoring the trinomial: $4y^2-11y-3=(4y+1)(y-3)$
15. Factoring the trinomial: $20x^2-41x+20=(4x-5)(5x-4)$
17. Factoring the trinomial: $20a^2+48ab-5b^2=(10a-b)(2a+5b)$
19. Factoring the trinomial: $20x^2-21x-5=(4x-5)(5x+1)$
21. Factoring the trinomial: $12m^2+16m-3=(6m-1)(2m+3)$
23. Factoring the trinomial: $20x^2+37x+15=(4x+5)(5x+3)$
25. Factoring the trinomial: $12a^2-25ab+12b^2=(3a-4b)(4a-3b)$
27. Factoring the trinomial: $3x^2-xy-14y^2=(3x-7y)(x+2y)$
29. Factoring the trinomial: $14x^2+29x-15=(2x+5)(7x-3)$
31. Factoring the trinomial: $6x^2-43x+55=(3x-5)(2x-11)$
33. Factoring the trinomial: $15t^2-67t+38=(5t-19)(3t-2)$
35. Factoring the trinomial: $4x^2+2x-6=2\left(2x^2+x-3\right)=2(2x+3)(x-1)$
37. Factoring the trinomial: $24a^2-50a+24=2\left(12a^2-25a+12\right)=2(4a-3)(3a-4)$
39. Factoring the trinomial: $10x^3-23x^2+12x=x\left(10x^2-23x+12\right)=x(5x-4)(2x-3)$
41. Factoring the trinomial: $6x^4-11x^3-10x^2=x^2\left(6x^2-11x-10\right)=x^2(3x+2)(2x-5)$
43. Factoring the trinomial: $10a^3-6a^2-4a=2a\left(5a^2-3a-2\right)=2a(5a+2)(a-1)$
45. Factoring the trinomial: $15x^3-102x^2-21x=3x\left(5x^2-34x-7\right)=3x(5x+1)(x-7)$
47. Factoring the trinomial: $35y^3-60y^2-20y=5y\left(7y^2-12y-4\right)=5y(7y+2)(y-2)$
49. Factoring the trinomial: $15a^4-2a^3-a^2=a^2\left(15a^2-2a-1\right)=a^2(5a+1)(3a-1)$
51. Factoring the trinomial: $24x^2y-6xy-45y=3y\left(8x^2-2x-15\right)=3y(4x+5)(2x-3)$
53. Factoring the trinomial: $12x^2y-34xy^2+14y^3=2y\left(6x^2-17xy+7y^2\right)=2y(2x-y)(3x-7y)$
55. Evaluating each expression when $x=2$:

$$2x^2+7x+3=2(2)^2+7(2)+3=8+14+3=25$$
$$(2x+1)(x+3)=(2\cdot 2+1)(2+3)=(5)(5)=25$$

57. Multiplying using the difference of squares formula: $(2x+3)(2x-3)=(2x)^2-(3)^2=4x^2-9$
59. Multiplying using the difference of squares formula:

$$(x+3)(x-3)\left(x^2+9\right)=\left(x^2-9\right)\left(x^2+9\right)=\left(x^2\right)^2-(9)^2=x^4-81$$

61. Multiplying using the difference of squares formula: $(x+3)(x-3)=x^2-(3)^2=x^2-9$
63. Multiplying using the difference of squares formula: $(6a+1)(6a-1)=(6a)^2-(1)^2=36a^2-1$

65. Multiplying using the square of binomial formula: $(x+4)^2 = x^2 + 2(x)(4) + (4)^2 = x^2 + 8x + 16$
67. Multiplying using the square of binomial formula: $(2x+3)^2 = (2x)^2 + 2(2x)(3) + (3)^2 = 4x^2 + 12x + 9$

6.4 The Difference of Two Squares

1. Factoring the binomial: $x^2 - 9 = (x+3)(x-3)$
3. Factoring the binomial: $a^2 - 36 = (a+6)(a-6)$
5. Factoring the binomial: $x^2 - 49 = (x+7)(x-7)$
7. Factoring the binomial: $4a^2 - 16 = 4(a^2 - 4) = 4(a+2)(a-2)$
9. The expression $9x^2 + 25$ cannot be factored.
11. Factoring the binomial: $25x^2 - 169 = (5x+13)(5x-13)$
13. Factoring the binomial: $9a^2 - 16b^2 = (3a+4b)(3a-4b)$
15. Factoring the binomial: $9 - m^2 = (3+m)(3-m)$
17. Factoring the binomial: $25 - 4x^2 = (5+2x)(5-2x)$
19. Factoring the binomial: $2x^2 - 18 = 2(x^2 - 9) = 2(x+3)(x-3)$
21. Factoring the binomial: $32a^2 - 128 = 32(a^2 - 4) = 32(a+2)(a-2)$
23. Factoring the binomial: $8x^2y - 18y = 2y(4x^2 - 9) = 2y(2x+3)(2x-3)$
25. Factoring the binomial: $a^4 - b^4 = (a^2 + b^2)(a^2 - b^2) = (a^2 + b^2)(a+b)(a-b)$
27. Factoring the binomial: $16m^4 - 81 = (4m^2 + 9)(4m^2 - 9) = (4m^2 + 9)(2m+3)(2m-3)$
29. Factoring the binomial: $3x^3y - 75xy^3 = 3xy(x^2 - 25y^2) = 3xy(x+5y)(x-5y)$
31. Factoring the trinomial: $x^2 - 2x + 1 = (x-1)(x-1) = (x-1)^2$
33. Factoring the trinomial: $x^2 + 2x + 1 = (x+1)(x+1) = (x+1)^2$
35. Factoring the trinomial: $a^2 - 10a + 25 = (a-5)(a-5) = (a-5)^2$
37. Factoring the trinomial: $y^2 + 4y + 4 = (y+2)(y+2) = (y+2)^2$
39. Factoring the trinomial: $x^2 - 4x + 4 = (x-2)(x-2) = (x-2)^2$
41. Factoring the trinomial: $m^2 - 12m + 36 = (m-6)(m-6) = (m-6)^2$
43. Factoring the trinomial: $4a^2 + 12a + 9 = (2a+3)(2a+3) = (2a+3)^2$
45. Factoring the trinomial: $49x^2 - 14x + 1 = (7x-1)(7x-1) = (7x-1)^2$
47. Factoring the trinomial: $9y^2 - 30y + 25 = (3y-5)(3y-5) = (3y-5)^2$
49. Factoring the trinomial: $x^2 + 10xy + 25y^2 = (x+5y)(x+5y) = (x+5y)^2$
51. Factoring the trinomial: $9a^2 + 6ab + b^2 = (3a+b)(3a+b) = (3a+b)^2$
53. Factoring the trinomial: $3a^2 + 18a + 27 = 3(a^2 + 6a + 9) = 3(a+3)(a+3) = 3(a+3)^2$
55. Factoring the trinomial: $2x^2 + 20xy + 50y^2 = 2(x^2 + 10xy + 25y^2) = 2(x+5y)(x+5y) = 2(x+5y)^2$
57. Factoring the trinomial: $5x^3 + 30x^2y + 45xy^2 = 5x(x^2 + 6xy + 9y^2) = 5x(x+3y)(x+3y) = 5x(x+3y)^2$
59. Factoring by grouping: $x^2 + 6x + 9 - y^2 = (x+3)^2 - y^2 = (x+3+y)(x+3-y)$
61. Factoring by grouping: $x^2 + 2xy + y^2 - 9 = (x+y)^2 - 9 = (x+y+3)(x+y-3)$

63. Since $(x+7)^2 = x^2 + 14x + 49$, the value is $b = 14$.

65. Since $(x+5)^2 = x^2 + 10x + 25$, the value is $c = 25$.

67. Using long division:

$$\begin{array}{r} x-2 \\ x-3\overline{)x^2-5x+8} \\ \underline{x^2-3x} \\ -2x+8 \\ \underline{-2x+6} \\ 2 \end{array}$$

The quotient is $x - 2 + \frac{2}{x-3}$.

69. Using long division:

$$\begin{array}{r} 3x-2 \\ 2x+3\overline{)6x^2+5x+3} \\ \underline{6x^2+9x} \\ -4x+3 \\ \underline{-4x-6} \\ 9 \end{array}$$

The quotient is $3x - 2 + \frac{9}{2x+3}$.

6.5 Factoring: A General Review

1. Factoring the polynomial: $x^2 - 81 = (x+9)(x-9)$
3. Factoring the polynomial: $x^2 + 2x - 15 = (x+5)(x-3)$
5. Factoring the polynomial: $x^2 + 6x + 9 = (x+3)(x+3) = (x+3)^2$
7. Factoring the polynomial: $y^2 - 10y + 25 = (y-5)(y-5) = (y-5)^2$
9. Factoring the polynomial: $2a^3b + 6a^2b + 2ab = 2ab\left(a^2 + 3a + 1\right)$
11. The polynomial $x^2 + x + 1$ cannot be factored.
13. Factoring the polynomial: $12a^2 - 75 = 3\left(4a^2 - 25\right) = 3(2a+5)(2a-5)$
15. Factoring the polynomial: $9x^2 - 12xy + 4y^2 = (3x-2y)(3x-2y) = (3x-2y)^2$
17. Factoring the polynomial: $4x^3 + 16xy^2 = 4x\left(x^2 + 4y^2\right)$
19. Factoring the polynomial: $2y^3 + 20y^2 + 50y = 2y\left(y^2 + 10y + 25\right) = 2y(y+5)(y+5) = 2y(y+5)^2$
21. Factoring the polynomial: $a^6 + 4a^4b^2 = a^4\left(a^2 + 4b^2\right)$
23. Factoring the polynomial: $xy + 3x + 4y + 12 = x(y+3) + 4(y+3) = (y+3)(x+4)$
25. Factoring the polynomial: $x^4 - 16 = \left(x^2 + 4\right)\left(x^2 - 4\right) = \left(x^2 + 4\right)(x+2)(x-2)$
27. Factoring the polynomial: $xy - 5x + 2y - 10 = x(y-5) + 2(y-5) = (y-5)(x+2)$
29. Factoring the polynomial: $5a^2 + 10ab + 5b^2 = 5\left(a^2 + 2ab + b^2\right) = 5(a+b)(a+b) = 5(a+b)^2$
31. The polynomial $x^2 + 49$ cannot be factored.
33. Factoring the polynomial: $3x^2 + 15xy + 18y^2 = 3\left(x^2 + 5xy + 6y^2\right) = 3(x+2y)(x+3y)$
35. Factoring the polynomial: $2x^2 + 15x - 38 = (2x+19)(x-2)$

37. Factoring the polynomial: $100x^2-300x+200=100\left(x^2-3x+2\right)=100(x-2)(x-1)$
39. Factoring the polynomial: $x^2-64=(x+8)(x-8)$
41. Factoring the polynomial: $x^2+3x+ax+3a=x(x+3)+a(x+3)=(x+3)(x+a)$
43. Factoring the polynomial: $49a^7-9a^5=a^5\left(49a^2-9\right)=a^5(7a+3)(7a-3)$
45. The polynomial $49x^2+9y^2$ cannot be factored.
47. Factoring the polynomial: $25a^3+20a^2+3a=a\left(25a^2+20a+3\right)=a(5a+3)(5a+1)$
49. Factoring the polynomial: $xa-xb+ay-by=x(a-b)+y(a-b)=(a-b)(x+y)$
51. Factoring the polynomial: $48a^4b-3a^2b=3a^2b\left(16a^2-1\right)=3a^2b(4a+1)(4a-1)$
53. Factoring the polynomial: $20x^4-45x^2=5x^2\left(4x^2-9\right)=5x^2(2x+3)(2x-3)$
55. Factoring the polynomial: $3x^2+35xy-82y^2=(3x+41y)(x-2y)$
57. Factoring the polynomial: $16x^5-44x^4+30x^3=2x^3\left(8x^2-22x+15\right)=2x^3(2x-3)(4x-5)$
59. Factoring the polynomial: $2x^2+2ax+3x+3a=2x(x+a)+3(x+a)=(x+a)(2x+3)$
61. Factoring the polynomial: $y^4-1=\left(y^2+1\right)\left(y^2-1\right)=\left(y^2+1\right)(y+1)(y-1)$
63. Factoring the polynomial:

$$12x^4y^2+36x^3y^3+27x^2y^4=3x^2y^2\left(4x^2+12xy+9y^2\right)=3x^2y^2(2x+3y)(2x+3y)=3x^2y^2(2x+3y)^2$$

65. Solving the equation:

$$\begin{aligned}3x-6&=9\\3x&=15\\x&=5\end{aligned}$$

67. Solving the equation:

$$\begin{aligned}2x+3&=0\\2x&=-3\\x&=-\frac{3}{2}\end{aligned}$$

69. Solving the equation:

$$\begin{aligned}4x+3&=0\\4x&=-3\\x&=-\frac{3}{4}\end{aligned}$$

71. Simplifying the expression: $x^8\bullet x^7=x^{8+7}=x^{15}$
73. Simplifying the expression: $\left(3x^3\right)^2\left(2x^4\right)^3=9x^6\bullet 8x^{12}=72x^{18}$
75. Writing in scientific notation: $57,600=5.76\times 10^4$

6.6 Solving Equations by Factoring

1. Setting each factor equal to 0:

$$\begin{aligned}x+2&=0 & x-1&=0\\x&=-2 & x&=1\end{aligned}$$

The solutions are –2 and 1.

3. Setting each factor equal to 0:

$$\begin{aligned}a-4&=0 & a-5&=0\\a&=4 & a&=5\end{aligned}$$

The solutions are 4 and 5.

5. Setting each factor equal to 0:

$$x = 0 \qquad x + 1 = 0 \Rightarrow x = -1 \qquad x - 3 = 0 \Rightarrow x = 3$$

The solutions are 0, –1 and 3.

7. Setting each factor equal to 0:

$$\begin{aligned} 3x + 2 &= 0 \\ 3x &= -2 \\ x &= -\frac{2}{3} \end{aligned} \qquad \begin{aligned} 2x + 3 &= 0 \\ 2x &= -3 \\ x &= -\frac{3}{2} \end{aligned}$$

The solutions are $-\frac{2}{3}$ and $-\frac{3}{2}$.

9. Setting each factor equal to 0:

$$m = 0 \qquad \begin{aligned} 3m + 4 &= 0 \\ 3m &= -4 \\ m &= -\frac{4}{3} \end{aligned} \qquad \begin{aligned} 3m - 4 &= 0 \\ 3m &= 4 \\ m &= \frac{4}{3} \end{aligned}$$

The solutions are 0, $-\frac{4}{3}$ and $\frac{4}{3}$.

11. Setting each factor equal to 0:

$$\begin{aligned} 2y &= 0 \\ y &= 0 \end{aligned} \qquad \begin{aligned} 3y + 1 &= 0 \\ 3y &= -1 \\ y &= -\frac{1}{3} \end{aligned} \qquad \begin{aligned} 5y + 3 &= 0 \\ 5y &= -3 \\ y &= -\frac{3}{5} \end{aligned}$$

The solutions are 0, $-\frac{1}{3}$ and $-\frac{3}{5}$.

13. Solving by factoring:

$$\begin{aligned} x^2 + 3x + 2 &= 0 \\ (x+2)(x+1) &= 0 \\ x &= -2, -1 \end{aligned}$$

15. Solving by factoring:

$$\begin{aligned} x^2 - 9x + 20 &= 0 \\ (x-4)(x-5) &= 0 \\ x &= 4, 5 \end{aligned}$$

17. Solving by factoring:

$$\begin{aligned} a^2 - 2a - 24 &= 0 \\ (a-6)(a+4) &= 0 \\ a &= 6, -4 \end{aligned}$$

19. Solving by factoring:

$$\begin{aligned} 100x^2 - 500x + 600 &= 0 \\ 100\left(x^2 - 5x + 6\right) &= 0 \\ 100(x-2)(x-3) &= 0 \\ x &= 2, 3 \end{aligned}$$

21. Solving by factoring:

$$\begin{aligned} x^2 &= -6x - 9 \\ x^2 + 6x + 9 &= 0 \\ (x+3)^2 &= 0 \\ x + 3 &= 0 \\ x &= -3 \end{aligned}$$

23. Solving by factoring:

$$\begin{aligned} a^2-16&=0\\ (a+4)(a-4)&=0\\ a&=-4,4 \end{aligned}$$

25. Solving by factoring:

$$\begin{aligned} 2x^2+5x-12&=0\\ (2x-3)(x+4)&=0\\ x&=\frac{3}{2},-4 \end{aligned}$$

27. Solving by factoring:

$$\begin{aligned} 9x^2+12x+4&=0\\ (3x+2)^2&=0\\ 3x+2&=0\\ x&=-\frac{2}{3} \end{aligned}$$

29. Solving by factoring:

$$\begin{aligned} a^2+25&=10a\\ a^2-10a+25&=0\\ (a-5)^2&=0\\ a-5&=0\\ a&=5 \end{aligned}$$

31. Solving by factoring:

$$\begin{aligned} 2x^2&=3x+20\\ 2x^2-3x-20&=0\\ (2x+5)(x-4)&=0\\ x&=-\frac{5}{2},4 \end{aligned}$$

33. Solving by factoring:

$$\begin{aligned} 3m^2&=20-7m\\ 3m^2+7m-20&=0\\ (3m-5)(m+4)&=0\\ m&=\frac{5}{3},-4 \end{aligned}$$

35. Solving by factoring:

$$\begin{aligned} 4x^2-49&=0\\ (2x+7)(2x-7)&=0\\ x&=-\frac{7}{2},\frac{7}{2} \end{aligned}$$

37. Solving by factoring:

$$\begin{aligned} x^2+6x&=0\\ x(x+6)&=0\\ x&=0,-6 \end{aligned}$$

39. Solving by factoring:

$$\begin{aligned} x^2-3x&=0\\ x(x-3)&=0\\ x&=0,3 \end{aligned}$$

41. Solving by factoring:

$$\begin{aligned} 2x^2 &= 8x \\ 2x^2 - 8x &= 0 \\ 2x(x-4) &= 0 \\ x &= 0, 4 \end{aligned}$$

43. Solving by factoring:

$$\begin{aligned} 3x^2 &= 15x \\ 3x^2 - 15x &= 0 \\ 3x(x-5) &= 0 \\ x &= 0, 5 \end{aligned}$$

45. Solving by factoring:

$$\begin{aligned} 1400 &= 400 + 700x - 100x^2 \\ 100x^2 - 700x + 1000 &= 0 \\ 100\left(x^2 - 7x + 10\right) &= 0 \\ 100(x-5)(x-2) &= 0 \\ x &= 2, 5 \end{aligned}$$

47. Solving by factoring:

$$\begin{aligned} 6x^2 &= -5x + 4 \\ 6x^2 + 5x - 4 &= 0 \\ (3x+4)(2x-1) &= 0 \\ x &= -\frac{4}{3}, \frac{1}{2} \end{aligned}$$

49. Solving by factoring:

$$\begin{aligned} x(2x-3) &= 20 \\ 2x^2 - 3x &= 20 \\ 2x^2 - 3x - 20 &= 0 \\ (2x+5)(x-4) &= 0 \\ x &= -\frac{5}{2}, 4 \end{aligned}$$

51. Solving by factoring:

$$\begin{aligned} t(t+2) &= 80 \\ t^2 + 2t &= 80 \\ t^2 + 2t - 80 &= 0 \\ (t+10)(t-8) &= 0 \\ t &= -10, 8 \end{aligned}$$

53. Solving by factoring:

$$\begin{aligned} 4000 &= (1300 - 100p)p \\ 4000 &= 1300p - 100p^2 \\ 100p^2 - 1300p + 4000 &= 0 \\ 100\left(p^2 - 13p + 40\right) &= 0 \\ 100(p-8)(p-5) &= 0 \\ p &= 5, 8 \end{aligned}$$

55. Solving by factoring:

$$\begin{aligned} x(14-x) &= 48 \\ 14x - x^2 &= 48 \\ -x^2 + 14x - 48 &= 0 \\ x^2 - 14x + 48 &= 0 \\ (x-6)(x-8) &= 0 \\ x &= 6, 8 \end{aligned}$$

57. Solving by factoring:

$$\begin{aligned} (x+5)^2 &= 2x+9 \\ x^2 + 10x + 25 &= 2x + 9 \\ x^2 + 8x + 16 &= 0 \\ (x+4)^2 &= 0 \\ x + 4 &= 0 \\ x &= -4 \end{aligned}$$

59. Solving by factoring:

$$\begin{aligned} (y-6)^2 &= y-4 \\ y^2 - 12y + 36 &= y - 4 \\ y^2 - 13y + 40 &= 0 \\ (y-5)(y-8) &= 0 \\ y &= 5, 8 \end{aligned}$$

61. Solving by factoring:

$$\begin{aligned} 10^2 &= (x+2)^2 + x^2 \\ 100 &= x^2 + 4x + 4 + x^2 \\ 100 &= 2x^2 + 4x + 4 \\ 0 &= 2x^2 + 4x - 96 \\ 0 &= 2\left(x^2 + 2x - 48\right) \\ 0 &= 2(x+8)(x-6) \\ x &= -8, 6 \end{aligned}$$

63. Solving by factoring:

$$\begin{aligned} 2x^3 + 11x^2 + 12x &= 0 \\ x\left(2x^2 + 11x + 12\right) &= 0 \\ x(2x+3)(x+4) &= 0 \\ x &= 0, -\frac{3}{2}, -4 \end{aligned}$$

65. Solving by factoring:

$$\begin{aligned} 4y^3 - 2y^2 - 30y &= 0 \\ 2y\left(2y^2 - y - 15\right) &= 0 \\ 2y(2y+5)(y-3) &= 0 \\ y &= 0, -\frac{5}{2}, 3 \end{aligned}$$

67. Solving by factoring:

$$\begin{aligned} 8x^3+16x^2&=10x \\ 8x^3+16x^2-10x&=0 \\ 2x\left(4x^2+8x-5\right)&=0 \\ 2x(2x-1)(2x+5)&=0 \\ x&=0,\frac{1}{2},-\frac{5}{2} \end{aligned}$$

69. Solving by factoring:

$$\begin{aligned} 20a^3&=-18a^2+18a \\ 20a^3+18a^2-18a&=0 \\ 2a\left(10a^2+9a-9\right)&=0 \\ 2a(5a-3)(2a+3)&=0 \\ a&=0,\frac{3}{5},-\frac{3}{2} \end{aligned}$$

71. Solving by factoring:

$$\begin{aligned} x^3+3x^2-4x-12&=0 \\ x^2(x+3)-4(x+3)&=0 \\ (x+3)\left(x^2-4\right)&=0 \\ (x+3)(x+2)(x-2)&=0 \\ x&=-3,-2,2 \end{aligned}$$

73. Solving by factoring:

$$\begin{aligned} x^3+x^2-16x-16&=0 \\ x^2(x+1)-16(x+1)&=0 \\ (x+1)\left(x^2-16\right)&=0 \\ (x+1)(x+4)(x-4)&=0 \\ x&=-1,-4,4 \end{aligned}$$

75. Let x represent the cost of the suit and $5x$ represent the cost of the bicycle. The equation is:

$$\begin{aligned} x+5x&=90 \\ 6x&=90 \\ x&=15 \\ 5x&=75 \end{aligned}$$

The suit costs $15 and the bicycle costs $75.

77. Let x represent the cost of the lot and $4x$ represent the cost of the house. The equation is:

$$\begin{aligned} x+4x&=3000 \\ 5x&=3000 \\ x&=600 \\ 4x&=2400 \end{aligned}$$

The lot cost $600 and the house cost $2,400.

79. Simplifying using the properties of exponents: $2^{-3}=\frac{1}{2^3}=\frac{1}{8}$

81. Simplifying using the properties of exponents: $\frac{x^5}{x^{-3}}=x^{5-(-3)}=x^{5+3}=x^8$

83. Simplifying using the properties of exponents: $\frac{\left(x^2\right)^3}{\left(x^{-3}\right)^4}=\frac{x^6}{x^{-12}}=x^{6-(-12)}=x^{6+12}=x^{18}$

85. Writing in scientific notation: $0.0056=5.6\times10^{-3}$

6.7 Applications

1. Let x and $x + 2$ represent the two integers. The equation is:

$$x(x+2)=80$$
$$x^2+2x=80$$
$$x^2+2x-80=0$$
$$(x+10)(x-8)=0$$
$$x=-10,8$$
$$x+2=-8,10$$

The two numbers are either –10 and –8, or 8 and 10.

3. Let x and $x + 2$ represent the two integers. The equation is:

$$x(x+2)=99$$
$$x^2+2x=99$$
$$x^2+2x-99=0$$
$$(x+11)(x-9)=0$$
$$x=-11,9$$
$$x+2=-9,11$$

The two numbers are either –11 and –9, or 9 and 11.

5. Let x and $x + 2$ represent the two integers. The equation is:

$$x(x+2)=5(x+x+2)-10$$
$$x^2+2x=5(2x+2)-10$$
$$x^2+2x=10x+10-10$$
$$x^2+2x=10x$$
$$x^2-8x=0$$
$$x(x-8)=0$$
$$x=0,8$$
$$x+2=2,10$$

The two numbers are either 0 and 2, or 8 and 10.

7. Let x and $14 - x$ represent the two numbers. The equation is:

$$x(14-x)=48$$
$$14x-x^2=48$$
$$0=x^2-14x+48$$
$$0=(x-8)(x-6)$$
$$x=8,6$$
$$14-x=6,8$$

The two numbers are 6 and 8.

9. Let x and $5x + 2$ represent the two numbers. The equation is:

$$x(5x+2)=24$$
$$5x^2+2x=24$$
$$5x^2+2x-24=0$$
$$(5x+12)(x-2)=0$$
$$x=-\frac{12}{5},2$$
$$5x+2=-10,12$$

The two numbers are either $-\frac{12}{5}$ and –10, or 2 and 12.

11. Let x and $4x$ represent the two numbers. The equation is:

$$\begin{aligned} x(4x) &= 4(x+4x) \\ 4x^2 &= 4(5x) \\ 4x^2 &= 20x \\ 4x^2 - 20x &= 0 \\ 4x(x-5) &= 0 \\ x &= 0, 5 \\ 4x &= 0, 20 \end{aligned}$$

The two numbers are either 0 and 0, or 5 and 20.

13. Let w represent the width and $w + 1$ represent the length. The equation is:

$$\begin{aligned} w(w+1) &= 12 \\ w^2 + w &= 12 \\ w^2 + w - 12 &= 0 \\ (w+4)(w-3) &= 0 \\ w &= 3 \quad (w = -4 \text{ is impossible}) \\ w+1 &= 4 \end{aligned}$$

The width is 3 inches and the length is 4 inches.

15. Let b represent the base and $2b$ represent the height. The equation is:

$$\begin{aligned} \frac{1}{2}b(2b) &= 9 \\ b^2 &= 9 \\ b^2 - 9 &= 0 \\ (b+3)(b-3) &= 0 \\ b &= 3 \quad (b = -3 \text{ is impossible}) \end{aligned}$$

The base is 3 inches.

17. Let x and $x + 2$ represent the two legs. The equation is:

$$\begin{aligned} x^2 + (x+2)^2 &= 10^2 \\ x^2 + x^2 + 4x + 4 &= 100 \\ 2x^2 + 4x + 4 &= 100 \\ 2x^2 + 4x - 96 &= 0 \\ 2\left(x^2 + 2x - 48\right) &= 0 \\ 2(x+8)(x-6) &= 0 \\ x &= 6 \quad (x = -8 \text{ is impossible}) \\ x+2 &= 8 \end{aligned}$$

The legs are 6 inches and 8 inches.

19. Let x represent the longer leg and $x + 1$ represent the hypotenuse. The equation is:

$$\begin{aligned} 5^2 + x^2 &= (x+1)^2 \\ 25 + x^2 &= x^2 + 2x + 1 \\ 25 &= 2x + 1 \\ 24 &= 2x \\ x &= 12 \end{aligned}$$

The longer leg is 12 meters.

21. Setting $C = \$1{,}400$:

$$\begin{aligned} 1400 &= 400 + 700x - 100x^2 \\ 100x^2 - 700x + 1000 &= 0 \\ 100\left(x^2 - 7x + 10\right) &= 0 \\ 100(x-5)(x-2) &= 0 \\ x &= 2, 5 \end{aligned}$$

The company can manufacture either 200 items or 500 items.

23. Setting $C = \$2{,}200$:

$$\begin{aligned} 2200 &= 600 + 1000x - 100x^2 \\ 100x^2 - 1000x + 1600 &= 0 \\ 100\left(x^2 - 10x + 16\right) &= 0 \\ 100(x-2)(x-8) &= 0 \\ x &= 2, 8 \end{aligned}$$

The company can manufacture either 200 videotapes or 800 videotapes.

25. The revenue is given by: $R = xp = (1200 - 100p)p$

Setting $R = \$3{,}200$:

$$\begin{aligned} 3200 &= (1200 - 100p)p \\ 3200 &= 1200p - 100p^2 \\ 100p^2 - 1200p + 3200 &= 0 \\ 100\left(p^2 - 12p + 32\right) &= 0 \\ 100(p-4)(p-8) &= 0 \\ p &= 4, 8 \end{aligned}$$

The company should sell the ribbons for either \$4 or \$8.

27. The revenue is given by: $R = xp = (1700 - 100p)p$

Setting $R = \$7{,}000$:

$$\begin{aligned} 7000 &= (1700 - 100p)p \\ 7000 &= 1700p - 100p^2 \\ 100p^2 - 1700p + 7000 &= 0 \\ 100\left(p^2 - 17p + 70\right) &= 0 \\ 100(p-7)(p-10) &= 0 \\ p &= 7, 10 \end{aligned}$$

The calculators should be sold for either \$7 or \$10.

29. Simplifying the expression: $\left(5x^3\right)^2\left(2x^6\right)^3 = 25x^6 \bullet 8x^{18} = 200x^{24}$

31. Simplifying the expression: $\dfrac{x^4}{x^{-3}} = x^{4-(-3)} = x^{4+3} = x^7$

33. Simplifying the expression: $\left(2\times10^{-4}\right)\left(4\times10^5\right) = 8\times10^1 = 80$

35. Simplifying the expression: $20ab^2 - 16ab^2 + 6ab^2 = 10ab^2$

37. Multiplying using the distributive property:

$$2x^2\left(3x^2 + 3x - 1\right) = 2x^2\left(3x^2\right) + 2x^2(3x) - 2x^2(1) = 6x^4 + 6x^3 - 2x^2$$

39. Multiplying using the square of binomial formula: $(3y-5)^2 = (3y)^2 - 2(3y)(5) + (5)^2 = 9y^2 - 30y + 25$

41. Multiplying using the difference of squares formula: $\left(2a^2+7\right)\left(2a^2-7\right) = \left(2a^2\right)^2 - (7)^2 = 4a^4 - 49$

Chapter 6 Review

1. Factoring the polynomial: $10x-20=10(x-2)$
3. Factoring the polynomial: $5x-5y=5(x-y)$
5. Factoring the polynomial: $8x+4=4(2x+1)$
7. Factoring the polynomial: $24y^2-40y+48=8\left(3y^2-5y+6\right)$
9. Factoring the polynomial: $49a^3-14b^3=7\left(7a^3-2b^3\right)$
11. Factoring the polynomial: $xy+bx+ay+ab=x(y+b)+a(y+b)=(y+b)(x+a)$
13. Factoring the polynomial: $2xy+10x-3y-15=2x(y+5)-3(y+5)=(y+5)(2x-3)$
15. Factoring the polynomial: $y^2+9y+14=(y+7)(y+2)$
17. Factoring the polynomial: $a^2-14a+48=(a-8)(a-6)$
19. Factoring the polynomial: $y^2+20y+99=(y+9)(y+11)$
21. Factoring the polynomial: $2x^2+13x+15=(2x+3)(x+5)$
23. Factoring the polynomial: $5y^2+11y+6=(5y+6)(y+1)$
25. Factoring the polynomial: $6r^2+5rt-6t^2=(3r-2t)(2r+3t)$
27. Factoring the polynomial: $n^2-81=(n+9)(n-9)$
29. The expression x^2+49 cannot be factored.
31. Factoring the polynomial: $64a^2-121b^2=(8a+11b)(8a-11b)$
33. Factoring the polynomial: $y^2+20y+100=(y+10)(y+10)=(y+10)^2$
35. Factoring the polynomial: $64t^2+16t+1=(8t+1)(8t+1)=(8t+1)^2$
37. Factoring the polynomial: $4r^2-12rt+9t^2=(2r-3t)(2r-3t)=(2r-3t)^2$
39. Factoring the polynomial: $2x^2+20x+48=2\left(x^2+10x+24\right)=2(x+4)(x+6)$
41. Factoring the polynomial: $3m^3-18m^2-21m=3m\left(m^2-6m-7\right)=3m(m-7)(m+1)$
43. Factoring the polynomial: $8x^2+16x+6=2\left(4x^2+8x+3\right)=2(2x+1)(2x+3)$
45. Factoring the polynomial: $20m^3-34m^2+6m=2m\left(10m^2-17m+3\right)=2m(5m-1)(2m-3)$
47. Factoring the polynomial: $4x^2+40x+100=4\left(x^2+10x+25\right)=4(x+5)(x+5)=4(x+5)^2$
49. Factoring the polynomial: $5x^2-45=5\left(x^2-9\right)=5(x+3)(x-3)$
51. Factoring the polynomial: $6a^3b+33a^2b^2+15ab^3=3ab\left(2a^2+11ab+5b^2\right)=3ab(2a+b)(a+5b)$
53. Factoring the polynomial: $4y^6+9y^4=y^4\left(4y^2+9\right)$
55. Factoring the polynomial: $30a^4b+35a^3b^2-15a^2b^3=5a^2b\left(6a^2+7ab-3b^2\right)=5a^2b(3a-b)(2a+3b)$
57. Setting each factor equal to 0:

$$\begin{aligned} x-5&=0 \qquad & x+2&=0 \\ x&=5 & x&=-2 \end{aligned}$$

The solutions are –2, 5.

59. Solving the equation by factoring:

$$\begin{aligned} m^2+3m&=10 \\ m^2+3m-10&=0 \\ (m+5)(m-2)&=0 \\ m&=-5,2 \end{aligned}$$

61. Solving the equation by factoring:

$$m^2 - 9m = 0$$
$$m(m-9) = 0$$
$$m = 0, 9$$

63. Solving the equation by factoring:

$$9x^4 + 9x^3 = 10x^2$$
$$9x^4 + 9x^3 - 10x^2 = 0$$
$$x^2\left(9x^2 + 9x - 10\right) = 0$$
$$x^2(3x-2)(3x+5) = 0$$
$$x = 0, \frac{2}{3}, -\frac{5}{3}$$

65. Let x and $x + 1$ represent the two integers. The equation is:

$$x(x+1) = 110$$
$$x^2 + x = 110$$
$$x^2 + x - 110 = 0$$
$$(x+11)(x-10) = 0$$
$$x = -11, 10$$
$$x+1 = -10, 11$$

The two integers are either –11 and –10, or 10 and 11.

67. Let x and $20 - x$ represent the two numbers. The equation is:

$$x(20-x) = 75$$
$$20x - x^2 = 75$$
$$0 = x^2 - 20x + 75$$
$$0 = (x-15)(x-5)$$
$$x = 15, 5$$
$$20 - x = 5, 15$$

The two numbers are 5 and 15.

69. Let b represent the base, and $8b$ represent the height. The equation is:

$$\frac{1}{2}(b)(8b) = 16$$
$$4b^2 = 16$$
$$4b^2 - 16 = 0$$
$$4\left(b^2 - 4\right) = 0$$
$$4(b+2)(b-2) = 0$$
$$b = 2 \quad (b = -2 \text{ is impossible})$$

The base is 2 inches.

Cumulative Review: Chapters 1-6

1. Simplifying the expression: $-|-9| = -9$

3. Simplifying the expression: $20 - (-9) = 20 + 9 = 29$

5. Simplifying the expression: $\frac{9(-2)}{-2} = \frac{-18}{-2} = 9$

7. Simplifying the expression: $\frac{-3(4-7)-5(7-2)}{-5-2-1} = \frac{-3(-3)-5(5)}{-8} = \frac{9-25}{-8} = \frac{-16}{-8} = 2$

9. Simplifying the expression: $6 - 2(4a+2) - 5 = 6 - 8a - 4 - 5 = -8a - 3$

11. Simplifying using the rules of exponents: $(9xy)^0 = 1$

13. Simplifying using the rules of exponents: $\frac{50x^8y^8}{25x^4y^2}+\frac{28x^7y^7}{14x^3y}=2x^4y^6+2x^4y^6=4x^4y^6$

15. Solving the equation:

$$3x=-18$$
$$\frac{1}{3}(3x)=\frac{1}{3}(-18)$$
$$x=-6$$

17. Solving the equation:

$$-\frac{x}{3}=7$$
$$-3\left(-\frac{x}{3}\right)=-3(7)$$
$$x=-21$$

19. Setting each factor equal to 0:

$$4m=0 \qquad m-7=0 \qquad 2m-7=0$$
$$m=0 \qquad m=7 \qquad 2m=7$$
$$m=\frac{7}{2}$$

The solutions are $0, 7$, and $\frac{7}{2}$.

21. Solving the inequality:

$$-2x>-8$$
$$-\frac{1}{2}(-2x)<-\frac{1}{2}(-8)$$
$$x<4$$

23. Graphing the line:

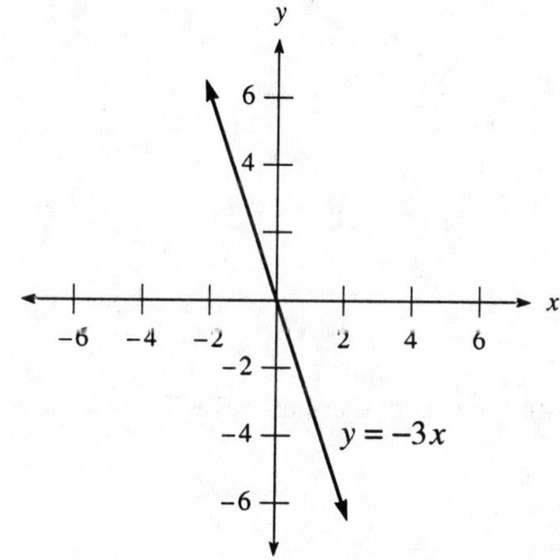

25. Checking the point $(0,0)$: $2(0)+3(0)=0\geq 6$ (false)
Graphing the linear inequality:

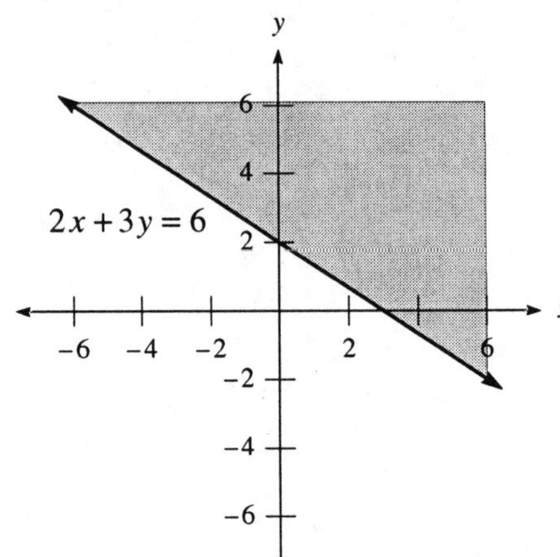

27. Finding the slope: $m=\dfrac{-4-3}{-2-7}=\dfrac{-7}{-9}=\dfrac{7}{9}$

29. The slope-intercept form is $y=-\dfrac{2}{5}x-\dfrac{2}{3}$.

31. Graphing both lines:

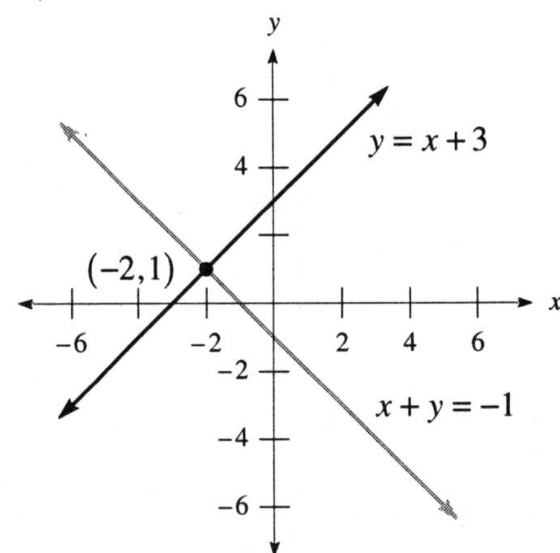

The intersection point is $(-2,1)$.

33. Multiplying the first equation by –3 and the second equation by 7:

$$-15x-21y=54$$
$$56x+21y=28$$

Adding the two equations:

$$41x=82$$
$$x=2$$

Substituting into the second equation:

$$8(2)+3y=4$$
$$16+3y=4$$
$$3y=-12$$
$$y=-4$$

The solution is $(2,-4)$.

35. Multiplying the first equation by –0.04:
$$-0.04x - 0.04y = -200$$
$$0.04x + 0.06y = 270$$
Adding the two equations:
$$0.02y = 70$$
$$y = 3500$$
Substituting into the first equation:
$$x + 3500 = 5000$$
$$x = 1500$$
The solution is $(1500, 3500)$.

37. Factoring the polynomial: $n^2 - 5n - 36 = (n-9)(n+4)$

39. Factoring the polynomial: $16 - a^2 = (4+a)(4-a)$

41. Factoring the polynomial: $45x^2y - 30xy^2 + 5y^3 = 5y(9x^2 - 6xy + y^2) = 5y(3x-y)(3x-y) = 5y(3x-y)^2$

43. Factoring the polynomial: $3xy + 15x - 2y - 10 = 3x(y+5) - 2(y+5) = (y+5)(3x-2)$

45. Commutative property of addition

47. Dividing by the monomial: $\dfrac{28x^4y^4 - 14x^2y^3 + 21xy^2}{-7xy^2} = \dfrac{28x^4y^4}{-7xy^2} + \dfrac{-14x^2y^3}{-7xy^2} + \dfrac{21xy^2}{-7xy^2} = -4x^3y^2 + 2xy - 3$

49. Let x and $x + 4$ represent the length of each piece. The equation is:
$$x + x + 4 = 72$$
$$2x + 4 = 72$$
$$2x = 68$$
$$x = 34$$
$$x + 4 = 38$$
The pieces are 34 inches and 38 inches in length.

Chapter 6 Test

1. Factoring the polynomial: $5x - 10 = 5(x-2)$
2. Factoring the polynomial: $18x^2y - 9xy - 36xy^2 = 9xy(2x - 1 - 4y)$
3. Factoring the polynomial: $x^2 + 2ax - 3bx - 6ab = x(x+2a) - 3b(x+2a) = (x+2a)(x-3b)$
4. Factoring the polynomial: $xy + 4x - 7y - 28 = x(y+4) - 7(y+4) = (y+4)(x-7)$
5. Factoring the polynomial: $x^2 - 5x + 6 = (x-2)(x-3)$
6. Factoring the polynomial: $x^2 - x - 6 = (x-3)(x+2)$
7. Factoring the polynomial: $a^2 - 16 = (a+4)(a-4)$
8. The polynomial $x^2 + 25$ cannot be factored.
9. Factoring the polynomial: $x^4 - 81 = (x^2+9)(x^2-9) = (x^2+9)(x+3)(x-3)$
10. Factoring the polynomial: $27x^2 - 75y^2 = 3(9x^2 - 25y^2) = 3(3x+5y)(3x-5y)$
11. Factoring the polynomial: $x^3 + 5x^2 - 9x - 45 = x^2(x+5) - 9(x+5) = (x+5)(x^2-9) = (x+5)(x+3)(x-3)$
12. Factoring the polynomial: $x^2 - bx + 5x - 5b = x(x-b) + 5(x-b) = (x-b)(x+5)$
13. Factoring the polynomial: $4a^2 + 22a + 10 = 2(2a^2 + 11a + 5) = 2(2a+1)(a+5)$
14. Factoring the polynomial: $3m^2 - 3m - 18 = 3(m^2 - m - 6) = 3(m-3)(m+2)$
15. Factoring the polynomial: $6y^2 + 7y - 5 = (3y+5)(2y-1)$
16. Factoring the polynomial: $12x^3 - 14x^2 - 10x = 2x(6x^2 - 7x - 5) = 2x(3x-5)(2x+1)$

17. Solving the equation by factoring:

$$x^2+7x+12=0$$
$$(x+4)(x+3)=0$$
$$x=-4,-3$$

18. Solving the equation by factoring:

$$x^2-4x+4=0$$
$$(x-2)^2=0$$
$$x-2=0$$
$$x=2$$

19. Solving the equation by factoring:

$$x^2-36=0$$
$$(x+6)(x-6)=0$$
$$x=-6,6$$

20. Solving the equation by factoring:

$$x^2=x+20$$
$$x^2-x-20=0$$
$$(x+4)(x-5)=0$$
$$x=-4,5$$

21. Solving the equation by factoring:

$$x^2-11x=-30$$
$$x^2-11x+30=0$$
$$(x-6)(x-5)=0$$
$$x=5,6$$

22. Solving the equation by factoring:

$$y^3=16y$$
$$y^3-16y=0$$
$$y\left(y^2-16\right)=0$$
$$y(y+4)(y-4)=0$$
$$y=0,-4,4$$

23. Solving the equation by factoring:

$$2a^2=a+15$$
$$2a^2-a-15=0$$
$$(2a+5)(a-3)=0$$
$$a=-\frac{5}{2},3$$

24. Solving the equation by factoring:

$$30x^3-20x^2=10x$$
$$30x^3-20x^2-10x=0$$
$$10x\left(3x^2-2x-1\right)=0$$
$$10x(3x+1)(x-1)=0$$
$$x=0,-\frac{1}{3},1$$

25. Let x and $20 - x$ represent the numbers. The equation is:

$$\begin{aligned} x(20-x) &= 64 \\ 20x - x^2 &= 64 \\ 0 &= x^2 - 20x + 64 \\ 0 &= (x-16)(x-4) \\ x &= 4,16 \\ 20 - x &= 16,4 \end{aligned}$$

The numbers are 4 and 16.

26. Let x and $x + 2$ represent the two integers. The equation is:

$$\begin{aligned} x(x+2) &= x + x + 2 + 7 \\ x^2 + 2x &= 2x + 9 \\ x^2 - 9 &= 0 \\ (x+3)(x-3) &= 0 \\ x &= -3,3 \\ x+2 &= -1,5 \end{aligned}$$

The integers are either –3 and –1, or 3 and 5.

27. Let w represent the width and $3w + 5$ represent the length. The equation is:

$$\begin{aligned} w(3w+5) &= 42 \\ 3w^2 + 5w &= 42 \\ 3w^2 + 5w - 42 &= 0 \\ (3w+14)(w-3) &= 0 \\ w &= 3 \quad \left(w = \frac{14}{3} \text{ is impossible}\right) \\ 3w + 5 &= 14 \end{aligned}$$

The width is 3 feet and the length is 14 feet.

28. Let x and $2x + 2$ represent the two legs. The equation is:

$$\begin{aligned} x^2 + (2x+2)^2 &= 13^2 \\ x^2 + 4x^2 + 8x + 4 &= 169 \\ 5x^2 + 8x - 165 &= 0 \\ (5x+33)(x-5) &= 0 \\ x &= 5 \quad \left(x = -\frac{33}{5} \text{ is impossible}\right) \\ 2x + 2 &= 12 \end{aligned}$$

The two legs are 5 meters and 12 meters in length.

29. Setting $C = \$800$:

$$\begin{aligned} 800 &= 200 + 500x - 100x^2 \\ 100x^2 - 500x + 600 &= 0 \\ 100\left(x^2 - 5x + 6\right) &= 0 \\ 100(x-2)(x-3) &= 0 \\ x &= 2,3 \end{aligned}$$

The company can manufacture either 200 items or 300 items.

30. The revenue is given by: $R = xp = (900 - 100p)p$

Setting $R = \$1{,}800$:

$$1800 = (900 - 100p)p$$
$$1800 = 900p - 100p^2$$
$$100p^2 - 900p + 1800 = 0$$
$$100(p^2 - 9p + 18) = 0$$
$$100(p-6)(p-3) = 0$$
$$p = 3, 6$$

The manufacturer should sell the items at either \$3 or \$6.

Chapter 7
Rational Expressions

7.1 Reducing Rational Expressions to Lowest Terms

1. Reducing the rational expression: $\frac{5}{5x-10}=\frac{5}{5(x-2)}=\frac{1}{x-2}$

   The variable restriction is $x \neq 2$.

3. Reducing the rational expression: $\frac{a-3}{a^2-9}=\frac{1(a-3)}{(a+3)(a-3)}=\frac{1}{a+3}$

   The variable restriction is $a \neq -3, 3$.

5. Reducing the rational expression: $\frac{x+5}{x^2-25}=\frac{1(x+5)}{(x+5)(x-5)}=\frac{1}{x-5}$

   The variable restriction is $x \neq -5, 5$.

7. Reducing the rational expression: $\frac{2x^2-8}{4}=\frac{2\left(x^2-4\right)}{4}=\frac{2(x+2)(x-2)}{4}=\frac{(x+2)(x-2)}{2}$

   There are no variable restrictions.

9. Reducing the rational expression: $\frac{2x-10}{3x-6}=\frac{2(x-5)}{3(x-2)}$

   The variable restriction is $x \neq 2$.

11. Reducing the rational expression: $\frac{10a+20}{5a+10}=\frac{10(a+2)}{5(a+2)}=\frac{2}{1}=2$

13. Reducing the rational expression: $\frac{5x^2-5}{4x+4}=\frac{5\left(x^2-1\right)}{4(x+1)}=\frac{5(x+1)(x-1)}{4(x+1)}=\frac{5(x-1)}{4}$

15. Reducing the rational expression: $\frac{x-3}{x^2-6x+9}=\frac{1(x-3)}{(x-3)^2}=\frac{1}{x-3}$

17. Reducing the rational expression: $\frac{3x+15}{3x^2+24x+45}=\frac{3(x+5)}{3\left(x^2+8x+15\right)}=\frac{3(x+5)}{3(x+5)(x+3)}=\frac{1}{x+3}$

19. Reducing the rational expression: $\frac{a^2-3a}{a^3-8a^2+15a}=\frac{a(a-3)}{a\left(a^2-8a+15\right)}=\frac{a(a-3)}{a(a-3)(a-5)}=\frac{1}{a-5}$

21. Reducing the rational expression: $\frac{3x-2}{9x^2-4}=\frac{1(3x-2)}{(3x+2)(3x-2)}=\frac{1}{3x+2}$

23. Reducing the rational expression: $\frac{x^2+8x+15}{x^2+5x+6}=\frac{(x+5)(x+3)}{(x+2)(x+3)}=\frac{x+5}{x+2}$

25. Reducing the rational expression: $\frac{2m^3-2m^2-12m}{m^2-5m+6}=\frac{2m(m^2-m-6)}{(m-2)(m-3)}=\frac{2m(m-3)(m+2)}{(m-2)(m-3)}=\frac{2m(m+2)}{m-2}$

27. Reducing the rational expression: $\frac{x^3+3x^2-4x}{x^3-16x}=\frac{x(x^2+3x-4)}{x(x^2-16)}=\frac{x(x+4)(x-1)}{x(x+4)(x-4)}=\frac{x-1}{x-4}$

29. Reducing the rational expression: $\frac{4x^3-10x^2+6x}{2x^3+x^2-3x}=\frac{2x(2x^2-5x+3)}{x(2x^2+x-3)}=\frac{2x(2x-3)(x-1)}{x(2x+3)(x-1)}=\frac{2(2x-3)}{2x+3}$

31. Reducing the rational expression: $\frac{4x^2-12x+9}{4x^2-9}=\frac{(2x-3)^2}{(2x+3)(2x-3)}=\frac{2x-3}{2x+3}$

33. Reducing the rational expression: $\frac{x+3}{x^4-81}=\frac{x+3}{(x^2+9)(x^2-9)}=\frac{x+3}{(x^2+9)(x+3)(x-3)}=\frac{1}{(x^2+9)(x-3)}$

35. Reducing the rational expression: $\frac{3x^2+x-10}{x^4-16}=\frac{(3x-5)(x+2)}{(x^2+4)(x^2-4)}=\frac{(3x-5)(x+2)}{(x^2+4)(x+2)(x-2)}=\frac{3x-5}{(x^2+4)(x-2)}$

37. Reducing the rational expression: $\frac{42x^3-20x^2-48x}{6x^2-5x-4}=\frac{2x(21x^2-10x-24)}{(3x-4)(2x+1)}=\frac{2x(7x+6)(3x-4)}{(3x-4)(2x+1)}=\frac{2x(7x+6)}{2x+1}$

39. Reducing the rational expression: $\frac{xy+3x+2y+6}{xy+3x+5y+15}=\frac{x(y+3)+2(y+3)}{x(y+3)+5(y+3)}=\frac{(y+3)(x+2)}{(y+3)(x+5)}=\frac{x+2}{x+5}$

41. Reducing the rational expression: $\frac{x^2-3x+ax-3a}{x^2-3x+bx-3b}=\frac{x(x-3)+a(x-3)}{x(x-3)+b(x-3)}=\frac{(x-3)(x+a)}{(x-3)(x+b)}=\frac{x+a}{x+b}$

43. Reducing the rational expression: $\frac{xy+bx+ay+ab}{xy+bx+3y+3b}=\frac{x(y+b)+a(y+b)}{x(y+b)+3(y+b)}=\frac{(y+b)(x+a)}{(y+b)(x+3)}=\frac{x+a}{x+3}$

45. Writing as a ratio: $\frac{8}{6}=\frac{4}{3}$

47. Writing as a ratio: $\frac{200}{250}=\frac{4}{5}$

49. Writing as a ratio: $\frac{32}{4}=\frac{8}{1}$

51. Writing as a ratio: $\frac{15x^2}{3x^3}=\frac{5}{x}$

53. The average speed is: $\frac{122 \text{ miles}}{3 \text{ hours}}\approx 40.7 \text{ miles / hour}$

55. The average speed is: $\frac{785 \text{ feet}}{20 \text{ minutes}}=39.25 \text{ feet / minute}$

57. The average speed is: $\frac{518 \text{ feet}}{40 \text{ seconds}}=12.95 \text{ feet / second}$

59. Her average speed on level ground is: $\frac{20 \text{ minutes}}{2 \text{ miles}}=10 \text{ minutes / mile}$, or $\frac{2 \text{ miles}}{20 \text{ minutes}}=0.1 \text{ miles / minute}$

Her average speed downhill is: $\frac{40 \text{ minutes}}{6 \text{ miles}}=\frac{20}{3} \text{ minutes / mile}$, or $\frac{6 \text{ miles}}{40 \text{ minutes}}=\frac{3}{20} \text{ miles / minute}$

61. The average fuel consumption is: $\frac{168 \text{ miles}}{3.5 \text{ gallons}}=48 \text{ miles / gallon}$

63. Substituting $x = 5$ and $y = 4$: $\frac{x^2 - y^2}{x - y} = \frac{5^2 - 4^2}{5 - 4} = \frac{25 - 16}{5 - 4} = \frac{9}{1} = 9$

The result is equal to $5 + 4 = 9$.

65. Substituting $a = 10$: $\frac{a - 7}{7 - a} = \frac{10 - 7}{7 - 10} = \frac{3}{-3} = -1$

67. Substituting $x = 3$ and $y = 1$: $\frac{x^4 - y^4}{x^2 + y^2} = \frac{3^4 - 1^4}{3^2 + 1^2} = \frac{81 - 1}{9 + 1} = \frac{80}{10} = 8$

69. Simplifying the expression: $\frac{27x^5}{9x^2} - \frac{45x^8}{15x^5} = 3x^3 - 3x^3 = 0$

71. Simplifying the expression: $\frac{72a^3b^7}{9ab^5} + \frac{64a^5b^3}{8a^3b} = 8a^2b^2 + 8a^2b^2 = 16a^2b^2$

73. Dividing by the monomial: $\frac{38x^7 + 42x^5 - 84x^3}{2x^3} = \frac{38x^7}{2x^3} + \frac{42x^5}{2x^3} - \frac{84x^3}{2x^3} = 19x^4 + 21x^2 - 42$

75. Dividing by the monomial: $\frac{28a^5b^5 + 36ab^4 - 44a^4b}{4ab} = \frac{28a^5b^5}{4ab} + \frac{36ab^4}{4ab} - \frac{44a^4b}{4ab} = 7a^4b^4 + 9b^3 - 11a^3$

7.2 Multiplication and Division of Rational Expressions

1. Simplifying the expression: $\frac{x + y}{3} \bullet \frac{6}{x + y} = \frac{6(x + y)}{3(x + y)} = 2$

3. Simplifying the expression: $\frac{2x + 10}{x^2} \bullet \frac{x^3}{4x + 20} = \frac{2(x + 5)}{x^2} \bullet \frac{x^3}{4(x + 5)} = \frac{2x^3(x + 5)}{4x^2(x + 5)} = \frac{x}{2}$

5. Simplifying the expression: $\frac{9}{2a - 8} \div \frac{3}{a - 4} = \frac{9}{2a - 8} \bullet \frac{a - 4}{3} = \frac{9}{2(a - 4)} \bullet \frac{a - 4}{3} = \frac{9(a - 4)}{6(a - 4)} = \frac{3}{2}$

7. Simplifying the expression:

$$\frac{x + 1}{x^2 - 9} \div \frac{2x + 2}{x + 3} = \frac{x + 1}{x^2 - 9} \bullet \frac{x + 3}{2x + 2} = \frac{x + 1}{(x + 3)(x - 3)} \bullet \frac{x + 3}{2(x + 1)} = \frac{(x + 1)(x + 3)}{2(x + 3)(x - 3)(x + 1)} = \frac{1}{2(x - 3)}$$

9. Simplifying the expression: $\frac{a^2 + 5a}{7a} \bullet \frac{4a^2}{a^2 + 4a} = \frac{a(a + 5)}{7a} \bullet \frac{4a^2}{a(a + 4)} = \frac{4a^3(a + 5)}{7a^2(a + 4)} = \frac{4a(a + 5)}{7(a + 4)}$

11. Simplifying the expression:

$$\frac{y^2 - 5y + 6}{2y + 4} \div \frac{2y - 6}{y + 2} = \frac{y^2 - 5y + 6}{2y + 4} \bullet \frac{y + 2}{2y - 6} = \frac{(y - 2)(y - 3)}{2(y + 2)} \bullet \frac{y + 2}{2(y - 3)} = \frac{(y - 2)(y - 3)(y + 2)}{4(y + 2)(y - 3)} = \frac{y - 2}{4}$$

13. Simplifying the expression:

$$\frac{2x - 8}{x^2 - 4} \bullet \frac{x^2 + 6x + 8}{x - 4} = \frac{2(x - 4)}{(x + 2)(x - 2)} \bullet \frac{(x + 4)(x + 2)}{x - 4} = \frac{2(x - 4)(x + 4)(x + 2)}{(x + 2)(x - 2)(x - 4)} = \frac{2(x + 4)}{x - 2}$$

15. Simplifying the expression:

$$\frac{x - 1}{x^2 - x - 6} \bullet \frac{x^2 + 5x + 6}{x^2 - 1} = \frac{x - 1}{(x - 3)(x + 2)} \bullet \frac{(x + 2)(x + 3)}{(x + 1)(x - 1)} = \frac{(x - 1)(x + 2)(x + 3)}{(x - 3)(x + 2)(x + 1)(x - 1)} = \frac{x + 3}{(x - 3)(x + 1)}$$

17. Simplifying the expression:

$$\frac{a^2 + 10a + 25}{a + 5} \div \frac{a^2 - 25}{a - 5} = \frac{a^2 + 10a + 25}{a + 5} \bullet \frac{a - 5}{a^2 - 25} = \frac{(a + 5)^2}{a + 5} \bullet \frac{a - 5}{(a + 5)(a - 5)} = \frac{(a + 5)^2(a - 5)}{(a + 5)^2(a - 5)} = 1$$

19. Simplifying the expression:

$$\frac{y^3-5y^2}{y^4+3y^3+2y^2} \div \frac{y^2-5y+6}{y^2-2y-3} = \frac{y^3-5y^2}{y^4+3y^3+2y^2} \bullet \frac{y^2-2y-3}{y^2-5y+6}$$
$$= \frac{y^2(y-5)}{y^2(y+2)(y+1)} \bullet \frac{(y-3)(y+1)}{(y-2)(y-3)}$$
$$= \frac{y^2(y-5)(y-3)(y+1)}{y^2(y+2)(y+1)(y-2)(y-3)}$$
$$= \frac{y-5}{(y+2)(y-2)}$$

21. Simplifying the expression:

$$\frac{2x^2+17x+21}{x^2+2x-35} \bullet \frac{x^2-25}{2x^2-7x-15} = \frac{(2x+3)(x+7)}{(x+7)(x-5)} \bullet \frac{(x+5)(x-5)}{(2x+3)(x-5)} = \frac{(2x+3)(x+7)(x+5)(x-5)}{(x+7)(x-5)^2(2x+3)} = \frac{x+5}{x-5}$$

23. Simplifying the expression:

$$\frac{2x^2+10x+12}{4x^2+24x+32} \bullet \frac{2x^2+18x+40}{x^2+8x+15} = \frac{2\left(x^2+5x+6\right)}{4\left(x^2+6x+8\right)} \bullet \frac{2\left(x^2+9x+20\right)}{x^2+8x+15}$$
$$= \frac{2(x+2)(x+3)}{4(x+4)(x+2)} \bullet \frac{2(x+5)(x+4)}{(x+5)(x+3)}$$
$$= \frac{4(x+2)(x+3)(x+4)(x+5)}{4(x+2)(x+3)(x+4)(x+5)}$$
$$= 1$$

25. Simplifying the expression:

$$\frac{2a^2+7a+3}{a^2-16} \div \frac{4a^2+8a+3}{2a^2-5a-12} = \frac{2a^2+7a+3}{a^2-16} \bullet \frac{2a^2-5a-12}{4a^2+8a+3}$$
$$= \frac{(2a+1)(a+3)}{(a+4)(a-4)} \bullet \frac{(2a+3)(a-4)}{(2a+1)(2a+3)}$$
$$= \frac{(2a+1)(a+3)(2a+3)(a-4)}{(a+4)(a-4)(2a+1)(2a+3)}$$
$$= \frac{a+3}{a+4}$$

27. Simplifying the expression:

$$\frac{4y^2-12y+9}{y^2-36} \div \frac{2y^2-5y+3}{y^2+5y-6} = \frac{4y^2-12y+9}{y^2-36} \bullet \frac{y^2+5y-6}{2y^2-5y+3}$$
$$= \frac{(2y-3)^2}{(y+6)(y-6)} \bullet \frac{(y+6)(y-1)}{(2y-3)(y-1)}$$
$$= \frac{(2y-3)^2(y+6)(y-1)}{(y+6)(y-6)(2y-3)(y-1)}$$
$$= \frac{2y-3}{y-6}$$

29. Simplifying the expression:

$$\frac{x^2-1}{6x^2+42x+60} \bullet \frac{7x^2+17x+6}{x+1} \bullet \frac{6x+30}{7x^2-11x-6} = \frac{(x+1)(x-1)}{6(x+5)(x+2)} \bullet \frac{(7x+3)(x+2)}{x+1} \bullet \frac{6(x+5)}{(7x+3)(x-2)}$$
$$= \frac{6(x+1)(x-1)(7x+3)(x+2)(x+5)}{6(x+5)(x+2)(x+1)(7x+3)(x-2)}$$
$$= \frac{x-1}{x-2}$$

31. Simplifying the expression:

$$\frac{18x^3+21x^2-60x}{21x^2-25x-4}\bullet\frac{28x^2-17x-3}{16x^3+28x^2-30x}=\frac{3x\left(6x^2+7x-20\right)}{21x^2-25x-4}\bullet\frac{28x^2-17x-3}{2x\left(8x^2+14x-15\right)}$$
$$=\frac{3x(3x-4)(2x+5)}{(7x+1)(3x-4)}\bullet\frac{(7x+1)(4x-3)}{2x(4x-3)(2x+5)}$$
$$=\frac{3x(3x-4)(2x+5)(7x+1)(4x-3)}{2x(7x+1)(3x-4)(4x-3)(2x+5)}$$
$$=\frac{3}{2}$$

33. Simplifying the expression: $\left(x^2-9\right)\left(\frac{2}{x+3}\right)=\frac{(x+3)(x-3)}{1}\bullet\frac{2}{x+3}=\frac{2(x+3)(x-3)}{x+3}=2(x-3)$

35. Simplifying the expression: $a(a+5)(a-5)\left(\frac{2}{a^2-25}\right)=\frac{a(a+5)(a-5)}{1}\bullet\frac{2}{(a+5)(a-5)}=\frac{2a(a+5)(a-5)}{(a+5)(a-5)}=2a$

37. Simplifying the expression: $\left(x^2-x-6\right)\left(\frac{x+1}{x-3}\right)=\frac{(x-3)(x+2)}{1}\bullet\frac{x+1}{x-3}=\frac{(x-3)(x+2)(x+1)}{x-3}=(x+2)(x+1)$

39. Simplifying the expression: $\left(x^2-4x-5\right)\left(\frac{-2x}{x+1}\right)=\frac{(x-5)(x+1)}{1}\bullet\frac{-2x}{x+1}=\frac{-2x(x-5)(x+1)}{x+1}=-2x(x-5)$

41. Simplifying the expression:

$$\frac{x^2-9}{x^2-3x}\bullet\frac{2x+10}{xy+5x+3y+15}=\frac{(x+3)(x-3)}{x(x-3)}\bullet\frac{2(x+5)}{x(y+5)+3(y+5)}=\frac{2(x+3)(x-3)(x+5)}{x(x-3)(y+5)(x+3)}=\frac{2(x+5)}{x(y+5)}$$

43. Simplifying the expression:

$$\frac{2x^2+4x}{x^2-y^2}\bullet\frac{x^2+3x+xy+3y}{x^2+5x+6}=\frac{2x(x+2)}{(x+y)(x-y)}\bullet\frac{x(x+3)+y(x+3)}{(x+2)(x+3)}=\frac{2x(x+2)(x+3)(x+y)}{(x+y)(x-y)(x+2)(x+3)}=\frac{2x}{x-y}$$

45. Simplifying the expression:

$$\frac{x^3-3x^2+4x-12}{x^4-16}\bullet\frac{3x^2+5x-2}{3x^2-10x+3}=\frac{x^2(x-3)+4(x-3)}{\left(x^2+4\right)\left(x^2-4\right)}\bullet\frac{(3x-1)(x+2)}{(3x-1)(x-3)}$$
$$=\frac{(x-3)\left(x^2+4\right)}{\left(x^2+4\right)(x+2)(x-2)}\bullet\frac{(3x-1)(x+2)}{(3x-1)(x-3)}$$
$$=\frac{(x-3)\left(x^2+4\right)(3x-1)(x+2)}{\left(x^2+4\right)(x+2)(x-2)(3x-1)(x-3)}$$
$$=\frac{1}{x-2}$$

47. Simplifying the expression: $\left(1-\frac{1}{2}\right)\left(1-\frac{1}{3}\right)\left(1-\frac{1}{4}\right)\left(1-\frac{1}{5}\right)=\left(\frac{2}{2}-\frac{1}{2}\right)\left(\frac{3}{3}-\frac{1}{3}\right)\left(\frac{4}{4}-\frac{1}{4}\right)\left(\frac{5}{5}-\frac{1}{5}\right)=\frac{1}{2}\bullet\frac{2}{3}\bullet\frac{3}{4}\bullet\frac{4}{5}=\frac{1}{5}$

49. Simplifying the expression: $\left(1-\frac{1}{2}\right)\left(1-\frac{1}{3}\right)\left(1-\frac{1}{4}\right)\cdots\left(1-\frac{1}{99}\right)\left(1-\frac{1}{100}\right)=\frac{1}{2}\bullet\frac{2}{3}\bullet\frac{3}{4}\bullet\cdots\frac{98}{99}\bullet\frac{99}{100}=\frac{1}{100}$

51. Since 5,280 feet = 1 mile, the height is: $\frac{14{,}494 \text{ feet}}{5{,}280 \text{ feet / mile}}\approx 2.7$ miles

53. Coverting to miles per hour: $\frac{1088 \text{ feet}}{1 \text{ second}}\bullet\frac{1 \text{ mile}}{5280 \text{ feet}}\bullet\frac{60 \text{ seconds}}{1 \text{ minute}}\bullet\frac{60 \text{ minutes}}{1 \text{ hour}}\approx 742$ miles / hour

55. Coverting to miles per hour: $\frac{785 \text{ feet}}{20 \text{ minutes}}\bullet\frac{60 \text{ minutes}}{1 \text{ hour}}\bullet\frac{1 \text{ mile}}{5280 \text{ feet}}\approx 0.45$ miles / hour

57. Coverting to miles per hour: $\dfrac{518\text{ feet}}{40\text{ seconds}}\bullet\dfrac{60\text{ seconds}}{1\text{ minute}}\bullet\dfrac{60\text{ minutes}}{1\text{ hour}}\bullet\dfrac{1\text{ mile}}{5280\text{ feet}}\approx 8.8\text{ miles / hour}$

59. Her average speed on level ground is: $\dfrac{2\text{ miles}}{1/3\text{ hour}}=6\text{ miles / hour}$

Her average speed downhill is: $\dfrac{6\text{ miles}}{2/3\text{ hour}}=9\text{ miles / hour}$

61. Adding the fractions: $\dfrac{1}{2}+\dfrac{5}{2}=\dfrac{6}{2}=3$

63. Adding the fractions: $2+\dfrac{3}{4}=\dfrac{2\bullet 4}{1\bullet 4}+\dfrac{3}{4}=\dfrac{8}{4}+\dfrac{3}{4}=\dfrac{11}{4}$

65. Adding the fractions: $\dfrac{1}{10}+\dfrac{3}{14}=\dfrac{1\bullet 7}{10\bullet 7}+\dfrac{3\bullet 5}{14\bullet 5}=\dfrac{7}{70}+\dfrac{15}{70}=\dfrac{22}{70}=\dfrac{11}{35}$

67. Simplifying the expression: $\dfrac{10x^4}{2x^2}+\dfrac{12x^6}{3x^4}=5x^2+4x^2=9x^2$

69. Simplifying the expression: $\dfrac{12a^2b^5}{3ab^3}+\dfrac{14a^4b^7}{7a^3b^5}=4ab^2+2ab^2=6ab^2$

7.3 Addition and Subtraction of Rational Expressions

1. Combining the fractions: $\dfrac{3}{x}+\dfrac{4}{x}=\dfrac{7}{x}$

3. Combining the fractions: $\dfrac{9}{a}-\dfrac{5}{a}=\dfrac{4}{a}$

5. Combining the fractions: $\dfrac{1}{x+1}+\dfrac{x}{x+1}=\dfrac{1+x}{x+1}=1$

7. Combining the fractions: $\dfrac{y^2}{y-1}-\dfrac{1}{y-1}=\dfrac{y^2-1}{y-1}=\dfrac{(y+1)(y-1)}{y-1}=y+1$

9. Combining the fractions: $\dfrac{x^2}{x+2}+\dfrac{4x+4}{x+2}=\dfrac{x^2+4x+4}{x+2}=\dfrac{(x+2)^2}{x+2}=x+2$

11. Combining the fractions: $\dfrac{x^2}{x-2}-\dfrac{4x-4}{x-2}=\dfrac{x^2-4x+4}{x-2}=\dfrac{(x-2)^2}{x-2}=x-2$

13. Combining the fractions: $\dfrac{x+2}{x+6}-\dfrac{x-4}{x+6}=\dfrac{x+2-x+4}{x+6}=\dfrac{6}{x+6}$

15. Combining the fractions: $\dfrac{y}{2}-\dfrac{2}{y}=\dfrac{y\bullet y}{2\bullet y}-\dfrac{2\bullet 2}{y\bullet 2}=\dfrac{y^2}{2y}-\dfrac{4}{2y}=\dfrac{y^2-4}{2y}=\dfrac{(y+2)(y-2)}{2y}$

17. Combining the fractions: $\dfrac{1}{2}+\dfrac{a}{3}=\dfrac{1\bullet 3}{2\bullet 3}+\dfrac{a\bullet 2}{3\bullet 2}=\dfrac{3}{6}+\dfrac{2a}{6}=\dfrac{2a+3}{6}$

19. Combining the fractions: $\dfrac{x}{x+1}+\dfrac{3}{4}=\dfrac{x\bullet 4}{(x+1)\bullet 4}+\dfrac{3\bullet(x+1)}{4\bullet(x+1)}=\dfrac{4x}{4(x+1)}+\dfrac{3x+3}{4(x+1)}=\dfrac{4x+3x+3}{4(x+1)}=\dfrac{7x+3}{4(x+1)}$

21. Combining the fractions:

$$\dfrac{x+1}{x-2}-\dfrac{4x+7}{5x-10}=\dfrac{(x+1)\bullet 5}{(x-2)\bullet 5}-\dfrac{4x+7}{5(x-2)}=\dfrac{5x+5}{5(x-2)}-\dfrac{4x+7}{5(x-2)}=\dfrac{5x+5-4x-7}{5(x-2)}=\dfrac{x-2}{5(x-2)}=\dfrac{1}{5}$$

23. Combining the fractions:

$$\dfrac{4x-2}{3x+12}-\dfrac{x-2}{x+4}=\dfrac{4x-2}{3(x+4)}-\dfrac{(x-2)\bullet 3}{(x+4)\bullet 3}=\dfrac{4x-2}{3(x+4)}-\dfrac{3x-6}{3(x+4)}=\dfrac{4x-2-3x+6}{3(x+4)}=\dfrac{x+4}{3(x+4)}=\dfrac{1}{3}$$

25. Combining the fractions:

$$\dfrac{6}{x(x-2)}+\dfrac{3}{x}=\dfrac{6}{x(x-2)}+\dfrac{3\bullet(x-2)}{x\bullet(x-2)}=\dfrac{6}{x(x-2)}+\dfrac{3x-6}{x(x-2)}=\dfrac{6+3x-6}{x(x-2)}=\dfrac{3x}{x(x-2)}=\dfrac{3}{x-2}$$

27. Combining the fractions:

$$\frac{4}{a}-\frac{12}{a^2+3a}=\frac{4\cdot(a+3)}{a\cdot(a+3)}-\frac{12}{a(a+3)}=\frac{4a+12}{a(a+3)}-\frac{12}{a(a+3)}=\frac{4a+12-12}{a(a+3)}=\frac{4a}{a(a+3)}=\frac{4}{a+3}$$

29. Combining the fractions:

$$\begin{aligned}\frac{2}{x+5}-\frac{10}{x^2-25}&=\frac{2\cdot(x-5)}{(x+5)\cdot(x-5)}-\frac{10}{(x+5)(x-5)}\\&=\frac{2x-10}{(x+5)(x-5)}-\frac{10}{(x+5)(x-5)}\\&=\frac{2x-10-10}{(x+5)(x-5)}\\&=\frac{2x-20}{(x+5)(x-5)}\\&=\frac{2(x-10)}{(x+5)(x-5)}\end{aligned}$$

31. Combining the fractions:

$$\begin{aligned}\frac{x-4}{x-3}+\frac{6}{x^2-9}&=\frac{(x-4)\cdot(x+3)}{(x-3)\cdot(x+3)}+\frac{6}{(x+3)(x-3)}\\&=\frac{x^2-x-12}{(x+3)(x-3)}+\frac{6}{(x+3)(x-3)}\\&=\frac{x^2-x-12+6}{(x+3)(x-3)}\\&=\frac{x^2-x-6}{(x+3)(x-3)}\\&=\frac{(x-3)(x+2)}{(x+3)(x-3)}\\&=\frac{x+2}{x+3}\end{aligned}$$

33. Combining the fractions:

$$\begin{aligned}\frac{a-4}{a-3}+\frac{5}{a^2-a-6}&=\frac{(a-4)\cdot(a+2)}{(a-3)\cdot(a+2)}+\frac{5}{(a-3)(a+2)}\\&=\frac{a^2-2a-8}{(a-3)(a+2)}+\frac{5}{(a-3)(a+2)}\\&=\frac{a^2-2a-8+5}{(a-3)(a+2)}\\&=\frac{a^2-2a-3}{(a-3)(a+2)}\\&=\frac{(a-3)(a+1)}{(a-3)(a+2)}\\&=\frac{a+1}{a+2}\end{aligned}$$

35. Combining the fractions:

$$\frac{8}{x^2-16}-\frac{7}{x^2-x-12}=\frac{8}{(x+4)(x-4)}-\frac{7}{(x-4)(x+3)}$$
$$=\frac{8(x+3)}{(x+4)(x-4)(x+3)}-\frac{7(x+4)}{(x+4)(x-4)(x+3)}$$
$$=\frac{8x+24}{(x+4)(x-4)(x+3)}-\frac{7x+28}{(x+4)(x-4)(x+3)}$$
$$=\frac{8x+24-7x-28}{(x+4)(x-4)(x+3)}$$
$$=\frac{x-4}{(x+4)(x-4)(x+3)}$$
$$=\frac{1}{(x+4)(x+3)}$$

37. Combining the fractions:

$$\frac{4y}{y^2+6y+5}-\frac{3y}{y^2+5y+4}=\frac{4y}{(y+5)(y+1)}-\frac{3y}{(y+4)(y+1)}$$
$$=\frac{4y(y+4)}{(y+5)(y+1)(y+4)}-\frac{3y(y+5)}{(y+5)(y+1)(y+4)}$$
$$=\frac{4y^2+16y}{(y+5)(y+1)(y+4)}-\frac{3y^2+15y}{(y+5)(y+1)(y+4)}$$
$$=\frac{4y^2+16y-3y^2-15y}{(y+5)(y+1)(y+4)}$$
$$=\frac{y^2+y}{(y+5)(y+1)(y+4)}$$
$$=\frac{y(y+1)}{(y+5)(y+1)(y+4)}$$
$$=\frac{y}{(y+5)(y+4)}$$

39. Combining the fractions:

$$\frac{4x+1}{x^2+5x+4}-\frac{x+3}{x^2+4x+3}=\frac{4x+1}{(x+4)(x+1)}-\frac{x+3}{(x+3)(x+1)}$$
$$=\frac{(4x+1)(x+3)}{(x+4)(x+1)(x+3)}-\frac{(x+3)(x+4)}{(x+4)(x+1)(x+3)}$$
$$=\frac{4x^2+13x+3}{(x+4)(x+1)(x+3)}-\frac{x^2+7x+12}{(x+4)(x+1)(x+3)}$$
$$=\frac{4x^2+13x+3-x^2-7x-12}{(x+4)(x+1)(x+3)}$$
$$=\frac{3x^2+6x-9}{(x+4)(x+1)(x+3)}$$
$$=\frac{3(x+3)(x-1)}{(x+4)(x+1)(x+3)}$$
$$=\frac{3(x-1)}{(x+4)(x+1)}$$

41. Combining the fractions:

$$\begin{aligned}\frac{1}{x}+\frac{x}{3x+9}-\frac{3}{x^2+3x}&=\frac{1}{x}+\frac{x}{3(x+3)}-\frac{3}{x(x+3)}\\&=\frac{1\cdot 3(x+3)}{x\cdot 3(x+3)}+\frac{x\cdot x}{3(x+3)\cdot x}-\frac{3\cdot 3}{x(x+3)\cdot 3}\\&=\frac{3x+9}{3x(x+3)}+\frac{x^2}{3x(x+3)}-\frac{9}{3x(x+3)}\\&=\frac{3x+9+x^2-9}{3x(x+3)}\\&=\frac{x^2+3x}{3x(x+3)}\\&=\frac{x(x+3)}{3x(x+3)}\\&=\frac{1}{3}\end{aligned}$$

43. Combining the fractions: $1+\frac{1}{x}=\frac{1\cdot x}{1\cdot x}+\frac{1}{x}=\frac{x}{x}+\frac{1}{x}=\frac{x+1}{x}$

45. Combining the fractions: $1-\frac{1}{x+1}=\frac{1\cdot(x+1)}{1\cdot(x+1)}-\frac{1}{x+1}=\frac{x+1}{x+1}-\frac{1}{x+1}=\frac{x+1-1}{x+1}=\frac{x}{x+1}$

47. Combining the fractions: $1+\frac{1}{x+2}=\frac{1\cdot(x+2)}{1\cdot(x+2)}+\frac{1}{x+2}=\frac{x+2}{x+2}+\frac{1}{x+2}=\frac{x+2+1}{x+2}=\frac{x+3}{x+2}$

49. Combining the fractions: $1-\frac{1}{x+3}=\frac{1\cdot(x+3)}{1\cdot(x+3)}-\frac{1}{x+3}=\frac{x+3}{x+3}-\frac{1}{x+3}=\frac{x+3-1}{x+3}=\frac{x+2}{x+3}$

51. The expression is: $x+2\left(\frac{1}{x}\right)=x+\frac{2}{x}=\frac{x\cdot x}{1\cdot x}+\frac{2}{x}=\frac{x^2}{x}+\frac{2}{x}=\frac{x^2+2}{x}$

53. Represent the two numbers as x and $2x$. The expression is: $\frac{1}{x}+\frac{1}{2x}=\frac{1\cdot 2}{x\cdot 2}+\frac{1}{2x}=\frac{2}{2x}+\frac{1}{2x}=\frac{3}{2x}$

55. Solving the equation:

$$\begin{aligned}2x+3(x-3)&=6\\2x+3x-9&=6\\5x-9&=6\\5x&=15\\x&=3\end{aligned}$$

57. Solving the equation:

$$\begin{aligned}x-3(x+3)&=x-3\\x-3x-9&=x-3\\-2x-9&=x-3\\-3x-9&=-3\\-3x&=6\\x&=-2\end{aligned}$$

59. Solving the equation:

$$\begin{aligned}7-2(3x+1)&=4x+3\\7-6x-2&=4x+3\\-6x+5&=4x+3\\-10x+5&=3\\-10x&=-2\\x&=\frac{1}{5}\end{aligned}$$

61. Solving the quadratic equation:

$$\begin{aligned} x^2+5x+6&=0\\ (x+2)(x+3)&=0\\ x&=-2,-3 \end{aligned}$$

63. Solving the quadratic equation:

$$\begin{aligned} x^2-x&=6\\ x^2-x-6&=0\\ (x-3)(x+2)&=0\\ x&=-2,3 \end{aligned}$$

65. Solving the quadratic equation:

$$\begin{aligned} x^2-5x&=0\\ x(x-5)&=0\\ x&=0,5 \end{aligned}$$

7.4 Equations Involving Rational Expressions

1. Multiplying both sides of the equation by 6:

$$\begin{aligned} 6\left(\frac{x}{3}+\frac{1}{2}\right)&=6\left(-\frac{1}{2}\right)\\ 2x+3&=-3\\ 2x&=-6\\ x&=-3 \end{aligned}$$

Since $x=-3$ checks in the original equation, the solution is $x=-3$.

3. Multiplying both sides of the equation by $5a$:

$$\begin{aligned} 5a\left(\frac{4}{a}\right)&=5a\left(\frac{1}{5}\right)\\ 20&=a \end{aligned}$$

Since $a=20$ checks in the original equation, the solution is $a=20$.

5. Multiplying both sides of the equation by x:

$$\begin{aligned} x\left(\frac{3}{x}+1\right)&=x\left(\frac{2}{x}\right)\\ 3+x&=2\\ x&=-1 \end{aligned}$$

Since $x=-1$ checks in the original equation, the solution is $x=-1$.

7. Multiplying both sides of the equation by $5a$:

$$\begin{aligned} 5a\left(\frac{3}{a}-\frac{2}{a}\right)&=5a\left(\frac{1}{5}\right)\\ 15-10&=a\\ a&=5 \end{aligned}$$

Since $a=5$ checks in the original equation, the solution is $a=5$.

9. Multiplying both sides of the equation by $2x$:

$$\begin{aligned} 2x\left(\frac{3}{x}+2\right)&=2x\left(\frac{1}{2}\right)\\ 6+4x&=x\\ 6&=-3x\\ x&=-2 \end{aligned}$$

Since $x=-2$ checks in the original equation, the solution is $x=-2$.

11. Multiplying both sides of the equation by $4y$:

$$4y\left(\frac{1}{y}-\frac{1}{2}\right)=4y\left(-\frac{1}{4}\right)$$
$$4-2y=-y$$
$$4=y$$

Since $y=4$ checks in the original equation, the solution is $y=4$.

13. Multiplying both sides of the equation by x^2:

$$x^2\left(1-\frac{8}{x}\right)=x^2\left(-\frac{15}{x^2}\right)$$
$$x^2-8x=-15$$
$$x^2-8x+15=0$$
$$(x-3)(x-5)=0$$
$$x=3,5$$

Both $x=3$ and $x=5$ check in the original equation.

15. Multiplying both sides of the equation by $2x$:

$$2x\left(\frac{x}{2}-\frac{4}{x}\right)=2x\left(-\frac{7}{2}\right)$$
$$x^2-8=-7x$$
$$x^2+7x-8=0$$
$$(x+8)(x-1)=0$$
$$x=-8,1$$

Both $x=-8$ and $x=1$ check in the original equation.

17. Multiplying both sides of the equation by 6:

$$6\left(\frac{x-3}{2}+\frac{2x}{3}\right)=6\left(\frac{5}{6}\right)$$
$$3(x-3)+2(2x)=5$$
$$3x-9+4x=5$$
$$7x-9=5$$
$$7x=14$$
$$x=2$$

Since $x=2$ checks in the original equation, the solution is $x=2$.

19. Multiplying both sides of the equation by 12:

$$12\left(\frac{x+1}{3}+\frac{x-3}{4}\right)=12\left(\frac{1}{6}\right)$$
$$4(x+1)+3(x-3)=2$$
$$4x+4+3x-9=2$$
$$7x-5=2$$
$$7x=7$$
$$x=1$$

Since $x=1$ checks in the original equation, the solution is $x=1$.

21. Multiplying both sides of the equation by $5(x+2)$:

$$5(x+2)\bullet\frac{6}{x+2}=5(x+2)\bullet\frac{3}{5}$$
$$30=3x+6$$
$$24=3x$$
$$x=8$$

Since $x=8$ checks in the original equation, the solution is $x=8$.

23. Multiplying both sides of the equation by $(y-2)(y-3)$:

$$\begin{aligned}(y-2)(y-3)\cdot\frac{3}{y-2}&=(y-2)(y-3)\cdot\frac{2}{y-3}\\3(y-3)&=2(y-2)\\3y-9&=2y-4\\y&=5\end{aligned}$$

Since $y=5$ checks in the original equation, the solution is $y=5$.

25. Multiplying both sides of the equation by $3(x-2)$:

$$\begin{aligned}3(x-2)\left(\frac{x}{x-2}+\frac{2}{3}\right)&=3(x-2)\left(\frac{2}{x-2}\right)\\3x+2(x-2)&=6\\3x+2x-4&=6\\5x-4&=6\\5x&=10\\x&=2\end{aligned}$$

Since $x=2$ does not check in the original equation, there is no solution.

27. Multiplying both sides of the equation by $2(x-2)$:

$$\begin{aligned}2(x-2)\left(\frac{x}{x-2}+\frac{3}{2}\right)&=2(x-2)\cdot\frac{9}{2(x-2)}\\2x+3(x-2)&=9\\2x+3x-6&=9\\5x-6&=9\\5x&=15\\x&=3\end{aligned}$$

Since $x=3$ checks in the original equation, the solution is $x=3$.

29. Multiplying both sides of the equation by $x^2+5x+6=(x+2)(x+3)$:

$$\begin{aligned}(x+2)(x+3)\left(\frac{5}{x+2}+\frac{1}{x+3}\right)&=(x+2)(x+3)\cdot\frac{-1}{(x+2)(x+3)}\\5(x+3)+1(x+2)&=-1\\5x+15+x+2&=-1\\6x+17&=-1\\6x&=-18\\x&=-3\end{aligned}$$

Since $x=-3$ does not check in the original equation, there is no solution.

31. Multiplying both sides of the equation by $x^2-4=(x+2)(x-2)$:

$$\begin{aligned}(x+2)(x-2)\left(\frac{8}{x^2-4}+\frac{3}{x+2}\right)&=(x+2)(x-2)\cdot\frac{1}{x-2}\\8+3(x-2)&=1(x+2)\\8+3x-6&=x+2\\3x+2&=x+2\\2x&=0\\x&=0\end{aligned}$$

Since $x=0$ checks in the original equation, the solution is $x=0$.

33. Multiplying both sides of the equation by $2(a-3)$:

$$2(a-3)\left(\frac{a}{2}+\frac{3}{a-3}\right)=2(a-3)\cdot\frac{a}{a-3}$$
$$a(a-3)+6=2a$$
$$a^2-3a+6=2a$$
$$a^2-5a+6=0$$
$$(a-2)(a-3)=0$$
$$a=2,3$$

Since $a=3$ does not check in the original equation, the solution is $a=2$.

35. Since $y^2-4=(y+2)(y-2)$ and $y^2+2y=y(y+2)$, the LCD is $y(y+2)(y-2)$. Multiplying by the LCD:

$$y(y+2)(y-2)\cdot\frac{6}{(y+2)(y-2)}=y(y+2)(y-2)\cdot\frac{4}{y(y+2)}$$
$$6y=4(y-2)$$
$$6y=4y-8$$
$$2y=-8$$
$$y=-4$$

Since $y=-4$ checks in the original equation, the solution is $y=-4$.

37. Since $a^2-9=(a+3)(a-3)$ and $a^2+a-12=(a+4)(a-3)$, the LCD is $(a+3)(a-3)(a+4)$. Multiplying by the LCD:

$$(a+3)(a-3)(a+4)\cdot\frac{2}{(a+3)(a-3)}=(a+3)(a-3)(a+4)\cdot\frac{3}{(a+4)(a-3)}$$
$$2(a+4)=3(a+3)$$
$$2a+8=3a+9$$
$$-a+8=9$$
$$-a=1$$
$$a=-1$$

Since $a=-1$ checks in the original equation, the solution is $a=-1$.

39. Multiplying both sides of the equation by $x^2-4x-5=(x-5)(x+1)$:

$$(x-5)(x+1)\left(\frac{3x}{x-5}-\frac{2x}{x+1}\right)=(x-5)(x+1)\cdot\frac{-42}{(x-5)(x+1)}$$
$$3x(x+1)-2x(x-5)=-42$$
$$3x^2+3x-2x^2+10x=-42$$
$$x^2+13x+42=0$$
$$(x+7)(x+6)=0$$
$$x=-7,-6$$

Both $x=-7$ and $x=-6$ check in the original equation.

41. Multiplying both sides of the equation by $x^2+5x+6=(x+2)(x+3)$:

$$(x+2)(x+3)\left(\frac{2x}{x+2}\right)=(x+2)(x+3)\left(\frac{x}{x+3}-\frac{3}{x^2+5x+6}\right)$$
$$2x(x+3)=x(x+2)-3$$
$$2x^2+6x=x^2+2x-3$$
$$x^2+4x+3=0$$
$$(x+3)(x+1)=0$$
$$x=-3,-1$$

Since $x=-3$ does not check in the original equation, the solution is $x=-1$.

43. Let x represent the number. The equation is:

$$\begin{aligned} 2(x-3)-5&=3\\ 2x-6-5&=3\\ 2x-11&=3\\ 2x&=14\\ x&=7 \end{aligned}$$

The number is 7.

45. Let w represent the width, and $2w+5$ represent the length. Using the perimeter formula:

$$\begin{aligned} 2w+2(2w+5)&=34\\ 2w+4w+10&=34\\ 6w+10&=34\\ 6w&=24\\ w&=4\\ 2w+5&=13 \end{aligned}$$

The length is 13 inches and the width is 4 inches.

47. Let x and $x+2$ represent the two integers. The equation is:

$$\begin{aligned} x(x+2)&=48\\ x^2+2x&=48\\ x^2+2x-48&=0\\ (x+8)(x-6)&=0\\ x&=-8,6\\ x+2&=-6,8 \end{aligned}$$

The two integers are either –8 and –6, or 6 and 8.

49. Let x and $x+2$ represent the two legs. The equation is:

$$\begin{aligned} x^2+(x+2)^2&=10^2\\ x^2+x^2+4x+4&=100\\ 2x^2+4x-96&=0\\ x^2+2x-48&=0\\ (x+8)(x-6)&=0\\ x&=6 \quad (x=-8 \text{ is impossible})\\ x+2&=8 \end{aligned}$$

The legs are 6 inches and 8 inches.

7.5 Applications

1. Let x and $3x$ represent the two numbers. The equation is:

$$\begin{aligned} \frac{1}{x}+\frac{1}{3x}&=\frac{16}{3}\\ 3x\left(\frac{1}{x}+\frac{1}{3x}\right)&=3x\left(\frac{16}{3}\right)\\ 3+1&=16x\\ 16x&=4\\ x&=\frac{1}{4}\\ 3x&=\frac{3}{4} \end{aligned}$$

The numbers are $\frac{1}{4}$ and $\frac{3}{4}$.

3. Let x represent the number. The equation is:

$$x+\frac{1}{x}=\frac{13}{6}$$
$$6x\left(x+\frac{1}{x}\right)=6x\left(\frac{13}{6}\right)$$
$$6x^2+6=13x$$
$$6x^2-13x+6=0$$
$$(3x-2)(2x-3)=0$$
$$x=\frac{2}{3},\frac{3}{2}$$

The number is either $\frac{2}{3}$ or $\frac{3}{2}$.

5. Let x represent the number. The equation is:

$$\frac{7+x}{9+x}=\frac{5}{7}$$
$$7(9+x)\bullet\frac{7+x}{9+x}=7(9+x)\bullet\frac{5}{7}$$
$$7(7+x)=5(9+x)$$
$$49+7x=45+5x$$
$$49+2x=45$$
$$2x=-4$$
$$x=-2$$

The number is –2.

7. Let x and $x+2$ represent the two integers. The equation is:

$$\frac{1}{x}+\frac{1}{x+2}=\frac{5}{12}$$
$$12x(x+2)\left(\frac{1}{x}+\frac{1}{x+2}\right)=12x(x+2)\left(\frac{5}{12}\right)$$
$$12(x+2)+12x=5x(x+2)$$
$$12x+24+12x=5x^2+10x$$
$$0=5x^2-14x-24$$
$$(5x+6)(x-4)=0$$
$$x=4 \quad \left(x=-\frac{6}{5}\text{ is impossible}\right)$$
$$x+2=6$$

The integers are 4 and 6.

9. Let x represent the rate of the boat in still water. Completing the table:

	d	r	t
Upstream	26	$x-3$	$\frac{26}{x-3}$
Downstream	38	$x+3$	$\frac{38}{x+3}$

The equation is:

$$\frac{26}{x-3}=\frac{38}{x+3}$$
$$(x+3)(x-3)\bullet\frac{26}{x-3}=(x+3)(x-3)\bullet\frac{38}{x+3}$$
$$26(x+3)=38(x-3)$$
$$26x+78=38x-114$$
$$-12x+78=-114$$
$$-12x=-192$$
$$x=16$$

The speed of the boat in still water is 16 mph.

11. Let x represent the plane speed in still air. Completing the table:

	d	r	t
Against Wind	140	$x-20$	$\frac{140}{x-20}$
With Wind	160	$x+20$	$\frac{160}{x+20}$

The equation is:

$$\frac{140}{x-20}=\frac{160}{x+20}$$
$$(x+20)(x-20)\bullet\frac{140}{x-20}=(x+20)(x-20)\bullet\frac{160}{x+20}$$
$$140(x+20)=160(x-20)$$
$$140x+2800=160x-3200$$
$$-20x+2800=-3200$$
$$-20x=-6000$$
$$x=300$$

The plane speed in still air is 300 mph.

13. Let x and $x + 20$ represent the rates of each plane. Completing the table:

	d	r	t
Plane 1	285	$x+20$	$\frac{285}{x+20}$
Plane 2	255	x	$\frac{255}{x}$

The equation is:

$$\frac{285}{x+20}=\frac{255}{x}$$
$$x(x+20)\bullet\frac{285}{x+20}=x(x+20)\bullet\frac{255}{x}$$
$$285x=255(x+20)$$
$$285x=255x+5100$$
$$30x=5100$$
$$x=170$$
$$x+20=190$$

The plane speeds are 170 mph and 190 mph.

15. Let x represent her rate downhill. Completing the table:

	d	r	t
Level Ground	2	$x-3$	$\frac{2}{x-3}$
Downhill	6	x	$\frac{6}{x}$

The equation is:

$$\frac{2}{x-3}+\frac{6}{x}=1$$
$$x(x-3)\left(\frac{2}{x-3}+\frac{6}{x}\right)=x(x-3)\bullet 1$$
$$2x+6(x-3)=x(x-3)$$
$$2x+6x-18=x^2-3x$$
$$8x-18=x^2-3x$$
$$0=x^2-11x+18$$
$$0=(x-2)(x-9)$$
$$x=9 \qquad (x=2 \text{ is impossible})$$

Tina runs 9 mph on the downhill part of the course.

17. Let x represent her rate on level ground. Completing the table:

	d	r	t
Level Ground	4	x	$\frac{4}{x}$
Downhill	5	$x+2$	$\frac{5}{x+2}$

The equation is:

$$\begin{aligned}\frac{4}{x}+\frac{5}{x+2}&=1\\ x(x+2)\left(\frac{4}{x}+\frac{5}{x+2}\right)&=x(x+2)\bullet 1\\ 4(x+2)+5x&=x(x+2)\\ 4x+8+5x&=x^2+2x\\ 9x+8&=x^2+2x\\ 0&=x^2-7x-8\\ 0&=(x-8)(x+1)\\ x&=8 \quad (x=-1 \text{ is impossible})\end{aligned}$$

Jerri jogs 8 mph on level ground.

19. Let t represent the time to fill the pool with both pipes left open. The equation is:

$$\begin{aligned}\frac{1}{12}-\frac{1}{15}&=\frac{1}{t}\\ 60t\left(\frac{1}{12}-\frac{1}{15}\right)&=60t\bullet\frac{1}{t}\\ 5t-4t&=60\\ t&=60\end{aligned}$$

It will take 60 hours to fill the pool with both pipes left open.

21. Let t represent the time to fill the bathtub with both faucets open. The equation is:

$$\begin{aligned}\frac{1}{10}+\frac{1}{12}&=\frac{1}{t}\\ 60t\left(\frac{1}{10}+\frac{1}{12}\right)&=60t\bullet\frac{1}{t}\\ 6t+5t&=60\\ 11t&=60\\ t&=\frac{60}{11}=5\frac{5}{11}\end{aligned}$$

It will take $5\frac{5}{11}$ minutes to fill the tub with both faucets open.

23. Let t represent the time to fill the sink with both the faucet and the drain left open. The equation is:

$$\begin{aligned}\frac{1}{3}-\frac{1}{4}&=\frac{1}{t}\\ 12t\left(\frac{1}{3}-\frac{1}{4}\right)&=12t\bullet\frac{1}{t}\\ 4t-3t&=12\\ t&=12\end{aligned}$$

It will take 12 minutes for the sink to overflow with both the faucet and drain left open.

25. Factoring the polynomial: $15a^3b^3-20a^2b-35ab^2=5ab\left(3a^2b^2-4a-7b\right)$

27. Factoring the polynomial: $x^2-4x-12=(x-6)(x+2)$

29. Factoring the polynomial: $x^4-16=\left(x^2+4\right)\left(x^2-4\right)=\left(x^2+4\right)(x+2)(x-2)$

31. Factoring the polynomial: $5x^3-25x^2-30x=5x\left(x^2-5x-6\right)=5x(x-6)(x+1)$

33. Solving the equation by factoring:
$$\begin{aligned} x^2-6x&=0\\ x(x-6)&=0\\ x&=0,6 \end{aligned}$$

35. Solving the equation by factoring:
$$\begin{aligned} x(x+2)&=80\\ x^2+2x&=80\\ x^2+2x-80&=0\\ (x+10)(x-8)&=0\\ x&=-10,8 \end{aligned}$$

37. Let x and $x+3$ represent the two integers. The equation is:
$$\begin{aligned} x^2+(x+3)^2&=15^2\\ x^2+x^2+6x+9&=225\\ 2x^2+6x-216&=0\\ x^2+3x-108&=0\\ (x+12)(x-9)&=0\\ x&=9 \quad (x=-12 \text{ is impossible})\\ x+3&=12 \end{aligned}$$
The two legs are 9 inches and 12 inches.

7.6 Complex Fractions

1. Simplifying the complex fraction: $\dfrac{\frac{3}{4}}{\frac{7}{8}}=\dfrac{\frac{3}{4}\cdot 8}{\frac{7}{8}\cdot 8}=\dfrac{6}{1}=6$

3. Simplifying the complex fraction: $\dfrac{\frac{2}{3}}{4}=\dfrac{\frac{2}{3}\cdot 3}{4\cdot 3}=\dfrac{2}{12}=\dfrac{1}{6}$

5. Simplifying the complex fraction: $\dfrac{\frac{x^2}{y}}{\frac{x}{y^3}}=\dfrac{\frac{x^2}{y}\cdot y^3}{\frac{x}{y^3}\cdot y^3}=\dfrac{x^2y^2}{x}=xy^2$

7. Simplifying the complex fraction: $\dfrac{\frac{4x^3}{y^6}}{\frac{8x^2}{y^7}}=\dfrac{\frac{4x^3}{y^6}\cdot y^7}{\frac{8x^2}{y^7}\cdot y^7}=\dfrac{4x^3y}{8x^2}=\dfrac{xy}{2}$

9. Simplifying the complex fraction: $\dfrac{y+\frac{1}{x}}{x+\frac{1}{y}}=\dfrac{\left(y+\frac{1}{x}\right)\cdot xy}{\left(x+\frac{1}{y}\right)\cdot xy}=\dfrac{xy^2+y}{x^2y+x}=\dfrac{y(xy+1)}{x(xy+1)}=\dfrac{y}{x}$

11. Simplifying the complex fraction: $\dfrac{1+\frac{1}{a}}{1-\frac{1}{a}}=\dfrac{\left(1+\frac{1}{a}\right)\cdot a}{\left(1-\frac{1}{a}\right)\cdot a}=\dfrac{a+1}{a-1}$

13. Simplifying the complex fraction: $\dfrac{\dfrac{x+1}{x^2-9}}{\dfrac{2}{x+3}}=\dfrac{\dfrac{x+1}{(x+3)(x-3)}\bullet(x+3)(x-3)}{\dfrac{2}{x+3}\bullet(x+3)(x-3)}=\dfrac{x+1}{2(x-3)}$

15. Simplifying the complex fraction: $\dfrac{\dfrac{1}{a+2}}{\dfrac{1}{a^2-a-6}}=\dfrac{\dfrac{1}{a+2}\bullet(a-3)(a+2)}{\dfrac{1}{(a-3)(a+2)}\bullet(a-3)(a+2)}=\dfrac{a-3}{1}=a-3$

17. Simplifying the complex fraction: $\dfrac{1-\dfrac{9}{y^2}}{1-\dfrac{1}{y}-\dfrac{6}{y^2}}=\dfrac{\left(1-\dfrac{9}{y^2}\right)\bullet y^2}{\left(1-\dfrac{1}{y}-\dfrac{6}{y^2}\right)\bullet y^2}=\dfrac{y^2-9}{y^2-y-6}=\dfrac{(y+3)(y-3)}{(y+2)(y-3)}=\dfrac{y+3}{y+2}$

19. Simplifying the complex fraction: $\dfrac{\dfrac{1}{y}+\dfrac{1}{x}}{\dfrac{1}{xy}}=\dfrac{\left(\dfrac{1}{y}+\dfrac{1}{x}\right)\bullet xy}{\left(\dfrac{1}{xy}\right)\bullet xy}=\dfrac{x+y}{1}=x+y$

21. Simplifying the complex fraction: $\dfrac{1-\dfrac{1}{a^2}}{1-\dfrac{1}{a}}=\dfrac{\left(1-\dfrac{1}{a^2}\right)\bullet a^2}{\left(1-\dfrac{1}{a}\right)\bullet a^2}=\dfrac{a^2-1}{a^2-a}=\dfrac{(a+1)(a-1)}{a(a-1)}=\dfrac{a+1}{a}$

23. Simplifying the complex fraction: $\dfrac{\dfrac{1}{10x}-\dfrac{y}{10x^2}}{\dfrac{1}{10}-\dfrac{y}{10x}}=\dfrac{\left(\dfrac{1}{10x}-\dfrac{y}{10x^2}\right)\bullet 10x^2}{\left(\dfrac{1}{10}-\dfrac{y}{10x}\right)\bullet 10x^2}=\dfrac{x-y}{x^2-xy}=\dfrac{1(x-y)}{x(x-y)}=\dfrac{1}{x}$

25. Simplifying the complex fraction: $\dfrac{\dfrac{1}{a+1}+2}{\dfrac{1}{a+1}+3}=\dfrac{\left(\dfrac{1}{a+1}+2\right)\bullet(a+1)}{\left(\dfrac{1}{a+1}+3\right)\bullet(a+1)}=\dfrac{1+2(a+1)}{1+3(a+1)}=\dfrac{1+2a+2}{1+3a+3}=\dfrac{2a+3}{3a+4}$

27. Simplifying each parenthesis first:

$$1-\frac{1}{x}=\frac{x}{x}-\frac{1}{x}=\frac{x-1}{x}$$

$$1-\frac{1}{x+1}=\frac{x+1}{x+1}-\frac{1}{x+1}=\frac{x}{x+1}$$

$$1-\frac{1}{x+2}=\frac{x+2}{x+2}-\frac{1}{x+2}=\frac{x+1}{x+2}$$

Now performing the multiplication: $\left(1-\dfrac{1}{x}\right)\left(1-\dfrac{1}{x+1}\right)\left(1-\dfrac{1}{x+2}\right)=\dfrac{x-1}{x}\bullet\dfrac{x}{x+1}\bullet\dfrac{x+1}{x+2}=\dfrac{x-1}{x+2}$

29. Simplifying each parenthesis first:

$$1+\frac{1}{x+3}=\frac{x+3}{x+3}+\frac{1}{x+3}=\frac{x+4}{x+3}$$

$$1+\frac{1}{x+2}=\frac{x+2}{x+2}+\frac{1}{x+2}=\frac{x+3}{x+2}$$

$$1+\frac{1}{x+1}=\frac{x+1}{x+1}+\frac{1}{x+1}=\frac{x+2}{x+1}$$

Now performing the multiplication: $\left(1+\dfrac{1}{x+3}\right)\left(1+\dfrac{1}{x+2}\right)\left(1+\dfrac{1}{x+1}\right)=\dfrac{x+4}{x+3}\bullet\dfrac{x+3}{x+2}\bullet\dfrac{x+2}{x+1}=\dfrac{x+4}{x+1}$

31. Simplifying each term in the sequence:

$$2+\frac{1}{2+1}=2+\frac{1}{3}=\frac{6}{3}+\frac{1}{3}=\frac{7}{3}$$

$$2+\frac{1}{2+\frac{1}{2+1}}=2+\frac{1}{\frac{7}{3}}=2+\frac{3}{7}=\frac{14}{7}+\frac{3}{7}=\frac{17}{7}$$

$$2+\frac{2}{2+\frac{1}{2+\frac{1}{2+1}}}=2+\frac{1}{\frac{17}{7}}=2+\frac{7}{17}=\frac{34}{17}+\frac{7}{17}=\frac{41}{17}$$

33. Solving the inequality:

$$\begin{aligned} 2x+3&<5\\ 2x+3-3&<5-3\\ 2x&<2\\ \frac{1}{2}(2x)&<\frac{1}{2}(2)\\ x&<1 \end{aligned}$$

35. Solving the inequality:

$$\begin{aligned} -3x&\le 21\\ -\frac{1}{3}(-3x)&\ge -\frac{1}{3}(21)\\ x&\ge -7 \end{aligned}$$

37. Solving the inequality:

$$\begin{aligned} -2x+8&>-4\\ -2x+8-8&>-4-8\\ -2x&>-12\\ -\frac{1}{2}(-2x)&<-\frac{1}{2}(-12)\\ x&<6 \end{aligned}$$

39. Solving the inequality:

$$\begin{aligned} 4-2(x+1)&\ge -2\\ 4-2x-2&\ge -2\\ -2x+2&\ge -2\\ -2x+2-2&\ge -2-2\\ -2x&\ge -4\\ -\frac{1}{2}(-2x)&\le -\frac{1}{2}(-4)\\ x&\le 2 \end{aligned}$$

7.7 Proportions

1. Solving the proportion:

$$\begin{aligned} \frac{x}{2}&=\frac{6}{12}\\ 12x&=12\\ x&=1 \end{aligned}$$

3. Solving the proportion:

$$\begin{aligned} \frac{2}{5}&=\frac{4}{x}\\ 2x&=20\\ x&=10 \end{aligned}$$

5. Solving the proportion:

$$\frac{10}{20}=\frac{20}{x}$$
$$10x=400$$
$$x=40$$

7. Solving the proportion:

$$\frac{a}{3}=\frac{5}{12}$$
$$12a=15$$
$$a=\frac{15}{12}=\frac{5}{4}$$

9. Solving the proportion:

$$\frac{2}{x}=\frac{6}{7}$$
$$6x=14$$
$$x=\frac{14}{6}=\frac{7}{3}$$

11. Solving the proportion:

$$\frac{x+1}{3}=\frac{4}{x}$$
$$x^2+x=12$$
$$x^2+x-12=0$$
$$(x+4)(x-3)=0$$
$$x=-4,3$$

13. Solving the proportion:

$$\frac{x}{2}=\frac{8}{x}$$
$$x^2=16$$
$$x^2-16=0$$
$$(x+4)(x-4)=0$$
$$x=-4,4$$

15. Solving the proportion:

$$\frac{4}{a+2}=\frac{a}{2}$$
$$a^2+2a=8$$
$$a^2+2a-8=0$$
$$(a+4)(a-2)=0$$
$$a=-4,2$$

17. Solving the proportion:

$$\frac{1}{x}=\frac{x-5}{6}$$
$$x^2-5x=6$$
$$x^2-5x-6=0$$
$$(x-6)(x+1)=0$$
$$x=-1,6$$

19. Comparing hits to games, the proportion is:

$$\frac{6}{18}=\frac{x}{45}$$
$$18x=270$$
$$x=15$$

He will get 15 hits in 45 games.

21. Comparing ml alcohol to ml water, the proportion is:

$$\frac{12}{16}=\frac{x}{28}$$
$$16x=336$$
$$x=21$$

The solution will have 21 ml of alcohol.

23. Comparing grams of fat to total grams, the proportion is:

$$\frac{13}{100}=\frac{x}{350}$$
$$100x=4550$$
$$x=45.5$$

There are 45.5 grams of fat in 350 grams of ice cream.

25. Comparing inches on the map to actual miles, the proportion is:

$$\frac{3.5}{100}=\frac{x}{420}$$
$$100x=1470$$
$$x=14.7$$

They are 14.7 inches apart on the map.

27. Comparing miles to hours, the proportion is:

$$\frac{245}{5}=\frac{x}{7}$$
$$5x=1715$$
$$x=343$$

He will travel 343 miles.

29. Reducing the fraction: $\frac{x^2-x-6}{x^2-9}=\frac{(x-3)(x+2)}{(x+3)(x-3)}=\frac{x+2}{x+3}$

31. Multiplying the fractions:

$$\frac{x^2-25}{x+4}\bullet\frac{2x+8}{x^2-9x+20}=\frac{(x+5)(x-5)}{x+4}\bullet\frac{2(x+4)}{(x-5)(x-4)}=\frac{2(x+5)(x-5)(x+4)}{(x+4)(x-5)(x-4)}=\frac{2(x+5)}{x-4}$$

33. Adding the fractions: $\frac{x}{x^2-16}+\frac{4}{x^2-16}=\frac{x+4}{x^2-16}=\frac{1(x+4)}{(x+4)(x-4)}=\frac{1}{x-4}$

7.8 Variation

1. The variation equation is $y=Kx$. Finding K:

$$10=K\bullet5$$
$$K=2$$

So $y=2x$. Substituting $x=4$: $y=2\bullet4=8$

3. The variation equation is $y=Kx$. Finding K:

$$39=K\bullet3$$
$$K=13$$

So $y=13x$. Substituting $x=10$: $y=13\bullet10=130$

5. The variation equation is $y=Kx$. Finding K:

$$-24=K\bullet4$$
$$K=-6$$

So $y=-6x$. Substituting $y=-30$:

$$-6x=-30$$
$$x=5$$

7. The variation equation is $y = Kx$. Finding K:

$$-7 = K \cdot (-1)$$
$$K = 7$$

So $y = 7x$. Substituting $y = -21$:

$$7x = -21$$
$$x = -3$$

9. The variation equation is $y = Kx^2$. Finding K:

$$75 = K \cdot 5^2$$
$$75 = 25K$$
$$K = 3$$

So $y = 3x^2$. Substituting $x = 1$: $y = 3 \cdot 1^2 = 3 \cdot 1 = 3$

11. The variation equation is $y = Kx^2$. Finding K:

$$48 = K \cdot 4^2$$
$$48 = 16K$$
$$K = 3$$

So $y = 3x^2$. Substituting $x = 9$: $y = 3 \cdot 9^2 = 3 \cdot 81 = 243$

13. The variation equation is $y = \frac{K}{x}$. Finding K:

$$5 = \frac{K}{2}$$
$$K = 10$$

So $y = \frac{10}{x}$. Substituting $x = 5$: $y = \frac{10}{5} = 2$

15. The variation equation is $y = \frac{K}{x}$. Finding K:

$$2 = \frac{K}{1}$$
$$K = 2$$

So $y = \frac{2}{x}$. Substituting $x = 4$: $y = \frac{2}{4} = \frac{1}{2}$

17. The variation equation is $y = \frac{K}{x}$. Finding K:

$$5 = \frac{K}{3}$$
$$K = 15$$

So $y = \frac{15}{x}$. Substituting $y = 15$:

$$\frac{15}{x} = 15$$
$$15 = 15x$$
$$x = 1$$

19. The variation equation is $y = \frac{K}{x}$. Finding K:

$$10 = \frac{K}{10}$$
$$K = 100$$

So $y = \frac{100}{x}$. Substituting $y = 100$:

$$\frac{100}{x} = 20$$
$$100 = 20x$$
$$x = 5$$

21. The variation equation is $y=\frac{K}{x^2}$. Finding K:

$$4=\frac{K}{5^2}$$
$$4=\frac{K}{25}$$
$$K=100$$

So $y=\frac{100}{x^2}$. Substituting $x=2$: $y=\frac{100}{2^2}=\frac{100}{4}=25$

23. The variation equation is $y=\frac{K}{x^2}$. Finding K:

$$4=\frac{K}{3^2}$$
$$4=\frac{K}{9}$$
$$K=36$$

So $y=\frac{36}{x^2}$. Substituting $x=2$: $y=\frac{36}{2^2}=\frac{36}{4}=9$

25. The variation equation is $t=Kd$. Finding K:

$$42=K\cdot 2$$
$$K=21$$

So $t=21d$. Substituting $d=4$: $t=21\cdot 4=84$ pounds

27. The variation equation is $P=KI^2$. Finding K:

$$30=K\cdot 2^2$$
$$30=4K$$
$$K=\frac{15}{2}$$

So $P=\frac{15}{2}I^2$. Substituting $I=7$: $P=\frac{15}{2}\cdot 7^2=\frac{15}{2}\cdot 49=367.5$

29. The variation equation is $M=Kh$. Finding K:

$$157=K\cdot 20$$
$$K=7.85$$

So $M=7.85h$. Substituting $h=30$: $M=7.85\cdot 30=\$235.50$

31. The variation equation is $F=\frac{K}{d^2}$. Finding K:

$$150=\frac{K}{4000^2}$$
$$150=\frac{K}{1.6\times 10^7}$$
$$K=2.4\times 10^9$$

So $F=\frac{2.4\times 10^9}{d^2}$. Substituting $d=5000$: $F=\frac{2.4\times 10^9}{(5000)^2}=\frac{2.4\times 10^9}{2.5\times 10^7}=96$ pounds

33. The variation equation is $I=\frac{K}{R}$. Finding K:

$$30=\frac{K}{2}$$
$$K=60$$

So $I=\frac{60}{R}$. Substituting $R=5$: $I=\frac{60}{5}=12$ amps

35. Adding the two equations:
$$5x = 10$$
$$x = 2$$
Substituting into the first equation:
$$2(2) + y = 3$$
$$4 + y = 3$$
$$y = -1$$
The solution is $(2,-1)$.

37. Multiplying the second equation by –4:
$$4x - 5y = 1$$
$$-4x + 8y = 8$$
Adding the two equations:
$$3y = 9$$
$$y = 3$$
Substituting into the second equation:
$$x - 2(3) = -2$$
$$x - 6 = -2$$
$$x = 4$$
The solution is $(4,3)$.

39. Substituting into the first equation:
$$5x + 2(3x - 2) = 7$$
$$5x + 6x - 4 = 7$$
$$11x - 4 = 7$$
$$11x = 11$$
$$x = 1$$
Substituting into the second equation: $y = 3(1) - 2 = 3 - 2 = 1$
The solution is $(1,1)$.

41. Substituting into the first equation:
$$2(2y + 1) - 3y = 4$$
$$4y + 2 - 3y = 4$$
$$y + 2 = 4$$
$$y = 2$$
Substituting into the second equation: $x = 2(2) + 1 = 4 + 1 = 5$
The solution is $(5,2)$.

Chapter 7 Review

1. Reducing the rational expression: $\dfrac{7}{14x-28} = \dfrac{7}{14(x-2)} = \dfrac{1}{2(x-2)}$

   The variable restriction is $x \neq 2$.

3. Reducing the rational expression: $\dfrac{8x-4}{4x+12} = \dfrac{4(2x-1)}{4(x+3)} = \dfrac{2x-1}{x+3}$

   The variable restriction is $x \neq -3$.

5. Reducing the rational expression: $\dfrac{3x^3+16x^2-12x}{2x^3+9x^2-18x} = \dfrac{x\left(3x^2+16x-12\right)}{x\left(2x^2+9x-18\right)} = \dfrac{x(3x-2)(x+6)}{x(2x-3)(x+6)} = \dfrac{3x-2}{2x-3}$

7. Reducing the rational expression: $\dfrac{x^2+5x-14}{x+7} = \dfrac{(x+7)(x-2)}{x+7} = x-2$

9. Reducing the rational expression: $\dfrac{xy+bx+ay+ab}{xy+5x+ay+5a} = \dfrac{x(y+b)+a(y+b)}{x(y+5)+a(y+5)} = \dfrac{(y+b)(x+a)}{(y+5)(x+a)} = \dfrac{y+b}{y+5}$

11. Performing the operations:

$$\frac{x^2+8x+16}{x^2+x-12} \div \frac{x^2-16}{x^2-x-6} = \frac{x^2+8x+16}{x^2+x-12} \bullet \frac{x^2-x-6}{x^2-16}$$
$$= \frac{(x+4)^2}{(x+4)(x-3)} \bullet \frac{(x+2)(x-3)}{(x+4)(x-4)}$$
$$= \frac{(x+4)^2(x+2)(x-3)}{(x+4)^2(x-3)(x-4)}$$
$$= \frac{x+2}{x-4}$$

13. Performing the operations:

$$\frac{3x^2-2x-1}{x^2+6x+8} \div \frac{3x^2+13x+4}{x^2+8x+16} = \frac{3x^2-2x-1}{x^2+6x+8} \bullet \frac{x^2+8x+16}{3x^2+13x+4}$$
$$= \frac{(3x+1)(x-1)}{(x+4)(x+2)} \bullet \frac{(x+4)^2}{(3x+1)(x+4)}$$
$$= \frac{(x+4)^2(3x+1)(x-1)}{(x+4)^2(x+2)(3x+1)}$$
$$= \frac{x-1}{x+2}$$

15. Performing the operations: $\frac{x^2}{x-9} - \frac{18x-81}{x-9} = \frac{x^2-18x+81}{x-9} = \frac{(x-9)^2}{x-9} = x-9$

17. Performing the operations: $\frac{x}{x+9} + \frac{5}{x} = \frac{x \bullet x}{(x+9) \bullet x} + \frac{5 \bullet (x+9)}{x \bullet (x+9)} = \frac{x^2}{x(x+9)} + \frac{5x+45}{x(x+9)} = \frac{x^2+5x+45}{x(x+9)}$

19. Performing the operations:

$$\frac{3}{x^2-36} - \frac{2}{x^2-4x-12} = \frac{3}{(x+6)(x-6)} - \frac{2}{(x-6)(x+2)}$$
$$= \frac{3(x+2)}{(x+6)(x-6)(x+2)} - \frac{2(x+6)}{(x+6)(x-6)(x+2)}$$
$$= \frac{3x+6}{(x+6)(x-6)(x+2)} - \frac{2x+12}{(x+6)(x-6)(x+2)}$$
$$= \frac{3x+6-2x-12}{(x+6)(x-6)(x+2)}$$
$$= \frac{x-6}{(x+6)(x-6)(x+2)}$$
$$= \frac{1}{(x+6)(x+2)}$$

21. Multiplying both sides of the equation by $2x$:

$$2x\left(\frac{3}{x}+\frac{1}{2}\right) = 2x\left(\frac{5}{x}\right)$$
$$6+x=10$$
$$x=4$$

Since $x=4$ checks in the original equation, the solution is $x=4$.

23. Multiplying both sides of the equation by x^2:

$$x^2\left(1-\frac{7}{x}\right)=x^2\left(\frac{-6}{x^2}\right)$$
$$x^2-7x=-6$$
$$x^2-7x+6=0$$
$$(x-6)(x-1)=0$$
$$x=1,6$$

Both $x=1$ and $x=6$ check in the original equation.

25. Since $y^2-16=(y+4)(y-4)$ and $y^2+4y=y(y+4)$, multiply each side of the equation by $y(y+4)(y-4)$:

$$y(y+4)(y-4)\bullet\frac{2}{(y+4)(y-4)}=y(y+4)(y-4)\bullet\frac{10}{y(y+4)}$$
$$2y=10(y-4)$$
$$2y=10y-40$$
$$-8y=-40$$
$$y=5$$

Since $y=5$ checks in the original equation, the solution is $y=5$.

27. Let x represent the speed of the boat in still water. Completing the table:

	d	r	t
Upstream	48	$x-3$	$\frac{48}{x-3}$
Downstream	72	$x+3$	$\frac{72}{x+3}$

The equation is:

$$\frac{48}{x-3}=\frac{72}{x+3}$$
$$(x+3)(x-3)\bullet\frac{48}{x-3}=(x+3)(x-3)\bullet\frac{72}{x+3}$$
$$48(x+3)=72(x-3)$$
$$48x+144=72x-216$$
$$-24x+144=-216$$
$$-24x=-360$$
$$x=15$$

The speed of the boat in still water is 15 mph.

29. Simplifying the complex fraction: $\dfrac{\frac{x+4}{x^2-16}}{\frac{2}{x-4}}=\dfrac{\frac{x+4}{(x+4)(x-4)}}{\frac{2}{x-4}}=\dfrac{\frac{1}{x-4}\bullet(x-4)}{\frac{2}{x-4}\bullet(x-4)}=\dfrac{1}{2}$

31. Simplifying the complex fraction: $\dfrac{\frac{1}{a-2}+4}{\frac{1}{a-2}+1}=\dfrac{\left(\frac{1}{a-2}+4\right)(a-2)}{\left(\frac{1}{a-2}+1\right)(a-2)}=\dfrac{1+4(a-2)}{1+1(a-2)}=\dfrac{1+4a-8}{1+a-2}=\dfrac{4a-7}{a-1}$

33. Writing as a fraction: $\dfrac{40\text{ seconds}}{3\text{ minutes}}=\dfrac{40\text{ seconds}}{180\text{ seconds}}=\dfrac{2}{9}$

35. Solving the proportion:

$$\frac{a}{3}=\frac{12}{a}$$
$$a^2=36$$
$$a^2-36=0$$
$$(a+6)(a-6)=0$$
$$a=-6,6$$

37. The variation equation is $y = Kx$. Finding K:

$$-20 = K \bullet 4$$
$$K = -5$$

So $y = -5x$. Substituting $x = 7$: $y = -5 \bullet 7 = -35$

Cumulative Review: Chapters 1-7

1. Simplifying the expression: $8 - 11 = 8 + (-11) = -3$

3. Simplifying the expression: $\frac{-48}{12} = -4$

5. Simplifying the expression: $5x - 4 - 9x = 5x - 9x - 4 = -4x - 4$

7. Simplifying the expression: $9^{-2} = \frac{1}{9^2} = \frac{1}{81}$

9. Simplifying the expression: $4^1 + 9^0 + (-7)^0 = 4 + 1 + 1 = 6$

11. Simplifying the expression:

$$\left(4a^3 - 10a^2 + 6\right) - \left(6a^3 + 5a - 7\right) = 4a^3 - 10a^2 + 6 - 6a^3 - 5a + 7 = -2a^3 - 10a^2 - 5a + 13$$

13. Simplifying the expression: $\frac{x^2}{x-7} - \frac{14x - 49}{x-7} = \frac{x^2 - 14x + 49}{x-7} = \frac{(x-7)^2}{x-7} = x - 7$

15. Simplifying the expression: $\frac{\frac{x-2}{x^2+6x+8}}{\frac{4}{x+4}} = \frac{\frac{x-2}{(x+4)(x+2)} \bullet (x+4)(x+2)}{\frac{4}{x+4} \bullet (x+4)(x+2)} = \frac{x-2}{4(x+2)}$

17. Solving the equation:

$$x - \frac{3}{4} = \frac{5}{6}$$
$$x = \frac{3}{4} + \frac{5}{6}$$
$$x = \frac{9}{12} + \frac{10}{12}$$
$$x = \frac{19}{12}$$

19. Solving the equation:

$$98r^2 - 18 = 0$$
$$2\left(49r^2 - 9\right) = 0$$
$$2(7r + 3)(7r - 3) = 0$$
$$r = -\frac{3}{7}, \frac{3}{7}$$

21. Multiplying each side of the equation by $3x$:

$$3x\left(\frac{5}{x} - \frac{1}{3}\right) = 3x\left(\frac{3}{x}\right)$$
$$15 - x = 9$$
$$-x = -6$$
$$x = 6$$

Since $x = 6$ checks in the original equation, the solution is $x = 6$.

23. Multiplying each side of the equation by $3(x-3)$:

$$\begin{aligned}3(x-3)\bullet\frac{x}{3}&=3(x-3)\bullet\frac{6}{x-3}\\x(x-3)&=18\\x^2-3x&=18\\x^2-3x-18&=0\\(x-6)(x+3)&=0\\x&=6,-3\end{aligned}$$

Both $x=-3$ and $x=6$ check in the original equation.

25. Multiplying the first equation by –5 and the second equation by 3:

$$\begin{aligned}-45x-70y&=20\\45x-24y&=27\end{aligned}$$

Adding the two equations:

$$\begin{aligned}-94y&=47\\y&=-\frac{1}{2}\end{aligned}$$

Substituting into the first equation:

$$\begin{aligned}9x+14\left(-\frac{1}{2}\right)&=-4\\9x-7&=-4\\9x&=3\\x&=\frac{1}{3}\end{aligned}$$

The solution is $\left(\frac{1}{3},-\frac{1}{2}\right)$.

27. To clear each equation of fractions, multiply the first equation by 6 and the second equation by 12:

$$\begin{aligned}6\left(\frac{1}{2}x+\frac{1}{3}y\right)&=6(-1)\\3x+2y&=-6\end{aligned}\qquad\qquad\begin{aligned}12\left(\frac{1}{3}x\right)&=12\left(\frac{1}{4}y+5\right)\\4x&=3y+60\\4x-3y&=60\end{aligned}$$

The system of equations is:

$$\begin{aligned}3x+2y&=-6\\4x-3y&=60\end{aligned}$$

Multiplying the first equation by 3 and the second equation by 2:

$$\begin{aligned}9x+6y&=-18\\8x-6y&=120\end{aligned}$$

Adding the two equations:

$$\begin{aligned}17x&=102\\x&=6\end{aligned}$$

Substituting into $3x+2y=-6$:

$$\begin{aligned}3(6)+2y&=-6\\18+2y&=-6\\2y&=-24\\y&=-12\end{aligned}$$

The solution is $(6,-12)$.

29. Graphing the inequality:

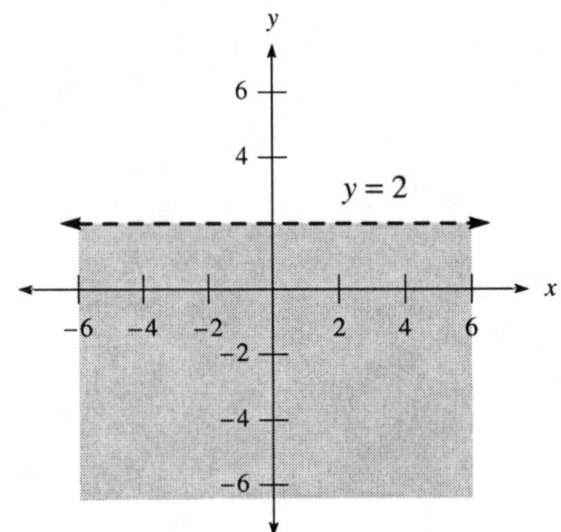

31. Factoring the polynomial: $xy+5x+ay+5a = x(y+5)+a(y+5) = (y+5)(x+a)$

33. Factoring the polynomial: $20y^2-27y+9 = (5y-3)(4y-3)$

35. Factoring the polynomial: $16x^2+72xy+81y^2 = (4x+9y)(4x+9y) = (4x+9y)^2$

37. Checking each ordered pair:

$(3,-1)$: $2(3)-5 = 6-5 = 1 \neq -1$

$(1,-3)$: $2(1)-5 = 2-5 = -3$

$(-2,9)$: $2(-2)-5 = -4-5 = -9 \neq 9$

The ordered pair $(1,-3)$ is a solution to the equation.

39. Solving for *y*:

$$\begin{aligned} 3x-y &= 4 \\ -y &= -3x+4 \\ y &= 3x-4 \end{aligned}$$

The slope is 3 and the *y*-intercept is –4.

41. Using the point-slope formula:

$$\begin{aligned} y-(-1) &= -\frac{2}{5}(x-(-2)) \\ y+1 &= -\frac{2}{5}(x+2) \\ y+1 &= -\frac{2}{5}x-\frac{4}{5} \\ y &= -\frac{2}{5}x-\frac{9}{5} \end{aligned}$$

43. associative property of addition

45. Simplifying the expression: $\frac{6x-12}{6x+12} \bullet \frac{3x+3}{12x-24} = \frac{6(x-2)}{6(x+2)} \bullet \frac{3(x+1)}{12(x-2)} = \frac{18(x-2)(x+1)}{72(x+2)(x-2)} = \frac{x+1}{4(x+2)}$

47. Simplifying the expression: $\frac{2xy+10x+3y+15}{3xy+15x+2y+10} = \frac{2x(y+5)+3(y+5)}{3x(y+5)+2(y+5)} = \frac{(y+5)(2x+3)}{(y+5)(3x+2)} = \frac{2x+3}{3x+2}$

49. Let *x* and *y* represent the two numbers. The system of equations is:

$$\begin{aligned} x+y &= 40 \\ x-y &= 18 \end{aligned}$$

Adding the two equations:

$$\begin{aligned} 2x &= 58 \\ x &= 29 \end{aligned}$$

Substituting into the first equation:

$$\begin{aligned} 29+y &= 40 \\ y &= 11 \end{aligned}$$

The two numbers are 29 and 11.

Chapter 7 Test

1. Reducing the rational expression: $\dfrac{x^2-16}{x^2-8x+16}=\dfrac{(x+4)(x-4)}{(x-4)^2}=\dfrac{x+4}{x-4}$

2. Reducing the rational expression: $\dfrac{10a+20}{5a^2+20a+20}=\dfrac{10(a+2)}{5\left(a^2+4a+4\right)}=\dfrac{10(a+2)}{5(a+2)^2}=\dfrac{2}{a+2}$

3. Reducing the rational expression: $\dfrac{xy+7x+5y+35}{x^2+ax+5x+5a}=\dfrac{x(y+7)+5(y+7)}{x(x+a)+5(x+a)}=\dfrac{(y+7)(x+5)}{(x+a)(x+5)}=\dfrac{y+7}{x+a}$

4. Performing the operations: $\dfrac{3x-12}{4}\bullet\dfrac{8}{2x-8}=\dfrac{3(x-4)}{4}\bullet\dfrac{8}{2(x-4)}=\dfrac{24(x-4)}{8(x-4)}=3$

5. Performing the operations:

$$\begin{aligned}\frac{x^2-49}{x+1}\div\frac{x+7}{x^2-1}&=\frac{x^2-49}{x+1}\bullet\frac{x^2-1}{x+7}\\&=\frac{(x+7)(x-7)}{x+1}\bullet\frac{(x+1)(x-1)}{x+7}\\&=\frac{(x+7)(x-7)(x+1)(x-1)}{(x+1)(x+7)}\\&=(x-7)(x-1)\end{aligned}$$

6. Performing the operations:

$$\begin{aligned}\frac{x^2-3x-10}{x^2-8x+15}\div\frac{3x^2+2x-8}{x^2+x-12}&=\frac{x^2-3x-10}{x^2-8x+15}\bullet\frac{x^2+x-12}{3x^2+2x-8}\\&=\frac{(x-5)(x+2)}{(x-5)(x-3)}\bullet\frac{(x+4)(x-3)}{(3x-4)(x+2)}\\&=\frac{(x-5)(x+2)(x+4)(x-3)}{(x-5)(x-3)(3x-4)(x+2)}\\&=\frac{x+4}{3x-4}\end{aligned}$$

7. Performing the operations: $\left(x^2-9\right)\left(\dfrac{x+2}{x+3}\right)=\dfrac{(x+3)(x-3)}{1}\bullet\dfrac{x+2}{x+3}=\dfrac{(x+3)(x-3)(x+2)}{x+3}=(x-3)(x+2)$

8. Performing the operations: $\dfrac{3}{x-2}-\dfrac{6}{x-2}=\dfrac{3-6}{x-2}=\dfrac{-3}{x-2}$

9. Performing the operations:

$$\begin{aligned}\frac{x}{x^2-9}+\frac{4}{4x-12}&=\frac{x}{(x+3)(x-3)}+\frac{4}{4(x-3)}\\&=\frac{x}{(x+3)(x-3)}+\frac{1\bullet(x+3)}{(x+3)(x-3)}\\&=\frac{x}{(x+3)(x-3)}+\frac{x+3}{(x+3)(x-3)}\\&=\frac{2x+3}{(x+3)(x-3)}\end{aligned}$$

10. Performing the operations:

$$\frac{2x}{x^2-1}+\frac{x}{x^2-3x+2}=\frac{2x}{(x+1)(x-1)}+\frac{x}{(x-1)(x-2)}$$
$$=\frac{2x\bullet(x-2)}{(x+1)(x-1)(x-2)}+\frac{x\bullet(x+1)}{(x+1)(x-1)(x-2)}$$
$$=\frac{2x^2-4x}{(x+1)(x-1)(x-2)}+\frac{x^2+x}{(x+1)(x-1)(x-2)}$$
$$=\frac{3x^2-3x}{(x+1)(x-1)(x-2)}$$
$$=\frac{3x(x-1)}{(x+1)(x-1)(x-2)}$$
$$=\frac{3x}{(x+1)(x-2)}$$

11. Multiplying both sides of the equation by 15:

$$15\bullet\frac{7}{5}=15\bullet\frac{x+2}{3}$$
$$21=5(x+2)$$
$$21=5x+10$$
$$11=5x$$
$$x=\frac{11}{5}$$

Since $x=\frac{11}{5}$ checks in the original equation, the solution is $x=\frac{11}{5}$.

12. Multiplying both sides of the equation by $x(x+4)$:

$$x(x+4)\bullet\frac{10}{x+4}=x(x+4)\bullet\left(\frac{6}{x}-\frac{4}{x}\right)$$
$$10x=6(x+4)-4(x+4)$$
$$10x=6x+24-4x-16$$
$$10x=2x+8$$
$$8x=8$$
$$x=1$$

Since $x=1$ checks in the original equation, the solution is $x=1$.

13. Multiplying both sides of the equation by $x^2-x-2=(x-2)(x+1)$:

$$(x-2)(x+1)\left(\frac{3}{x-2}-\frac{4}{x+1}\right)=(x-2)(x+1)\bullet\frac{5}{(x-2)(x+1)}$$
$$3(x+1)-4(x-2)=5$$
$$3x+3-4x+8=5$$
$$-x+11=5$$
$$-x=-6$$
$$x=6$$

Since $x=6$ checks in the original equation, the solution is $x=6$.

14. Let x represent the speed of the boat in still water. Completing the table:

	d	r	t
Upstream	26	$x-2$	$\frac{26}{x-2}$
Downstream	34	$x+2$	$\frac{34}{x+2}$

The equation is:

$$\frac{26}{x-2}=\frac{34}{x+2}$$
$$(x+2)(x-2)\bullet\frac{26}{x-2}=(x+2)(x-2)\bullet\frac{34}{x+2}$$
$$26(x+2)=34(x-2)$$
$$26x+52=34x-68$$
$$-8x+52=-68$$
$$-8x=-120$$
$$x=15$$

The speed of the boat in still water is 15 mph.

15. Let t represent the time to empty the pool with both pipes open. The equation is:

$$\frac{1}{12}-\frac{1}{15}=\frac{1}{t}$$
$$60t\left(\frac{1}{12}-\frac{1}{15}\right)=60t\bullet\frac{1}{t}$$
$$5t-4t=60$$
$$t=60$$

It will take 60 hours to empty the pool with both pipes open.

16. The ratio of alcohol to water is given by: $\frac{27 \text{ ml}}{54 \text{ ml}}=\frac{1}{2}$

The ratio of alcohol to total volume is given by: $\frac{27 \text{ ml}}{81 \text{ ml}}=\frac{1}{3}$

17. Comparing defective parts to total parts, the proportion is:

$$\frac{8}{100}=\frac{x}{1650}$$
$$100x=13200$$
$$x=132$$

The machine can be expected to produce 132 defective parts.

18. Simplifying the complex fraction: $\dfrac{1+\frac{1}{x}}{1-\frac{1}{x}}=\dfrac{\left(1+\frac{1}{x}\right)\bullet x}{\left(1-\frac{1}{x}\right)\bullet x}=\dfrac{x+1}{x-1}$

19. Simplifying the complex fraction: $\dfrac{1-\frac{16}{x^2}}{1-\frac{2}{x}-\frac{8}{x^2}}=\dfrac{\left(1-\frac{16}{x^2}\right)\bullet x^2}{\left(1-\frac{2}{x}-\frac{8}{x^2}\right)\bullet x^2}=\dfrac{x^2-16}{x^2-2x-8}=\dfrac{(x+4)(x-4)}{(x-4)(x+2)}=\dfrac{x+4}{x+2}$

20. The variation equation is $y=Kx^2$. Finding K:

$$36=K\bullet 3^2$$
$$36=9K$$
$$K=4$$

So $y=4x^2$. Substituting $x=5$: $y=4\bullet 5^2=4\bullet 25=100$

21. The variation equation is $y=\frac{K}{x}$. Finding K:

$$6=\frac{K}{3}$$
$$K=18$$

So $y=\frac{18}{x}$. Substituting $x=9$: $y=\frac{18}{9}=2$

Chapter 8
Roots and Radicals

8.1 Definitions and Common Roots

1. Finding the root: $\sqrt{9}=3$
3. Finding the root: $-\sqrt{9}=-3$
5. Finding the root: $\sqrt{-25}$ is not a real number
7. Finding the root: $-\sqrt{144}=-12$
9. Finding the root: $\sqrt{625}=25$
11. Finding the root: $\sqrt{-49}$ is not a real number
13. Finding the root: $-\sqrt{64}=-8$
15. Finding the root: $-\sqrt{100}=-10$
17. Finding the root: $\sqrt{1225}=35$
19. Finding the root: $\sqrt[4]{1}=1$
21. Finding the root: $\sqrt[3]{-8}=-2$
23. Finding the root: $-\sqrt[3]{125}=-5$
25. Finding the root: $\sqrt[3]{-1}=-1$
27. Finding the root: $\sqrt[3]{-27}=-3$
29. Finding the root: $-\sqrt[4]{16}=-2$
31. Simplifying the expression: $\sqrt{x^2}=x$
33. Simplifying the expression: $\sqrt{9x^2}=3x$
35. Simplifying the expression: $\sqrt{x^2y^2}=xy$
37. Simplifying the expression: $\sqrt{(a+b)^2}=a+b$
39. Simplifying the expression: $\sqrt{49x^2y^2}=7xy$
41. Simplifying the expression: $\sqrt[3]{x^3}=x$
43. Simplifying the expression: $\sqrt[3]{8x^3}=2x$
45. Simplifying the expression: $\sqrt{x^4}=x^2$
47. Simplifying the expression: $\sqrt{36a^6}=6a^3$
49. Simplifying the expression: $\sqrt{25a^8b^4}=5a^4b^2$
51. Simplifying the expression: $\sqrt[3]{x^6}=x^2$
53. Simplifying the expression: $\sqrt[3]{27a^{12}}=3a^4$
55. Simplifying the expression: $\sqrt[4]{x^8}=x^2$

57. Simplifying the expression: $\sqrt{9}+\sqrt{16}=3+4=7$

59. Simplifying the expression: $\sqrt{9+16}=\sqrt{25}=5$

61. Simplifying the expression: $\sqrt{144}+\sqrt{25}=12+5=17$

63. Simplifying the expression: $\sqrt{144+25}=\sqrt{169}=13$

65. Simplifying each expression:

$$\frac{5+\sqrt{49}}{2}=\frac{5+7}{2}=\frac{12}{2}=6$$

$$\frac{5-\sqrt{49}}{2}=\frac{5-7}{2}=\frac{-2}{2}=-1$$

67. Simplifying each expression:

$$\frac{2+\sqrt{16}}{2}=\frac{2+4}{2}=\frac{6}{2}=3$$

$$\frac{2-\sqrt{16}}{2}=\frac{2-4}{2}=\frac{-2}{2}=-1$$

69. Simplifying the expression: $\sqrt{x^2+6x+9}=\sqrt{(x+3)^2}=x+3$

71. Simplifying each expression when $x=4$:

$$\sqrt{x^2+9}=\sqrt{4^2+9}=\sqrt{16+9}=\sqrt{25}=5$$

$$x+3=4+3=7$$

73. Finding the annual rate of return: $r=\dfrac{\sqrt{65}-\sqrt{50}}{\sqrt{50}}\approx 0.140=14.0\%$

75. Finding the annual rate of return: $r=\dfrac{\sqrt{600}-\sqrt{500}}{\sqrt{500}}\approx 0.095=9.5\%$

77. Using the Pythagorean Theorem:

$$x^2=3^2+4^2=9+16=25$$

$$x=\sqrt{25}=5$$

79. Using the Pythagorean Theorem:

$$x^2=5^2+10^2=25+100=125$$

$$x=\sqrt{125}\approx 11.2$$

81. Let l represent the length of the wire. Using the Pythagorean Theorem:

$$l^2=24^2+18^2=576+324=900$$

$$l=\sqrt{900}=30$$

The length of the wire is 30 feet.

83. Reducing the rational expression: $\dfrac{x^2-16}{x+4}=\dfrac{(x+4)(x-4)}{x+4}=x-4$

85. Reducing the rational expression: $\dfrac{10a+20}{5a^2-20}=\dfrac{10(a+2)}{5\left(a^2-4\right)}=\dfrac{10(a+2)}{5(a+2)(a-2)}=\dfrac{2}{a-2}$

87. Reducing the rational expression: $\dfrac{2x^2-5x-3}{x^2-3x}=\dfrac{(2x+1)(x-3)}{x(x-3)}=\dfrac{2x+1}{x}$

89. Reducing the rational expression: $\dfrac{xy+3x+2y+6}{xy+3x+ay+3a}=\dfrac{x(y+3)+2(y+3)}{x(y+3)+a(y+3)}=\dfrac{(y+3)(x+2)}{(y+3)(x+a)}=\dfrac{x+2}{x+a}$

8.2 Properties of Radicals

1. Simplifying the radical expression: $\sqrt{8}=\sqrt{4\bullet 2}=\sqrt{4}\sqrt{2}=2\sqrt{2}$

3. Simplifying the radical expression: $\sqrt{12}=\sqrt{4\bullet 3}=\sqrt{4}\sqrt{3}=2\sqrt{3}$

5. Simplifying the radical expression: $\sqrt[3]{24}=\sqrt[3]{8\bullet 3}=\sqrt[3]{8}\sqrt[3]{3}=2\sqrt[3]{3}$

7. Simplifying the radical expression: $\sqrt{50x^2}=\sqrt{25x^2\bullet 2}=\sqrt{25x^2}\sqrt{2}=5x\sqrt{2}$

9. Simplifying the radical expression: $\sqrt{45a^2b^2}=\sqrt{9a^2b^2\bullet 5}=\sqrt{9a^2b^2}\sqrt{5}=3ab\sqrt{5}$

11. Simplifying the radical expression: $\sqrt[3]{54x^3}=\sqrt[3]{27x^3\bullet 2}=\sqrt[3]{27x^3}\sqrt[3]{2}=3x\sqrt[3]{2}$

13. Simplifying the radical expression: $\sqrt{32x^4}=\sqrt{16x^4\bullet 2}=\sqrt{16x^4}\sqrt{2}=4x^2\sqrt{2}$

15. Simplifying the radical expression: $5\sqrt{80}=5\sqrt{16\bullet 5}=5\sqrt{16}\sqrt{5}=5\bullet 4\sqrt{5}=20\sqrt{5}$

17. Simplifying the radical expression: $\frac{1}{2}\sqrt{28x^3}=\frac{1}{2}\sqrt{4x^2\bullet 7x}=\frac{1}{2}\sqrt{4x^2}\sqrt{7x}=\frac{1}{2}\bullet 2x\sqrt{7x}=x\sqrt{7x}$

19. Simplifying the radical expression: $x\sqrt[3]{8x^4}=x\sqrt[3]{8x^3\bullet x}=x\sqrt[3]{8x^3}\sqrt[3]{x}=x\bullet 2x\sqrt[3]{x}=2x^2\sqrt[3]{x}$

21. Simplifying the radical expression: $2a\sqrt[3]{27a^5}=2a\sqrt[3]{27a^3\bullet a^2}=2a\sqrt[3]{27a^3}\sqrt[3]{a^2}=2a\bullet 3a\sqrt[3]{a^2}=6a^2\sqrt[3]{a^2}$

23. Simplifying the radical expression: $\frac{4}{3}\sqrt{45a^3}=\frac{4}{3}\sqrt{9a^2\bullet 5a}=\frac{4}{3}\sqrt{9a^2}\sqrt{5a}=\frac{4}{3}\bullet 3a\sqrt{5a}=4a\sqrt{5a}$

25. Simplifying the radical expression: $3\sqrt{50xy^2}=3\sqrt{25y^2\bullet 2x}=3\sqrt{25y^2}\sqrt{2x}=3\bullet 5y\sqrt{2x}=15y\sqrt{2x}$

27. Simplifying the radical expression: $7\sqrt{12x^2y}=7\sqrt{4x^2\bullet 3y}=7\sqrt{4x^2}\sqrt{3y}=7\bullet 2x\sqrt{3y}=14x\sqrt{3y}$

29. Simplifying the radical expression: $\sqrt{\frac{16}{25}}=\frac{\sqrt{16}}{\sqrt{25}}=\frac{4}{5}$

31. Simplifying the radical expression: $\sqrt{\frac{4}{9}}=\frac{\sqrt{4}}{\sqrt{9}}=\frac{2}{3}$

33. Simplifying the radical expression: $\sqrt[3]{\frac{8}{27}}=\frac{\sqrt[3]{8}}{\sqrt[3]{27}}=\frac{2}{3}$

35. Simplifying the radical expression: $\sqrt[4]{\frac{16}{81}}=\frac{\sqrt[4]{16}}{\sqrt[4]{81}}=\frac{2}{3}$

37. Simplifying the radical expression: $\sqrt{\frac{100x^2}{25}}=\sqrt{4x^2}=2x$

39. Simplifying the radical expression: $\sqrt{\frac{81a^2b^2}{9}}=\sqrt{9a^2b^2}=3ab$

41. Simplifying the radical expression: $\sqrt[3]{\frac{27x^3}{8y^3}}=\frac{\sqrt[3]{27x^3}}{\sqrt[3]{8y^3}}=\frac{3x}{2y}$

43. Simplifying the radical expression: $\sqrt{\frac{50}{9}}=\frac{\sqrt{50}}{\sqrt{9}}=\frac{\sqrt{25\bullet 2}}{3}=\frac{5\sqrt{2}}{3}$

45. Simplifying the radical expression: $\sqrt{\frac{75}{25}}=\sqrt{3}$

47. Simplifying the radical expression: $\sqrt{\frac{128}{49}}=\frac{\sqrt{128}}{\sqrt{49}}=\frac{\sqrt{64\bullet 2}}{7}=\frac{8\sqrt{2}}{7}$

49. Simplifying the radical expression: $\sqrt{\frac{288x}{25}}=\frac{\sqrt{288x}}{\sqrt{25}}=\frac{\sqrt{144\bullet 2x}}{5}=\frac{12\sqrt{2x}}{5}$

51. Simplifying the radical expression: $\sqrt{\frac{54a^2}{25}}=\frac{\sqrt{54a^2}}{\sqrt{25}}=\frac{\sqrt{9a^2\bullet 6}}{5}=\frac{3a\sqrt{6}}{5}$

53. Simplifying the radical expression: $\frac{3\sqrt{50}}{2} = \frac{3\sqrt{25 \cdot 2}}{2} = \frac{3 \cdot 5\sqrt{2}}{2} = \frac{15\sqrt{2}}{2}$

55. Simplifying the radical expression: $\frac{7\sqrt{28y^2}}{3} = \frac{7\sqrt{4y^2 \cdot 7}}{3} = \frac{7 \cdot 2y\sqrt{7}}{3} = \frac{14y\sqrt{7}}{3}$

57. Simplifying the radical expression: $\frac{5\sqrt{72a^2b^2}}{\sqrt{36}} = \frac{5\sqrt{36a^2b^2 \cdot 2}}{6} = \frac{5 \cdot 6ab\sqrt{2}}{6} = \frac{30ab\sqrt{2}}{6} = 5ab\sqrt{2}$

59. Simplifying the radical expression: $\frac{6\sqrt{8x^2y}}{\sqrt{4}} = \frac{6\sqrt{4x^2 \cdot 2y}}{2} = \frac{6 \cdot 2x\sqrt{2y}}{2} = \frac{12x\sqrt{2y}}{2} = 6x\sqrt{2y}$

61. Simplifying the radical expression: $\frac{8\sqrt{12x^2y^3}}{\sqrt{100}} = \frac{8\sqrt{4x^2y^2 \cdot 3y}}{10} = \frac{8 \cdot 2xy\sqrt{3y}}{10} = \frac{16xy\sqrt{3y}}{10} = \frac{8xy\sqrt{3y}}{5}$

63. Approximating each value:

$\sqrt{12} \approx 3.464$

$2\sqrt{3} \approx 3.464$

65. Substituting $h = 25$ feet: $t = \sqrt{\frac{25}{16}} = \frac{\sqrt{25}}{\sqrt{16}} = \frac{5}{4} = 1\frac{1}{4}$ seconds

67. Performing the operations: $\frac{8x}{x^2 - 5x} \cdot \frac{x^2 - 25}{4x^2 + 4x} = \frac{8x}{x(x-5)} \cdot \frac{(x+5)(x-5)}{4x(x+1)} = \frac{8x(x+5)(x-5)}{4x^2(x-5)(x+1)} = \frac{2(x+5)}{x(x+1)}$

69. Performing the operations:

$$\begin{aligned} \frac{x^2+3x-4}{3x^2+7x-20} \div \frac{x^2-2x+1}{3x^2-2x-5} &= \frac{x^2+3x-4}{3x^2+7x-20} \cdot \frac{3x^2-2x-5}{x^2-2x+1} \\ &= \frac{(x+4)(x-1)}{(3x-5)(x+4)} \cdot \frac{(3x-5)(x+1)}{(x-1)^2} \\ &= \frac{(x+4)(x-1)(3x-5)(x+1)}{(3x-5)(x+4)(x-1)^2} \\ &= \frac{x+1}{x-1} \end{aligned}$$

71. Performing the operations: $(x^2 - 36)\left(\frac{x+3}{x-6}\right) = \frac{(x+6)(x-6)}{1} \cdot \frac{x+3}{x-6} = \frac{(x+6)(x-6)(x+3)}{x-6} = (x+6)(x+3)$

8.3 Simplified Form for Radicals

1. Simplifying the radical expression: $\sqrt{\frac{1}{2}} = \frac{\sqrt{1}}{\sqrt{2}} \cdot \frac{\sqrt{2}}{\sqrt{2}} = \frac{\sqrt{2}}{\sqrt{4}} = \frac{\sqrt{2}}{2}$

3. Simplifying the radical expression: $\sqrt{\frac{1}{3}} = \frac{\sqrt{1}}{\sqrt{3}} \cdot \frac{\sqrt{3}}{\sqrt{3}} = \frac{\sqrt{3}}{\sqrt{9}} = \frac{\sqrt{3}}{3}$

5. Simplifying the radical expression: $\sqrt{\frac{2}{5}} = \frac{\sqrt{2}}{\sqrt{5}} \cdot \frac{\sqrt{5}}{\sqrt{5}} = \frac{\sqrt{10}}{\sqrt{25}} = \frac{\sqrt{10}}{5}$

7. Simplifying the radical expression: $\sqrt{\frac{3}{2}} = \frac{\sqrt{3}}{\sqrt{2}} \cdot \frac{\sqrt{2}}{\sqrt{2}} = \frac{\sqrt{6}}{\sqrt{4}} = \frac{\sqrt{6}}{2}$

9. Simplifying the radical expression: $\sqrt{\frac{20}{3}} = \frac{\sqrt{20}}{\sqrt{3}} \cdot \frac{\sqrt{3}}{\sqrt{3}} = \frac{\sqrt{60}}{\sqrt{9}} = \frac{\sqrt{4 \cdot 15}}{3} = \frac{2\sqrt{15}}{3}$

11. Simplifying the radical expression: $\sqrt{\frac{45}{6}} = \sqrt{\frac{15}{2}} = \frac{\sqrt{15}}{\sqrt{2}} \cdot \frac{\sqrt{2}}{\sqrt{2}} = \frac{\sqrt{30}}{\sqrt{4}} = \frac{\sqrt{30}}{2}$

13. Simplifying the radical expression: $\sqrt{\frac{20}{5}} = \sqrt{4} = 2$

15. Simplifying the radical expression: $\frac{\sqrt{21}}{\sqrt{3}}=\sqrt{\frac{21}{3}}=\sqrt{7}$

17. Simplifying the radical expression: $\frac{\sqrt{35}}{\sqrt{7}}=\sqrt{\frac{35}{7}}=\sqrt{5}$

19. Simplifying the radical expression: $\frac{10\sqrt{15}}{5\sqrt{3}}=\frac{10}{5}\bullet\sqrt{\frac{15}{3}}=2\sqrt{5}$

21. Simplifying the radical expression: $\frac{6\sqrt{21}}{3\sqrt{7}}=\frac{6}{3}\bullet\sqrt{\frac{21}{7}}=2\sqrt{3}$

23. Simplifying the radical expression: $\frac{6\sqrt{35}}{12\sqrt{5}}=\frac{6}{12}\bullet\sqrt{\frac{35}{5}}=\frac{1}{2}\sqrt{7}=\frac{\sqrt{7}}{2}$

25. Simplifying the radical expression: $\sqrt{\frac{4x^2y^2}{2}}=\sqrt{2x^2y^2}=xy\sqrt{2}$

27. Simplifying the radical expression: $\sqrt{\frac{5x^2y}{3}}=\frac{\sqrt{5x^2y}}{\sqrt{3}}\bullet\frac{\sqrt{3}}{\sqrt{3}}=\frac{\sqrt{15x^2y}}{\sqrt{9}}=\frac{x\sqrt{15y}}{3}$

29. Simplifying the radical expression: $\sqrt{\frac{16a^4}{5}}=\frac{\sqrt{16a^4}}{\sqrt{5}}\bullet\frac{\sqrt{5}}{\sqrt{5}}=\frac{4a^2\sqrt{5}}{\sqrt{25}}=\frac{4a^2\sqrt{5}}{5}$

31. Simplifying the radical expression: $\sqrt{\frac{72a^5}{5}}=\frac{\sqrt{72a^5}}{\sqrt{5}}\bullet\frac{\sqrt{5}}{\sqrt{5}}=\frac{\sqrt{360a^5}}{\sqrt{25}}=\frac{\sqrt{36a^4\bullet 10a}}{5}=\frac{6a^2\sqrt{10a}}{5}$

33. Simplifying the radical expression: $\sqrt{\frac{20x^2y^3}{3}}=\frac{\sqrt{20x^2y^3}}{\sqrt{3}}\bullet\frac{\sqrt{3}}{\sqrt{3}}=\frac{\sqrt{60x^2y^3}}{\sqrt{9}}=\frac{\sqrt{4x^2y^2\bullet 15y}}{3}=\frac{2xy\sqrt{15y}}{3}$

35. Simplifying the radical expression: $\frac{2\sqrt{20x^2y^3}}{3}=\frac{2\sqrt{4x^2y^2\bullet 5y}}{3}=\frac{2\bullet 2xy\sqrt{5y}}{3}=\frac{4xy\sqrt{5y}}{3}$

37. Simplifying the radical expression: $\frac{6\sqrt{54a^2b^3}}{5}=\frac{6\sqrt{9a^2b^2\bullet 6b}}{5}=\frac{6\bullet 3ab\sqrt{6b}}{5}=\frac{18ab\sqrt{6b}}{5}$

39. Simplifying the radical expression: $\frac{3\sqrt{72x^4}}{\sqrt{2x}}=3\sqrt{\frac{72x^4}{2x}}=3\sqrt{36x^3}=3\sqrt{36x^2\bullet x}=3\bullet 6x\sqrt{x}=18x\sqrt{x}$

41. Simplifying the radical expression: $\sqrt[3]{\frac{1}{2}}=\frac{\sqrt[3]{1}}{\sqrt[3]{2}}\bullet\frac{\sqrt[3]{4}}{\sqrt[3]{4}}=\frac{\sqrt[3]{4}}{\sqrt[3]{8}}=\frac{\sqrt[3]{4}}{2}$

43. Simplifying the radical expression: $\sqrt[3]{\frac{1}{9}}=\frac{\sqrt[3]{1}}{\sqrt[3]{9}}\bullet\frac{\sqrt[3]{3}}{\sqrt[3]{3}}=\frac{\sqrt[3]{3}}{\sqrt[3]{27}}=\frac{\sqrt[3]{3}}{3}$

45. Simplifying the radical expression: $\sqrt[3]{\frac{3}{2}}=\frac{\sqrt[3]{3}}{\sqrt[3]{2}}\bullet\frac{\sqrt[3]{4}}{\sqrt[3]{4}}=\frac{\sqrt[3]{12}}{\sqrt[3]{8}}=\frac{\sqrt[3]{12}}{2}$

47. Simplifying the radical expression: $\sqrt[3]{\frac{3}{4}}=\frac{\sqrt[3]{3}}{\sqrt[3]{4}}\bullet\frac{\sqrt[3]{2}}{\sqrt[3]{2}}=\frac{\sqrt[3]{6}}{\sqrt[3]{8}}=\frac{\sqrt[3]{6}}{2}$

49. The approximate values are:

$$\frac{1}{\sqrt{2}}\approx 0.707$$

$$\frac{\sqrt{2}}{2}\approx 0.707$$

51. Substituting $h=24$: $d=\sqrt{\frac{3\bullet 24}{2}}=\sqrt{36}=6$ miles

53. Drawing the figure:

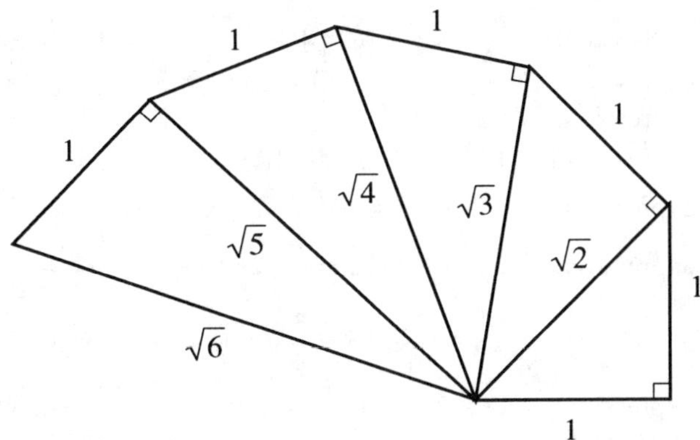

55. Simplifying the terms in the sequence:

$$\sqrt{1^2+1}=\sqrt{1+1}=\sqrt{2}$$

$$\sqrt{\left(\sqrt{2}\right)^2+1}=\sqrt{2+1}=\sqrt{3}$$

$$\sqrt{\left(\sqrt{3}\right)^2+1}=\sqrt{3+1}=\sqrt{4}=2$$

57. Combining the terms: $3x+7x=(3+7)x=10x$

59. Combining the terms: $15x+8x=(15+8)x=23x$

61. Combining the terms: $7a-3a+6a=(7-3+6)a=10a$

63. Combining the rational expressions: $\frac{x^2}{x+5}+\frac{10x+25}{x+5}=\frac{x^2+10x+25}{x+5}=\frac{(x+5)^2}{x+5}=x+5$

65. Combining the rational expressions: $\frac{a}{3}+\frac{2}{5}=\frac{a\cdot 5}{3\cdot 5}+\frac{2\cdot 3}{5\cdot 3}=\frac{5a}{15}+\frac{6}{15}=\frac{5a+6}{15}$

67. Combining the rational expressions:

$$\begin{aligned}\frac{6}{a^2-9}-\frac{5}{a^2-a-6}&=\frac{6}{(a+3)(a-3)}-\frac{5}{(a-3)(a+2)}\\&=\frac{6\cdot(a+2)}{(a+3)(a-3)(a+2)}-\frac{5\cdot(a+3)}{(a+3)(a-3)(a+2)}\\&=\frac{6a+12}{(a+3)(a-3)(a+2)}-\frac{5a+15}{(a+3)(a-3)(a+2)}\\&=\frac{6a+12-5a-15}{(a+3)(a-3)(a+2)}\\&=\frac{a-3}{(a+3)(a-3)(a+2)}\\&=\frac{1}{(a+3)(a+2)}\end{aligned}$$

8.4 Addition and Subtraction of Radical Expressions

1. Simplifying the radical expression: $3\sqrt{2}+4\sqrt{2}=7\sqrt{2}$

3. Simplifying the radical expression: $9\sqrt{5}-7\sqrt{5}=2\sqrt{5}$

5. Simplifying the radical expression: $\sqrt{3}+6\sqrt{3}=7\sqrt{3}$

7. Simplifying the radical expression: $\frac{5}{8}\sqrt{5}-\frac{3}{7}\sqrt{5}=\frac{35}{56}\sqrt{5}-\frac{24}{56}\sqrt{5}=\frac{11}{56}\sqrt{5}$

9. Simplifying the radical expression: $14\sqrt{13}-\sqrt{13}=13\sqrt{13}$

11. Simplifying the radical expression: $-3\sqrt{10}+9\sqrt{10}=6\sqrt{10}$

13. Simplifying the radical expression: $5\sqrt{5}+\sqrt{5}=6\sqrt{5}$
15. Simplifying the radical expression: $\sqrt{8}+2\sqrt{2}=\sqrt{4\bullet 2}+2\sqrt{2}=2\sqrt{2}+2\sqrt{2}=4\sqrt{2}$
17. Simplifying the radical expression: $3\sqrt{3}-\sqrt{27}=3\sqrt{3}-\sqrt{9\bullet 3}=3\sqrt{3}-3\sqrt{3}=0$
19. Simplifying the radical expression:

$$5\sqrt{12}-10\sqrt{48}=5\sqrt{4\bullet 3}-10\sqrt{16\bullet 3}=5\bullet 2\sqrt{3}-10\bullet 4\sqrt{3}=10\sqrt{3}-40\sqrt{3}=-30\sqrt{3}$$

21. Simplifying the radical expression: $-\sqrt{75}-\sqrt{3}=-\sqrt{25\bullet 3}-\sqrt{3}=-5\sqrt{3}-\sqrt{3}=-6\sqrt{3}$
23. Simplifying the radical expression: $\frac{1}{5}\sqrt{75}-\frac{1}{2}\sqrt{12}=\frac{1}{5}\sqrt{25\bullet 3}-\frac{1}{2}\sqrt{4\bullet 3}=\frac{1}{5}\bullet 5\sqrt{3}-\frac{1}{2}\bullet 2\sqrt{3}=\sqrt{3}-\sqrt{3}=0$
25. Simplifying the radical expression: $\frac{3}{4}\sqrt{8}+\frac{3}{10}\sqrt{75}=\frac{3}{4}\sqrt{4\bullet 2}+\frac{3}{10}\sqrt{25\bullet 3}=\frac{3}{4}\bullet 2\sqrt{2}+\frac{3}{10}\bullet 5\sqrt{3}=\frac{3}{2}\sqrt{2}+\frac{3}{2}\sqrt{3}$
27. Simplifying the radical expression: $\sqrt{27}-2\sqrt{12}+\sqrt{3}=\sqrt{9\bullet 3}-2\sqrt{4\bullet 3}+\sqrt{3}=3\sqrt{3}-4\sqrt{3}+\sqrt{3}=0$
29. Simplifying the radical expression:

$$\begin{aligned}\frac{5}{6}\sqrt{72}-\frac{3}{8}\sqrt{8}+\frac{3}{10}\sqrt{50}&=\frac{5}{6}\sqrt{36\bullet 2}-\frac{3}{8}\sqrt{4\bullet 2}+\frac{3}{10}\sqrt{25\bullet 2}\\&=\frac{5}{6}\bullet 6\sqrt{2}-\frac{3}{8}\bullet 2\sqrt{2}+\frac{3}{10}\bullet 5\sqrt{2}\\&=5\sqrt{2}-\frac{3}{4}\sqrt{2}+\frac{3}{2}\sqrt{2}\\&=\frac{20}{4}\sqrt{2}-\frac{3}{4}\sqrt{2}+\frac{6}{4}\sqrt{2}\\&=\frac{23}{4}\sqrt{2}\end{aligned}$$

31. Simplifying the radical expression:

$$5\sqrt{7}+2\sqrt{28}-4\sqrt{63}=5\sqrt{7}+2\sqrt{4\bullet 7}-4\sqrt{9\bullet 7}=5\sqrt{7}+2\bullet 2\sqrt{7}-4\bullet 3\sqrt{7}=5\sqrt{7}+4\sqrt{7}-12\sqrt{7}=-3\sqrt{7}$$

33. Simplifying the radical expression:

$$\begin{aligned}6\sqrt{48}-2\sqrt{12}+5\sqrt{27}&=6\sqrt{16\bullet 3}-2\sqrt{4\bullet 3}+5\sqrt{9\bullet 3}\\&=6\bullet 4\sqrt{3}-2\bullet 2\sqrt{3}+5\bullet 3\sqrt{3}\\&=24\sqrt{3}-4\sqrt{3}+15\sqrt{3}\\&=35\sqrt{3}\end{aligned}$$

35. Simplifying the radical expression:

$$\begin{aligned}6\sqrt{48}-\sqrt{72}-3\sqrt{300}&=6\sqrt{16\bullet 3}-\sqrt{36\bullet 2}-3\sqrt{100\bullet 3}\\&=6\bullet 4\sqrt{3}-6\sqrt{2}-3\bullet 10\sqrt{3}\\&=24\sqrt{3}-6\sqrt{2}-30\sqrt{3}\\&=-6\sqrt{3}-6\sqrt{2}\end{aligned}$$

37. Simplifying the radical expression: $\sqrt{x^3}+x\sqrt{x}=\sqrt{x^2\bullet x}+x\sqrt{x}=x\sqrt{x}+x\sqrt{x}=2x\sqrt{x}$
39. Simplifying the radical expression: $5\sqrt{3a^2}-a\sqrt{3}=5a\sqrt{3}-a\sqrt{3}=4a\sqrt{3}$
41. Simplifying the radical expression:

$$5\sqrt{8x^3}+x\sqrt{50x}=5\sqrt{4x^2\bullet 2x}+x\sqrt{25\bullet 2x}=5\bullet 2x\sqrt{2x}+x\bullet 5\sqrt{2x}=10x\sqrt{2x}+5x\sqrt{2x}=15x\sqrt{2x}$$

43. Simplifying the radical expression:

$$3\sqrt{75x^3y}-2x\sqrt{3xy}=3\sqrt{25x^2\bullet 3xy}-2x\sqrt{3xy}=3\bullet 5x\sqrt{3xy}-2x\sqrt{3xy}=15x\sqrt{3xy}-2x\sqrt{3xy}=13x\sqrt{3xy}$$

45. Simplifying the radical expression: $\sqrt{20ab^2}-b\sqrt{45a}=\sqrt{4b^2\bullet 5a}-b\sqrt{9\bullet 5a}=2b\sqrt{5a}-3b\sqrt{5a}=-b\sqrt{5a}$
47. Simplifying the radical expression:

$$9\sqrt{18x^3}-2x\sqrt{48x}=9\sqrt{9x^2\bullet 2x}-2x\sqrt{16\bullet 3x}=9\bullet 3x\sqrt{2x}-2x\bullet 4\sqrt{3x}=27x\sqrt{2x}-8x\sqrt{3x}$$

49. Simplifying the radical expression:

$$\begin{aligned} 7\sqrt{50x^2y}+8x\sqrt{8y}-7\sqrt{32x^2y} &= 7\sqrt{25x^2\bullet 2y}+8x\sqrt{4\bullet 2y}-7\sqrt{16x^2\bullet 2y} \\ &= 7\bullet 5x\sqrt{2y}+8x\bullet 2\sqrt{2y}-7\bullet 4x\sqrt{2y} \\ &= 35x\sqrt{2y}+16x\sqrt{2y}-28x\sqrt{2y} \\ &= 23x\sqrt{2y} \end{aligned}$$

51. Simplifying the expression: $\dfrac{8-\sqrt{24}}{6}=\dfrac{8-\sqrt{4\bullet 6}}{6}=\dfrac{8-2\sqrt{6}}{6}=\dfrac{2(4-\sqrt{6})}{6}=\dfrac{4-\sqrt{6}}{3}$

53. Simplifying the expression: $\dfrac{6+\sqrt{8}}{2}=\dfrac{6+\sqrt{4\bullet 2}}{2}=\dfrac{6+2\sqrt{2}}{2}=\dfrac{2(3+\sqrt{2})}{2}=3+\sqrt{2}$

55. Simplifying the expression: $\dfrac{-10+\sqrt{50}}{10}=\dfrac{-10+\sqrt{25\bullet 2}}{10}=\dfrac{-10+5\sqrt{2}}{10}=\dfrac{5(-2+\sqrt{2})}{10}=\dfrac{-2+\sqrt{2}}{2}$

57. Approximating each expression:

$\sqrt{5}+\sqrt{3}\approx 3.968$

$\sqrt{8}\approx 2.828$

No, the two values are not equal.

59. The correct statement is: $4\sqrt{3}+5\sqrt{3}=9\sqrt{3}$

61. Multiplying the expressions: $(3x+y)^2=(3x)^2+2(3x)(y)+y^2=9x^2+6xy+y^2$

63. Multiplying the expressions: $(3x-4y)(3x+4y)=(3x)^2-(4y)^2=9x^2-16y^2$

65. Multiplying each side of the equation by 6:

$$\begin{aligned} 6\left(\frac{x}{3}-\frac{1}{2}\right) &= 6\left(\frac{5}{2}\right) \\ 2x-3 &= 15 \\ 2x &= 18 \\ x &= 9 \end{aligned}$$

Since $x=9$ checks in the original equation, the solution is $x=9$.

67. Multiplying each side of the equation by x^2:

$$\begin{aligned} x^2\left(1-\frac{5}{x}\right) &= x^2\left(\frac{-6}{x^2}\right) \\ x^2-5x &= -6 \\ x^2-5x+6 &= 0 \\ (x-2)(x-3) &= 0 \\ x &= 2,3 \end{aligned}$$

Both $x=2$ and $x=3$ check in the original equation.

69. Multiplying each side of the equation by $2(a-4)$:

$$\begin{aligned} 2(a-4)\left(\frac{a}{a-4}-\frac{a}{2}\right) &= 2(a-4)\bullet\frac{4}{a-4} \\ 2a-a(a-4) &= 8 \\ 2a-a^2+4a &= 8 \\ -a^2+6a-8 &= 0 \\ a^2-6a+8 &= 0 \\ (a-4)(a-2) &= 0 \\ a &= 2,4 \end{aligned}$$

Since $a=4$ does not check in the original equation, the solution is $a=2$.

8.5 Multiplication and Division of Radicals

1. Multiplying the radicals: $\sqrt{3}\sqrt{2}=\sqrt{6}$
3. Multiplying the radicals: $\sqrt{6}\sqrt{2}=\sqrt{12}=\sqrt{4\bullet 3}=2\sqrt{3}$
5. Multiplying the radicals: $(2\sqrt{3})(5\sqrt{7})=10\sqrt{21}$
7. Multiplying the radicals: $(4\sqrt{3})(2\sqrt{6})=8\sqrt{18}=8\sqrt{9\bullet 2}=8\bullet 3\sqrt{2}=24\sqrt{2}$
9. Multiplying the radicals: $\sqrt{2}(\sqrt{3}-1)=\sqrt{6}-\sqrt{2}$
11. Multiplying the radicals: $\sqrt{2}(\sqrt{3}+\sqrt{2})=\sqrt{6}+\sqrt{4}=\sqrt{6}+2$
13. Multiplying the radicals: $\sqrt{3}(2\sqrt{2}+\sqrt{3})=2\sqrt{6}+\sqrt{9}=2\sqrt{6}+3$
15. Multiplying the radicals: $\sqrt{3}(2\sqrt{3}-\sqrt{5})=2\sqrt{9}-\sqrt{15}=2\bullet 3-\sqrt{15}=6-\sqrt{15}$
17. Multiplying the radicals: $2\sqrt{3}(\sqrt{2}+\sqrt{5})=2\sqrt{6}+2\sqrt{15}$
19. Simplifying the expression: $(\sqrt{2}+1)^2=(\sqrt{2})^2+2(\sqrt{2})(1)+(1)^2=2+2\sqrt{2}+1=3+2\sqrt{2}$
21. Simplifying the expression: $(\sqrt{x}+3)^2=(\sqrt{x})^2+2(\sqrt{x})(3)+(3)^2=x+6\sqrt{x}+9$
23. Simplifying the expression: $(5-\sqrt{2})^2=(5)^2-2(5)(\sqrt{2})+(\sqrt{2})^2=25-10\sqrt{2}+2=27-10\sqrt{2}$
25. Simplifying the expression: $\left(\sqrt{a}-\frac{1}{2}\right)^2=(\sqrt{a})^2-2(\sqrt{a})\left(\frac{1}{2}\right)+\left(\frac{1}{2}\right)^2=a-\sqrt{a}+\frac{1}{4}$
27. Simplifying the expression: $(3+\sqrt{7})^2=(3)^2+2(3)(\sqrt{7})+(\sqrt{7})^2=9+6\sqrt{7}+7=16+6\sqrt{7}$
29. Simplifying the expression: $(\sqrt{5}+3)(\sqrt{5}+2)=5+3\sqrt{5}+2\sqrt{5}+6=11+5\sqrt{5}$
31. Simplifying the expression: $(\sqrt{2}-5)(\sqrt{2}+6)=2-5\sqrt{2}+6\sqrt{2}-30=-28+\sqrt{2}$
33. Simplifying the expression: $\left(\sqrt{3}+\frac{1}{2}\right)\left(\sqrt{2}+\frac{1}{3}\right)=\sqrt{6}+\frac{1}{2}\sqrt{2}+\frac{1}{3}\sqrt{3}+\frac{1}{6}$
35. Simplifying the expression: $(\sqrt{x}+6)(\sqrt{x}-6)=(\sqrt{x})^2-(6)^2=x-36$
37. Simplifying the expression: $\left(\sqrt{a}+\frac{1}{3}\right)\left(\sqrt{a}+\frac{2}{3}\right)=a+\frac{1}{3}\sqrt{a}+\frac{2}{3}\sqrt{a}+\frac{2}{9}=a+\sqrt{a}+\frac{2}{9}$
39. Simplifying the expression: $(\sqrt{5}-2)(\sqrt{5}+2)=(\sqrt{5})^2-(2)^2=5-4=1$
41. Simplifying the expression: $(2\sqrt{7}+3)(3\sqrt{7}-4)=6\sqrt{49}+9\sqrt{7}-8\sqrt{7}-12=42+\sqrt{7}-12=30+\sqrt{7}$
43. Simplifying the expression: $(2\sqrt{x}+4)(3\sqrt{x}+2)=6x+12\sqrt{x}+4\sqrt{x}+8=6x+16\sqrt{x}+8$
45. Simplifying the expression: $(7\sqrt{a}+2\sqrt{b})(7\sqrt{a}-2\sqrt{b})=(7\sqrt{a})^2-(2\sqrt{b})^2=49a-4b$
47. Rationalizing the denominator: $\frac{\sqrt{3}}{\sqrt{5}-\sqrt{2}}\bullet\frac{\sqrt{5}+\sqrt{2}}{\sqrt{5}+\sqrt{2}}=\frac{\sqrt{15}+\sqrt{6}}{5-2}=\frac{\sqrt{15}+\sqrt{6}}{3}$
49. Rationalizing the denominator: $\frac{\sqrt{5}}{\sqrt{5}+\sqrt{2}}\bullet\frac{\sqrt{5}-\sqrt{2}}{\sqrt{5}-\sqrt{2}}=\frac{\sqrt{25}-\sqrt{10}}{5-2}=\frac{5-\sqrt{10}}{3}$
51. Rationalizing the denominator: $\frac{8}{3-\sqrt{5}}\bullet\frac{3+\sqrt{5}}{3+\sqrt{5}}=\frac{8(3+\sqrt{5})}{9-5}=\frac{8(3+\sqrt{5})}{4}=2(3+\sqrt{5})=6+2\sqrt{5}$
53. Rationalizing the denominator: $\frac{\sqrt{3}+\sqrt{2}}{\sqrt{3}-\sqrt{2}}\bullet\frac{\sqrt{3}+\sqrt{2}}{\sqrt{3}+\sqrt{2}}=\frac{3+\sqrt{6}+\sqrt{6}+2}{3-2}=\frac{5+2\sqrt{6}}{1}=5+2\sqrt{6}$

55. Rationalizing the denominator: $\dfrac{\sqrt{7}-\sqrt{3}}{\sqrt{7}+\sqrt{3}} \cdot \dfrac{\sqrt{7}-\sqrt{3}}{\sqrt{7}-\sqrt{3}} = \dfrac{7-\sqrt{21}-\sqrt{21}+3}{7-3} = \dfrac{10-2\sqrt{21}}{4} = \dfrac{2\left(5-\sqrt{21}\right)}{4} = \dfrac{5-\sqrt{21}}{2}$

57. Rationalizing the denominator: $\dfrac{\sqrt{x}+2}{\sqrt{x}-2} \cdot \dfrac{\sqrt{x}+2}{\sqrt{x}+2} = \dfrac{x+2\sqrt{x}+2\sqrt{x}+4}{x-4} = \dfrac{x+4\sqrt{x}+4}{x-4}$

59. Rationalizing the denominator: $\dfrac{\sqrt{5}-\sqrt{2}}{\sqrt{5}+\sqrt{3}} \cdot \dfrac{\sqrt{5}-\sqrt{3}}{\sqrt{5}-\sqrt{3}} = \dfrac{5-\sqrt{10}-\sqrt{15}+\sqrt{6}}{5-3} = \dfrac{5-\sqrt{10}-\sqrt{15}+\sqrt{6}}{2}$

61. The correct statement is: $2\left(3\sqrt{5}\right) = 6\sqrt{5}$

63. The correct statement is: $\left(\sqrt{3}+7\right)^2 = \left(\sqrt{3}\right)^2 + 2\left(\sqrt{3}\right)(7) + (7)^2 = 3+14\sqrt{3}+49 = 52+14\sqrt{3}$

65. Solving the equation by factoring:

$$x^2+5x-6=0$$
$$(x+6)(x-1)=0$$
$$x=-6,1$$

67. Solving the equation by factoring:

$$x^2-3x=0$$
$$x(x-3)=0$$
$$x=0,3$$

69. Solving the proportion:

$$\frac{x}{3}=\frac{27}{x}$$
$$x^2=81$$
$$x^2-81=0$$
$$(x+9)(x-9)=0$$
$$x=-9,9$$

71. Solving the proportion:

$$\frac{x}{5}=\frac{3}{x+2}$$
$$x^2+2x=15$$
$$x^2+2x-15=0$$
$$(x+5)(x-3)=0$$
$$x=-5,3$$

73. Comparing miles to hours, the proportion is:

$$\frac{375}{15}=\frac{x}{20}$$
$$15x=7500$$
$$x=500$$

You will drive 500 miles.

8.6 Equations Involving Radicals

1. Solving the equation by squaring:

$$\sqrt{x+1}=2$$
$$\left(\sqrt{x+1}\right)^2=(2)^2$$
$$x+1=4$$
$$x=3$$

This value checks in the original equation.

3. Solving the equation by squaring:

$$\begin{aligned} \sqrt{x+5} &= 7 \\ \left(\sqrt{x+5}\right)^2 &= (7)^2 \\ x+5 &= 49 \\ x &= 44 \end{aligned}$$

This value checks in the original equation.

5. Solving the equation by squaring:

$$\begin{aligned} \sqrt{x-9} &= -6 \\ \left(\sqrt{x-9}\right)^2 &= (-6)^2 \\ x-9 &= 36 \\ x &= 45 \end{aligned}$$

Since $\sqrt{45-9} = \sqrt{36} = 6 \neq -6$, there is no solution to the equation.

7. Solving the equation by squaring:

$$\begin{aligned} \sqrt{x-5} &= -4 \\ \left(\sqrt{x-5}\right)^2 &= (-4)^2 \\ x-5 &= 16 \\ x &= 21 \end{aligned}$$

Since $\sqrt{21-5} = \sqrt{16} = 4 \neq -4$, there is no solution to the equation.

9. Solving the equation by squaring:

$$\begin{aligned} \sqrt{x-8} &= 0 \\ \left(\sqrt{x-8}\right)^2 &= (0)^2 \\ x-8 &= 0 \\ x &= 8 \end{aligned}$$

This value checks in the original equation.

11. Solving the equation by squaring:

$$\begin{aligned} \sqrt{2x+1} &= 3 \\ \left(\sqrt{2x+1}\right)^2 &= (3)^2 \\ 2x+1 &= 9 \\ 2x &= 8 \\ x &= 4 \end{aligned}$$

This value checks in the original equation.

13. Solving the equation by squaring:

$$\begin{aligned} \sqrt{2x-3} &= -5 \\ \left(\sqrt{2x-3}\right)^2 &= (-5)^2 \\ 2x-3 &= 25 \\ 2x &= 28 \\ x &= 14 \end{aligned}$$

Since $\sqrt{2 \cdot 14 - 3} = \sqrt{28-3} = \sqrt{25} = 5 \neq -5$, there is no solution to the equation.

15. Solving the equation by squaring:

$$\begin{aligned} \sqrt{3x+6} &= 2 \\ \left(\sqrt{3x+6}\right)^2 &= (2)^2 \\ 3x+6 &= 4 \\ 3x &= -2 \\ x &= -\frac{2}{3} \end{aligned}$$

This value checks in the original equation.

17. Solving the equation by squaring:

$$\begin{aligned} 2\sqrt{x} &= 10 \\ \sqrt{x} &= 5 \\ \left(\sqrt{x}\right)^2 &= (5)^2 \\ x &= 25 \end{aligned}$$

This value checks in the original equation.

19. Solving the equation by squaring:

$$\begin{aligned} 3\sqrt{a} &= 6 \\ \sqrt{a} &= 2 \\ \left(\sqrt{a}\right)^2 &= (2)^2 \\ a &= 4 \end{aligned}$$

This value checks in the original equation.

21. Solving the equation by squaring:

$$\begin{aligned} \sqrt{3x+4} - 3 &= 2 \\ \sqrt{3x+4} &= 5 \\ \left(\sqrt{3x+4}\right)^2 &= (5)^2 \\ 3x+4 &= 25 \\ 3x &= 21 \\ x &= 7 \end{aligned}$$

This value checks in the original equation.

23. Solving the equation by squaring:

$$\begin{aligned} \sqrt{5y-4} - 2 &= 4 \\ \sqrt{5y-4} &= 6 \\ \left(\sqrt{5y-4}\right)^2 &= (6)^2 \\ 5y-4 &= 36 \\ 5y &= 40 \\ y &= 8 \end{aligned}$$

This value checks in the original equation.

25. Solving the equation by squaring:

$$\begin{aligned} \sqrt{2x+1} + 5 &= 2 \\ \sqrt{2x+1} &= -3 \\ \left(\sqrt{2x+1}\right)^2 &= (-3)^2 \\ 2x+1 &= 9 \\ 2x &= 8 \\ x &= 4 \end{aligned}$$

Since $\sqrt{2 \cdot 4+1} + 5 = \sqrt{9} + 5 = 3 + 5 = 8 \neq 2$, there is no solution to the equation.

27. Solving the equation by squaring:

$$\begin{aligned} \sqrt{x+3} &= x-3 \\ \left(\sqrt{x+3}\right)^2 &= (x-3)^2 \\ x+3 &= x^2 - 6x + 9 \\ 0 &= x^2 - 7x + 6 \\ 0 &= (x-6)(x-1) \\ x &= 6, 1 \end{aligned}$$

Since $x = 1$ does not check in the original equation, the solution is $x = 6$.

29. Solving the equation by squaring:

$$\begin{aligned}\sqrt{a+2} &= a+2\\ \left(\sqrt{a+2}\right)^2 &= (a+2)^2\\ a+2 &= a^2+4a+4\\ 0 &= a^2+3a+2\\ 0 &= (a+2)(a+1)\\ a &= -2,-1\end{aligned}$$

Both values check in the original equation.

31. Solving the equation by squaring:

$$\begin{aligned}\sqrt{2x+9} &= x+5\\ \left(\sqrt{2x+9}\right)^2 &= (x+5)^2\\ 2x+9 &= x^2+10x+25\\ 0 &= x^2+8x+16\\ 0 &= (x+4)^2\\ x &= -4\end{aligned}$$

This value checks in the original equation.

33. Solving the equation by squaring:

$$\begin{aligned}\sqrt{y-4} &= y-6\\ \left(\sqrt{y-4}\right)^2 &= (y-6)^2\\ y-4 &= y^2-12y+36\\ 0 &= y^2-13y+40\\ 0 &= (y-8)(y-5)\\ y &= 5,8\end{aligned}$$

Since $y=5$ does not check in the original equation, the solution is $y=8$.

35. Solving the equation by squaring:

$$\begin{aligned}\sqrt{5x+1} &= x+1\\ \left(\sqrt{5x+1}\right)^2 &= (x+1)^2\\ 5x+1 &= x^2+2x+1\\ 0 &= x^2-3x\\ 0 &= x(x-3)\\ x &= 0,3\end{aligned}$$

Both values check in the original equation.

37. Let x represent the number. The equation is:

$$\begin{aligned}x+2 &= \sqrt{8x}\\ (x+2)^2 &= \left(\sqrt{8x}\right)^2\\ x^2+4x+4 &= 8x\\ x^2-4x+4 &= 0\\ (x-2)^2 &= 0\\ x-2 &= 0\\ x &= 2\end{aligned}$$

The number is 2.

39. Let x represent the number. The equation is:

$$\begin{aligned} x-3&=2\sqrt{x} \\ (x-3)^2&=\left(2\sqrt{x}\right)^2 \\ x^2-6x+9&=4x \\ x^2-10x+9&=0 \\ (x-9)(x-1)&=0 \\ x&=9 \quad (x=1 \text{ does not check in the original equation}) \end{aligned}$$

The number is 9.

41. Substituting $T=2$:

$$\begin{aligned} 2&=\frac{11}{7}\sqrt{\frac{L}{2}} \\ 14&=11\sqrt{\frac{L}{2}} \\ \frac{14}{11}&=\sqrt{\frac{L}{2}} \\ \left(\frac{14}{11}\right)^2&=\left(\sqrt{\frac{L}{2}}\right)^2 \\ \frac{196}{121}&=\frac{L}{2} \\ L&=\frac{392}{121}\approx 3.24 \text{ feet} \end{aligned}$$

43. Reducing the rational expression: $\dfrac{x^2-x-6}{x^2-9}=\dfrac{(x-3)(x+2)}{(x+3)(x-3)}=\dfrac{x+2}{x+3}$

45. Performing the operations:

$$\frac{x^2-25}{x+4}\bullet\frac{2x+8}{x^2-9x+20}=\frac{(x+5)(x-5)}{x+4}\bullet\frac{2(x+4)}{(x-4)(x-5)}=\frac{2(x+5)(x-5)(x+4)}{(x+4)(x-4)(x-5)}=\frac{2(x+5)}{x-4}$$

47. Performing the operations: $\dfrac{x}{x^2-16}+\dfrac{4}{x^2-16}=\dfrac{x+4}{x^2-16}=\dfrac{x+4}{(x+4)(x-4)}=\dfrac{1}{x-4}$

49. Simplifying the complex fraction: $\dfrac{1-\frac{25}{x^2}}{1-\frac{8}{x}+\frac{15}{x^2}}\bullet\dfrac{x^2}{x^2}=\dfrac{x^2-25}{x^2-8x+15}=\dfrac{(x+5)(x-5)}{(x-5)(x-3)}=\dfrac{x+5}{x-3}$

51. Multiplying each side of the equation by $x^2-9=(x+3)(x-3)$:

$$\begin{aligned} (x+3)(x-3)\left(\frac{x}{x^2-9}-\frac{3}{x-3}\right)&=(x+3)(x-3)\bullet\frac{1}{x+3} \\ x-3(x+3)&=x-3 \\ x-3x-9&=x-3 \\ -2x-9&=x-3 \\ -3x&=6 \\ x&=-2 \end{aligned}$$

Since $x=-2$ checks in the original equation, the solution is $x=-2$.

53. Let t represent the time to fill the pool with both pipes open. The equation is:

$$\begin{aligned} \frac{1}{8}-\frac{1}{12}&=\frac{1}{t} \\ 24t\left(\frac{1}{8}-\frac{1}{12}\right)&=24t\cdot\frac{1}{t} \\ 3t-2t&=24 \\ t&=24 \end{aligned}$$

It will take 24 hours to fill the pool with both pipes left open.

55. The variation equation is $y=Kx$. Finding K:

$$\begin{aligned} 8&=K\cdot 12 \\ K&=\frac{2}{3} \end{aligned}$$

So $y=\frac{2}{3}x$. Substituting $x=36$: $y=\frac{2}{3}(36)=24$

Chapter 8 Review

1. Finding the root: $\sqrt{25}=5$
3. Finding the root: $\sqrt[3]{-1}=-1$
5. Finding the root: $\sqrt{100x^2y^4}=10xy^2$
7. Simplifying the expression: $\sqrt{24}=\sqrt{4\cdot 6}=2\sqrt{6}$
9. Simplifying the expression: $\sqrt{90x^3y^4}=\sqrt{9x^2y^4\cdot 10x}=3xy^2\sqrt{10x}$
11. Simplifying the expression: $3\sqrt{20x^3y}=3\sqrt{4x^2\cdot 5xy}=3\cdot 2x\sqrt{5xy}=6x\sqrt{5xy}$
13. Simplifying the expression: $\sqrt{\frac{8}{81}}=\frac{\sqrt{8}}{\sqrt{81}}=\frac{\sqrt{4\cdot 2}}{9}=\frac{2\sqrt{2}}{9}$
15. Simplifying the expression: $\sqrt{\frac{49a^2b^2}{16}}=\frac{\sqrt{49a^2b^2}}{\sqrt{16}}=\frac{7ab}{4}$
17. Simplifying the expression: $\sqrt{\frac{40a^2}{121}}=\frac{\sqrt{40a^2}}{\sqrt{121}}=\frac{\sqrt{4a^2\cdot 10}}{11}=\frac{2a\sqrt{10}}{11}$
19. Simplifying the expression: $\frac{3\sqrt{120a^2b^2}}{\sqrt{25}}=\frac{3\sqrt{4a^2b^2\cdot 30}}{5}=\frac{3\cdot 2ab\sqrt{30}}{5}=\frac{6ab\sqrt{30}}{5}$
21. Simplifying the radical expression: $\frac{2}{\sqrt{7}}\cdot\frac{\sqrt{7}}{\sqrt{7}}=\frac{2\sqrt{7}}{7}$
23. Simplifying the radical expression: $\sqrt{\frac{5}{48}}=\frac{\sqrt{5}}{\sqrt{48}}\cdot\frac{\sqrt{3}}{\sqrt{3}}=\frac{\sqrt{15}}{\sqrt{144}}=\frac{\sqrt{15}}{12}$
25. Simplifying the radical expression: $\sqrt{\frac{32ab^2}{3}}=\frac{\sqrt{32ab^2}}{\sqrt{3}}\cdot\frac{\sqrt{3}}{\sqrt{3}}=\frac{\sqrt{96ab^2}}{\sqrt{9}}=\frac{\sqrt{16b^2\cdot 6a}}{3}=\frac{4b\sqrt{6a}}{3}$
27. Rationalizing the denominator: $\frac{3}{\sqrt{3}-4}\cdot\frac{\sqrt{3}+4}{\sqrt{3}+4}=\frac{3(\sqrt{3}+4)}{3-16}=\frac{3\sqrt{3}+12}{-13}=\frac{-3\sqrt{3}-12}{13}$
29. Rationalizing the denominator: $\frac{3}{\sqrt{5}-\sqrt{2}}\cdot\frac{\sqrt{5}+\sqrt{2}}{\sqrt{5}+\sqrt{2}}=\frac{3(\sqrt{5}+\sqrt{2})}{5-2}=\frac{3(\sqrt{5}+\sqrt{2})}{3}=\sqrt{5}+\sqrt{2}$
31. Rationalizing the denominator: $\frac{\sqrt{5}-\sqrt{2}}{\sqrt{5}+\sqrt{2}}\cdot\frac{\sqrt{5}-\sqrt{2}}{\sqrt{5}-\sqrt{2}}=\frac{5-\sqrt{10}-\sqrt{10}+2}{5-2}=\frac{7-2\sqrt{10}}{3}$
33. Combining the expressions: $3\sqrt{5}-7\sqrt{5}=-4\sqrt{5}$

35. Combining the expressions:
$$\begin{aligned}-2\sqrt{45}-5\sqrt{80}+2\sqrt{20}&=-2\sqrt{9\cdot 5}-5\sqrt{16\cdot 5}+2\sqrt{4\cdot 5}\\&=-2\cdot 3\sqrt{5}-5\cdot 4\sqrt{5}+2\cdot 2\sqrt{5}\\&=-6\sqrt{5}-20\sqrt{5}+4\sqrt{5}\\&=-22\sqrt{5}\end{aligned}$$

37. Combining the expressions:
$$\sqrt{40a^3b^2}-a\sqrt{90ab^2}=\sqrt{4a^2b^2\cdot 10a}-a\sqrt{9b^2\cdot 10a}=2ab\sqrt{10a}-3ab\sqrt{10a}=-ab\sqrt{10a}$$

39. Multiplying the expressions: $4\sqrt{2}\left(\sqrt{3}+\sqrt{5}\right)=4\sqrt{6}+4\sqrt{10}$

41. Multiplying the expressions: $\left(2\sqrt{5}-4\right)\left(\sqrt{5}+3\right)=2\sqrt{25}-4\sqrt{5}+6\sqrt{5}-12=10+2\sqrt{5}-12=-2+2\sqrt{5}$

43. Solving the equation by squaring:
$$\begin{aligned}\sqrt{x-3}&=3\\\left(\sqrt{x-3}\right)^2&=(3)^2\\x-3&=9\\x&=12\end{aligned}$$
This value checks in the original equation.

45. Solving the equation by squaring:
$$\begin{aligned}5\sqrt{a}&=20\\\sqrt{a}&=4\\\left(\sqrt{a}\right)^2&=(4)^2\\a&=16\end{aligned}$$
This value checks in the original equation.

47. Solving the equation by squaring:
$$\begin{aligned}\sqrt{2x+1}+10&=8\\\sqrt{2x+1}&=-2\\\left(\sqrt{2x+1}\right)^2&=(-2)^2\\2x+1&=4\\2x&=3\\x&=\frac{3}{2}\end{aligned}$$
Since $\sqrt{2\cdot\frac{3}{2}+1}+10=\sqrt{3+1}+10=\sqrt{4}+10=2+10=12\neq 8$, there is no solution to the equation.

49. Using the Pythagorean Theorem:
$$\begin{aligned}x^2&=\left(\sqrt{2}\right)^2+(1)^2=2+1=3\\x&=\sqrt{3}\end{aligned}$$

Cumulative Review: Chapters 1-8

1. Simplifying the expression: $\left(\frac{4}{5}\right)\left(\frac{5}{4}\right)=\frac{20}{20}=1$

3. Simplifying the expression: $-\left|-\frac{1}{2}\right|=-\frac{1}{2}$

5. Simplifying the expression: $\frac{4^2-8^2}{(4-8)^2}=\frac{16-64}{(-4)^2}=\frac{-48}{16}=-3$

7. Simplifying the expression: $4x-7x=(4-7)x=-3x$

9. Simplifying the expression: $\dfrac{(b^6)^2(b^3)^4}{(b^{10})^3}=\dfrac{b^{12}\bullet b^{12}}{b^{30}}=\dfrac{b^{24}}{b^{30}}=b^{24-30}=b^{-6}=\dfrac{1}{b^6}$

11. Simplifying the expression: $\left(\frac{1}{2}y+2\right)\left(\frac{1}{2}y-2\right)=\left(\frac{1}{2}y\right)^2-(2)^2=\frac{1}{4}y^2-4$

13. Simplifying the expression: $\dfrac{\dfrac{1}{a+6}+3}{\dfrac{1}{a+6}+2}\bullet\dfrac{a+6}{a+6}=\dfrac{1+3(a+6)}{1+2(a+6)}=\dfrac{1+3a+18}{1+2a+12}=\dfrac{3a+19}{2a+13}$

15. Simplifying the expression: $\sqrt{\dfrac{90a^2}{169}}=\dfrac{\sqrt{90a^2}}{\sqrt{169}}=\dfrac{\sqrt{9a^2\bullet 10}}{13}=\dfrac{3a\sqrt{10}}{13}$

17. Simplifying the expression:

$$\begin{aligned}\frac{7a}{a^2-3a-54}+\frac{5}{a-9}&=\frac{7a}{(a-9)(a+6)}+\frac{5}{a-9}\bullet\frac{a+6}{a+6}\\&=\frac{7a}{(a-9)(a+6)}+\frac{5a+30}{(a-9)(a+6)}\\&=\frac{7a+5a+30}{(a-9)(a+6)}\\&=\frac{12a+30}{(a-9)(a+6)}\\&=\frac{6(2a+5)}{(a-9)(a+6)}\end{aligned}$$

19. Solving the equation:

$$\begin{aligned}3(5x-1)&=6(2x+3)-21\\15x-3&=12x+18-21\\15x-3&=12x-3\\3x-3&=-3\\3x&=0\\x&=0\end{aligned}$$

21. Setting each factor equal to 0 results in $x=0$, $x=-\frac{2}{3}$, or $x=4$.

23. Multiplying each side of the equation by $16a$:

$$\begin{aligned}16a\left(\frac{4}{a}\right)&=16a\left(\frac{a}{16}\right)\\64&=a^2\\a^2-64&=0\\(a+8)(a-8)&=0\\a&=-8,8\end{aligned}$$

Both values check in the original equation.

25. Adding the two equations:

$$\begin{aligned}2x&=4\\x&=2\end{aligned}$$

Substituting into the first equation:

$$\begin{aligned}2+y&=1\\y&=-1\end{aligned}$$

The solution is $(2,-1)$.

27. Substituting into the first equation:

$$x-(3x-1)=5$$
$$x-3x+1=5$$
$$-2x+1=5$$
$$-2x=4$$
$$x=-2$$

Substituting into the second equation: $y=3(-2)-1=-6-1=-7$

The solution is $(-2,-7)$. Note this system could also be solved by graphing:

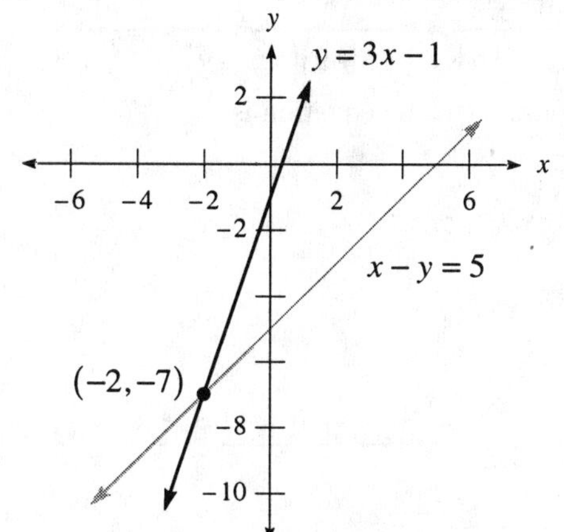

29. Graphing the interval on a number line:

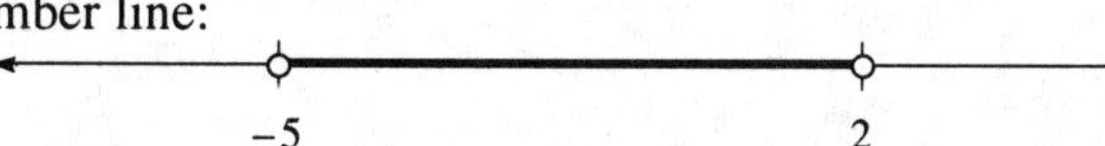

31. Graphing the line:

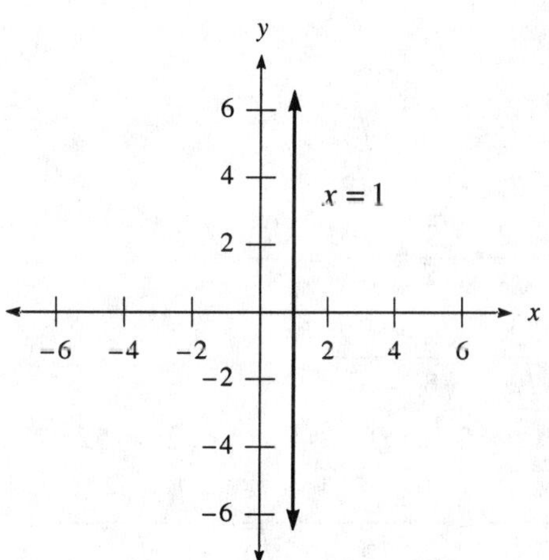

33. To find the x-intercept, let $y = 0$:

$$0=-x+7$$
$$x=7$$

To find the y-intercept, let $x = 0$: $y=-(0)+7=7$

35. Using the point-slope formula:

$$y-(-3)=--1(x-3)$$
$$y+3=--x+3$$
$$y=-x$$

37. Factoring the polynomial: $r^2+r-20=(r+5)(r-4)$

39. Factoring the polynomial: $x^5-x^4-30x^3=x^3\left(x^2-x-30\right)=x^3(x-6)(x+5)$

41. Simplifying the expression: $\dfrac{(6\times10^5)(6\times10^{-3})}{9\times10^{-4}}=\dfrac{36\times10^2}{9\times10^{-4}}=4\times10^6$

43. Rationalizing the denominator: $\dfrac{5}{\sqrt{3}}\bullet\dfrac{\sqrt{3}}{\sqrt{3}}=\dfrac{5\sqrt{3}}{\sqrt{9}}=\dfrac{5\sqrt{3}}{3}$

45. Combining the expressions:

$$5\sqrt{63x^2}-x\sqrt{28}=5\sqrt{9x^2\bullet7}-x\sqrt{4\bullet7}=5\bullet3x\sqrt{7}-x\bullet2\sqrt{7}=15x\sqrt{7}-2x\sqrt{7}=13x\sqrt{7}$$

47. Reducing the rational expression: $\dfrac{x^2-9}{x^4-81}=\dfrac{1(x^2-9)}{(x^2+9)(x^2-9)}=\dfrac{1}{x^2+9}$

49. Let x and $2x+5$ represent the two numbers. The equation is:

$$\begin{aligned} x+2x+5&=35\\ 3x+5&=35\\ 3x&=30\\ x&=10\\ 2x+5&=25 \end{aligned}$$

The two numbers are 10 and 25.

Chapter 8 Test

1. Finding the root: $\sqrt{16}=4$
2. Finding the root: $-\sqrt{36}=-6$
3. The roots are $\sqrt{49}=7$ and $-\sqrt{49}=-7$.
4. Finding the root: $\sqrt[3]{27}=3$
5. Finding the root: $\sqrt[3]{-8}=-2$
6. Finding the root: $-\sqrt[4]{81}=-3$
7. Simplifying the expression: $\sqrt{75}=\sqrt{25\bullet3}=5\sqrt{3}$
8. Simplifying the expression: $\sqrt{32}=\sqrt{16\bullet2}=4\sqrt{2}$
9. Simplifying the expression: $\sqrt{\dfrac{2}{3}}=\dfrac{\sqrt{2}}{\sqrt{3}}\bullet\dfrac{\sqrt{3}}{\sqrt{3}}=\dfrac{\sqrt{6}}{\sqrt{9}}=\dfrac{\sqrt{6}}{3}$
10. Simplifying the expression: $\dfrac{1}{\sqrt[3]{4}}\bullet\dfrac{\sqrt[3]{2}}{\sqrt[3]{2}}=\dfrac{\sqrt[3]{2}}{\sqrt[3]{8}}=\dfrac{\sqrt[3]{2}}{2}$
11. Simplifying the expression: $3\sqrt{50x^2}=3\sqrt{25x^2\bullet2}=3\bullet5x\sqrt{2}=15x\sqrt{2}$
12. Simplifying the expression: $\sqrt{\dfrac{12x^2y^3}{5}}=\dfrac{\sqrt{12x^2y^3}}{\sqrt{5}}\bullet\dfrac{\sqrt{5}}{\sqrt{5}}=\dfrac{\sqrt{60x^2y^3}}{\sqrt{25}}=\dfrac{\sqrt{4x^2y^2\bullet15y}}{5}=\dfrac{2xy\sqrt{15y}}{5}$
13. Combining the radicals: $5\sqrt{12}-2\sqrt{27}=5\sqrt{4\bullet3}-2\sqrt{9\bullet3}=5\bullet2\sqrt{3}-2\bullet3\sqrt{3}=10\sqrt{3}-6\sqrt{3}=4\sqrt{3}$
14. Combining the radicals: $2x\sqrt{18}+5\sqrt{2x^2}=2x\sqrt{9\bullet2}+5\sqrt{x^2\bullet2}=2x\bullet3\sqrt{2}+5\bullet x\sqrt{2}=6x\sqrt{2}+5x\sqrt{2}=11x\sqrt{2}$
15. Multiplying the expressions: $\sqrt{3}(\sqrt{5}-2)=\sqrt{15}-2\sqrt{3}$
16. Multiplying the expressions: $(\sqrt{5}+7)(\sqrt{5}-8)=\sqrt{25}+7\sqrt{5}-8\sqrt{5}-56=5-\sqrt{5}-56=-51-\sqrt{5}$
17. Multiplying the expressions: $(\sqrt{x}+6)(\sqrt{x}-6)=(\sqrt{x})^2-(6)^2=x-36$
18. Multiplying the expressions: $(\sqrt{5}-\sqrt{3})^2=(\sqrt{5})^2-2(\sqrt{5})(\sqrt{3})+(\sqrt{3})^2=5-2\sqrt{15}+3=8-2\sqrt{15}$
19. Rationalizing the denominator: $\dfrac{\sqrt{7}-\sqrt{3}}{\sqrt{7}+\sqrt{3}}\bullet\dfrac{\sqrt{7}-\sqrt{3}}{\sqrt{7}-\sqrt{3}}=\dfrac{7-\sqrt{21}-\sqrt{21}+3}{7-3}=\dfrac{10-2\sqrt{21}}{4}=\dfrac{2(5-\sqrt{21})}{4}=\dfrac{5-\sqrt{21}}{2}$

20. Rationalizing the denominator: $\frac{\sqrt{x}}{\sqrt{x}+5} \bullet \frac{\sqrt{x}-5}{\sqrt{x}-5} = \frac{x-5\sqrt{x}}{x-25}$

21. Solving the equation by squaring:

$$\begin{aligned} \sqrt{2x+1}+2 &= 7 \\ \sqrt{2x+1} &= 5 \\ \left(\sqrt{2x+1}\right)^2 &= (5)^2 \\ 2x+1 &= 25 \\ 2x &= 24 \\ x &= 12 \end{aligned}$$

This value checks in the original equation.

22. Solving the equation by squaring:

$$\begin{aligned} \sqrt{3x+1}+6 &= 2 \\ \sqrt{3x+1} &= -4 \\ \left(\sqrt{3x+1}\right)^2 &= (-4)^2 \\ 3x+1 &= 16 \\ 3x &= 15 \\ x &= 5 \end{aligned}$$

Since $\sqrt{3 \bullet 5+1}+6 = \sqrt{15+1}+6 = \sqrt{16}+6 = 4+6 = 10 \neq 2$, there is no solution to the equation.

23. Solving the equation by squaring:

$$\begin{aligned} \sqrt{2x-3} &= x-3 \\ \left(\sqrt{2x-3}\right)^2 &- (x-3)^2 \\ 2x-3 &= x^2-6x+9 \\ 0 &= x^2-8x+12 \\ 0 &= (x-6)(x-2) \\ x &= 2,6 \end{aligned}$$

Since $x = 2$ does not check in the original equation, the solution is $x = 6$.

24. Let x represent the number. The equation is:

$$\begin{aligned} x-4 &= 3\sqrt{x} \\ (x-4)^2 &= \left(3\sqrt{x}\right)^2 \\ x^2-8x+16 &= 9x \\ x^2-17x+16 &= 0 \\ (x-16)(x-1) &= 0 \\ x &= 1,16 \end{aligned}$$

Since $x = 1$ does not check in the original equation, $x = 16$. The number is 16.

25. Using the Pythagorean Theorem:

$$\begin{aligned} x^2 &= \left(\sqrt{5}\right)^2 + (1)^2 = 5+1 = 6 \\ x &= \sqrt{6} \end{aligned}$$

Chapter 9
More Quadratic Equations

9.1 More Quadratic Equations

1. Solving the equation:

$$x^2 = 9$$
$$x = \pm\sqrt{9}$$
$$x = \pm 3$$

3. Solving the equation:

$$a^2 = 25$$
$$a = \pm\sqrt{25}$$
$$a = \pm 5$$

5. Solving the equation:

$$y^2 = 8$$
$$y = \pm\sqrt{8}$$
$$y = \pm 2\sqrt{2}$$

7. Solving the equation:

$$2x^2 = 100$$
$$x^2 = 50$$
$$x = \pm\sqrt{50}$$
$$x = \pm 5\sqrt{2}$$

9. Solving the equation:

$$3a^2 = 54$$
$$a^2 = 18$$
$$a = \pm\sqrt{18}$$
$$a = \pm 3\sqrt{2}$$

11. Solving the equation:

$$(x+2)^2 = 4$$
$$x+2 = \pm\sqrt{4}$$
$$x+2 = \pm 2$$
$$x = -2 \pm 2$$
$$x = -4, 0$$

13. Solving the equation:

$$(x+1)^2 = 25$$
$$x+1 = \pm\sqrt{25}$$
$$x+1 = \pm 5$$
$$x = -1 \pm 5$$
$$x = -6, 4$$

15. Solving the equation:

$$\begin{aligned} (a-5)^2 &= 75 \\ a-5 &= \pm\sqrt{75} \\ a-5 &= \pm 5\sqrt{3} \\ a &= 5 \pm 5\sqrt{3} \end{aligned}$$

17. Solving the equation:

$$\begin{aligned} (y+1)^2 &= 50 \\ y+1 &= \pm\sqrt{50} \\ y+1 &= \pm 5\sqrt{2} \\ y &= -1 \pm 5\sqrt{2} \end{aligned}$$

19. Solving the equation:

$$\begin{aligned} (2x+1)^2 &= 25 \\ 2x+1 &= \pm\sqrt{25} \\ 2x+1 &= \pm 5 \\ 2x &= -1 \pm 5 \\ 2x &= -6, 4 \\ x &= -3, 2 \end{aligned}$$

21. Solving the equation:

$$\begin{aligned} (4a-5)^2 &= 36 \\ 4a-5 &= \pm\sqrt{36} \\ 4a-5 &= \pm 6 \\ 4a &= 5 \pm 6 \\ 4a &= -1, 11 \\ a &= -\frac{1}{4}, \frac{11}{4} \end{aligned}$$

23. Solving the equation:

$$\begin{aligned} (3y-1)^2 &= 12 \\ 3y-1 &= \pm\sqrt{12} \\ 3y-1 &= \pm 2\sqrt{3} \\ 3y &= 1 \pm 2\sqrt{3} \\ y &= \frac{1 \pm 2\sqrt{3}}{3} \end{aligned}$$

25. Solving the equation:

$$\begin{aligned} (6x+2)^2 &= 27 \\ 6x+2 &= \pm\sqrt{27} \\ 6x+2 &= \pm 3\sqrt{3} \\ 6x &= -2 \pm 3\sqrt{3} \\ x &= \frac{-2 \pm 3\sqrt{3}}{6} \end{aligned}$$

27. Solving the equation:

$$\begin{aligned} (3x-9)^2 &= 27 \\ 3x-9 &= \pm\sqrt{27} \\ 3x-9 &= \pm 3\sqrt{3} \\ 3x &= 9 \pm 3\sqrt{3} \\ x &= 3 \pm \sqrt{3} \end{aligned}$$

29 . Solving the equation:

$$
\begin{aligned}
(3x+6)^2 &= 45 \\
3x+6 &= \pm\sqrt{45} \\
3x+6 &= \pm 3\sqrt{5} \\
3x &= -6 \pm 3\sqrt{5} \\
x &= -2 \pm \sqrt{5}
\end{aligned}
$$

31 . Solving the equation:

$$
\begin{aligned}
(2y-4)^2 &= 8 \\
2y-4 &= \pm\sqrt{8} \\
2y-4 &= \pm 2\sqrt{2} \\
2y &= 4 \pm 2\sqrt{2} \\
y &= 2 \pm \sqrt{2}
\end{aligned}
$$

33 . Solving the equation:

$$
\begin{aligned}
\left(x-\frac{2}{3}\right)^2 &= \frac{25}{9} \\
x-\frac{2}{3} &= \pm\sqrt{\frac{25}{9}} \\
x-\frac{2}{3} &= \pm\frac{5}{3} \\
x &= \frac{2}{3} \pm \frac{5}{3} \\
x &= -1, \frac{7}{3}
\end{aligned}
$$

35 . Solving the equation:

$$
\begin{aligned}
\left(x+\frac{1}{2}\right)^2 &= \frac{7}{4} \\
x+\frac{1}{2} &= \pm\sqrt{\frac{7}{4}} \\
x+\frac{1}{2} &= \pm\frac{\sqrt{7}}{2} \\
x &= \frac{-1 \pm \sqrt{7}}{2}
\end{aligned}
$$

37 . Solving the equation:

$$
\begin{aligned}
\left(a-\frac{4}{5}\right)^2 &= \frac{12}{25} \\
a-\frac{4}{5} &= \pm\sqrt{\frac{12}{25}} \\
a-\frac{4}{5} &= \pm\frac{2\sqrt{3}}{5} \\
a &= \frac{4 \pm 2\sqrt{3}}{5}
\end{aligned}
$$

39 . Solving the equation:

$$
\begin{aligned}
x^2+10x+25 &= 7 \\
(x+5)^2 &= 7 \\
x+5 &= \pm\sqrt{7} \\
x &= -5 \pm \sqrt{7}
\end{aligned}
$$

41. Solving the equation:

$$\begin{aligned} x^2 - 2x + 1 &= 9 \\ (x-1)^2 &= 9 \\ x - 1 &= \pm\sqrt{9} \\ x - 1 &= \pm 3 \\ x &= 1 \pm 3 \\ x &= -2, 4 \end{aligned}$$

43. Solving the equation:

$$\begin{aligned} x^2 + 12x + 36 &= 8 \\ (x+6)^2 &= 8 \\ x + 6 &= \pm\sqrt{8} \\ x + 6 &= \pm 2\sqrt{2} \\ x &= -6 \pm 2\sqrt{2} \end{aligned}$$

45. Checking the solution: $(x+1)^2 = \left(-1 + 5\sqrt{2} + 1\right)^2 = \left(5\sqrt{2}\right)^2 = 25 \cdot 2 = 50$

47. Let x represent the number. The equation is:

$$\begin{aligned} (x+3)^2 &= 16 \\ x + 3 &= \pm\sqrt{16} \\ x + 3 &= \pm 4 \\ x &= -3 \pm 4 \\ x &= -7, 1 \end{aligned}$$

The number is either –7 or 1.

49. Solving for r:

$$\begin{aligned} 100(1+r)^2 &= A \\ (1+r)^2 &= \frac{A}{100} \\ 1 + r &= \sqrt{\frac{A}{100}} \qquad \text{(since } 1 + r > 0\text{)} \\ 1 + r &= \frac{\sqrt{A}}{10} \\ r &= -1 + \frac{\sqrt{A}}{10} \end{aligned}$$

51. Let x represent the height of the triangle. Draw the figure:

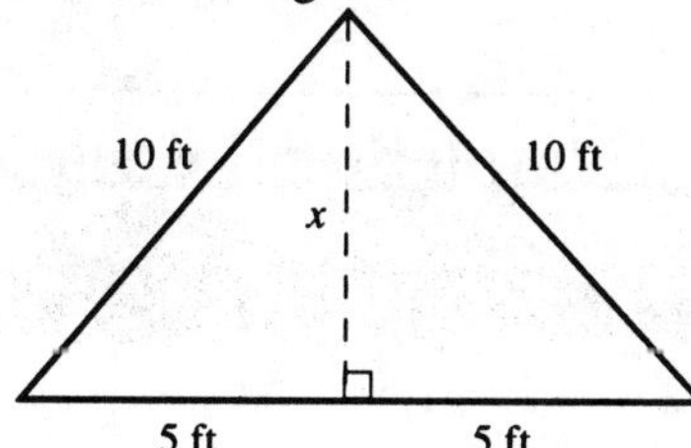

Using the Pythagorean theorem:

$$\begin{aligned} 5^2 + x^2 &= 10^2 \\ 25 + x^2 &= 100 \\ x^2 &= 75 \\ x &= \sqrt{75} \\ x &= 5\sqrt{3} \approx 8.66 \end{aligned}$$

The height is $5\sqrt{3} \approx 8.66$ feet.

53. Let x represent the height of the triangle. Draw the figure:

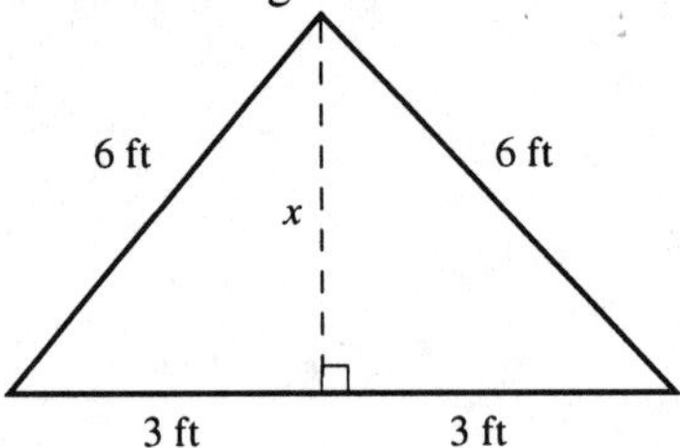

Using the Pythagorean theorem:

$$3^2 + x^2 = 6^2$$
$$9 + x^2 = 36$$
$$x^2 = 27$$
$$x = \sqrt{27}$$
$$x = 3\sqrt{3} \approx 5.2 \text{ feet}$$

No, a person 5 ft 8 in. tall cannot stand up inside the tent.

55. Let x represent the height of the triangle. Draw the figure:

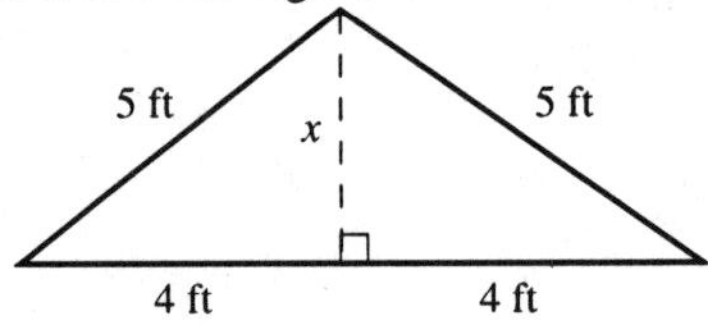

Using the Pythagorean theorem:

$$4^2 + x^2 = 5^2$$
$$16 + x^2 = 25$$
$$x^2 = 9$$
$$x = \sqrt{9} = 3$$

The height is 3 feet.

57. Multiplying using the square of binomial formula: $(x+3)^2 = x^2 + 2(x)(3) + (3)^2 = x^2 + 6x + 9$

59. Multiplying using the square of binomial formula: $(x-5)^2 = x^2 - 2(x)(5) + (5)^2 = x^2 - 10x + 25$

61. Factoring the polynomial: $x^2 - 12x + 36 = (x-6)(x-6) = (x-6)^2$

63. Factoring the polynomial: $x^2 + 4x + 4 = (x+2)(x+2) = (x+2)^2$

65. Finding the root: $\sqrt[3]{8} = 2$

67. Finding the root: $\sqrt[4]{16} = 2$

9.2 Completing the Square

1. The correct term is 9, since: $x^2 + 6x + 9 = (x+3)^2$

3. The correct term is 1, since: $x^2 + 2x + 1 = (x+1)^2$

5. The correct term is 16, since: $y^2 - 8y + 16 = (y-4)^2$

7. The correct term is 1, since: $y^2 - 2y + 1 = (y-1)^2$

9. The correct term is 64, since: $x^2 + 16x + 64 = (x+8)^2$

11. The correct term is $\frac{9}{4}$, since: $a^2 - 3a + \frac{9}{4} = \left(a - \frac{3}{2}\right)^2$

13. The correct term is $\frac{49}{4}$, since: $x^2 - 7x + \frac{49}{4} = \left(x - \frac{7}{2}\right)^2$

15. The correct term is $\frac{1}{4}$, since: $y^2+y+\frac{1}{4}=\left(y+\frac{1}{2}\right)^2$

17. The correct term is $\frac{9}{16}$, since: $x^2-\frac{3}{2}x+\frac{9}{16}=\left(x-\frac{3}{4}\right)^2$

19. Solve by completing the square:

$$\begin{aligned}x^2+4x&=12\\x^2+4x+4&=12+4\\(x+2)^2&=16\\x+2&=\pm4\\x&=-2\pm4\\x&=-6,2\end{aligned}$$

21. Solve by completing the square:

$$\begin{aligned}x^2-6x&=16\\x^2-6x+9&=16+9\\(x-3)^2&=25\\x-3&=\pm5\\x&=3\pm5\\x&=-2,8\end{aligned}$$

23. Solve by completing the square:

$$\begin{aligned}a^2+2a&=3\\a^2+2a+1&=3+1\\(a+1)^2&=4\\a+1&=\pm2\\a&=-1\pm2\\a&=-3,1\end{aligned}$$

25. Solve by completing the square:

$$\begin{aligned}x^2-10x&=0\\x^2-10x+25&=0+25\\(x-5)^2&=25\\x-5&=\pm5\\x&=5\pm5\\x&=0,10\end{aligned}$$

27. Solve by completing the square:

$$\begin{aligned}y^2+2y-15&=0\\y^2+2y&=15\\y^2+2y+1&=15+1\\(y+1)^2&=16\\y+1&=\pm4\\y&=-1\pm4\\y&=-5,3\end{aligned}$$

29. Solve by completing the square:

$$\begin{aligned}x^2+4x-3&=0\\x^2+4x&=3\\x^2+4x+4&=3+4\\(x+2)^2&=7\\x+2&=\pm\sqrt{7}\\x&=-2\pm\sqrt{7}\end{aligned}$$

31. Solve by completing the square:

$$x^2-4x=4$$
$$x^2-4x+4=4+4$$
$$(x-2)^2=8$$
$$x-2=\pm\sqrt{8}$$
$$x-2=\pm 2\sqrt{2}$$
$$x=2\pm 2\sqrt{2}$$

33. Solve by completing the square:

$$a^2=7a+8$$
$$a^2-7a=8$$
$$a^2-7a+\frac{49}{4}=8+\frac{49}{4}$$
$$\left(a-\frac{7}{2}\right)^2=\frac{81}{4}$$
$$a-\frac{7}{2}=\pm\sqrt{\frac{81}{4}}$$
$$a-\frac{7}{2}=\pm\frac{9}{2}$$
$$a=\frac{7}{2}\pm\frac{9}{2}$$
$$a=-1,8$$

35. Solve by completing the square:

$$4x^2+8x-4=0$$
$$x^2+2x-1=0$$
$$x^2+2x=1$$
$$x^2+2x+1=1+1$$
$$(x+1)^2=2$$
$$x+1=\pm\sqrt{2}$$
$$x=-1\pm\sqrt{2}$$

37. Solve by completing the square:

$$2x^2+2x-4=0$$
$$x^2+x-2=0$$
$$x^2+x=2$$
$$x^2+x+\frac{1}{4}=2+\frac{1}{4}$$
$$\left(x+\frac{1}{2}\right)^2=\frac{9}{4}$$
$$x+\frac{1}{2}=\pm\sqrt{\frac{9}{4}}$$
$$x+\frac{1}{2}=\pm\frac{3}{2}$$
$$x=-\frac{1}{2}\pm\frac{3}{2}$$
$$x=-2,1$$

39. Solve by completing the square:

$$
\begin{aligned}
4x^2 + 8x + 1 &= 0 \\
x^2 + 2x + \frac{1}{4} &= 0 \\
x^2 + 2x &= -\frac{1}{4} \\
x^2 + 2x + 1 &= -\frac{1}{4} + 1 \\
(x+1)^2 &= \frac{3}{4} \\
x + 1 &= \pm\sqrt{\frac{3}{4}} \\
x + 1 &= \pm\frac{\sqrt{3}}{2} \\
x &= -1 \pm \frac{\sqrt{3}}{2} \\
x &= \frac{-2 \pm \sqrt{3}}{2}
\end{aligned}
$$

41. Solve by completing the square:

$$
\begin{aligned}
2x^2 - 2x &= 1 \\
x^2 - x &= \frac{1}{2} \\
x^2 - x + \frac{1}{4} &= \frac{1}{2} + \frac{1}{4} \\
\left(x - \frac{1}{2}\right)^2 &= \frac{3}{4} \\
x - \frac{1}{2} &= \pm\sqrt{\frac{3}{4}} \\
x - \frac{1}{2} &= \pm\frac{\sqrt{3}}{2} \\
x &= \frac{1}{2} \pm \frac{\sqrt{3}}{2} \\
x &= \frac{1 \pm \sqrt{3}}{2}
\end{aligned}
$$

43. Solve by completing the square:

$$
\begin{aligned}
4a^2 - 4a + 1 &= 0 \\
a^2 - a + \frac{1}{4} &= 0 \\
\left(a - \frac{1}{2}\right)^2 &= 0 \\
a - \frac{1}{2} &= 0 \\
a &= \frac{1}{2}
\end{aligned}
$$

45. Solve by completing the square:

$$\begin{aligned}
3y^2 - 9y &= 2 \\
y^2 - 3y &= \frac{2}{3} \\
y^2 - 3y + \frac{9}{4} &= \frac{2}{3} + \frac{9}{4} \\
\left(y - \frac{3}{2}\right)^2 &= \frac{8}{12} + \frac{27}{12} = \frac{35}{12} \\
y - \frac{3}{2} &= \pm\sqrt{\frac{35}{12}} \bullet \frac{\sqrt{3}}{\sqrt{3}} \\
y - \frac{3}{2} &= \pm\frac{\sqrt{105}}{6} \\
y &= \frac{3}{2} \pm \frac{\sqrt{105}}{6} \\
y &= \frac{9 \pm \sqrt{105}}{6}
\end{aligned}$$

47. The values are:

$$-2 + \sqrt{7} \approx -2 + 2.646 = 0.646$$
$$-2 - \sqrt{7} \approx -2 - 2.646 = -4.646$$

49. Checking the solution:

$$\begin{aligned}
4\left(-1+\sqrt{2}\right)^2 + 8\left(-1+\sqrt{2}\right) - 4 &= 4\left(1 - 2\sqrt{2} + 2\right) - 8 + 8\sqrt{2} - 4 \\
&= 12 - 8\sqrt{2} - 8 + 8\sqrt{2} - 4 \\
&= 0
\end{aligned}$$

51. Adding the two solutions: $\left(-2+\sqrt{7}\right) + \left(-2-\sqrt{7}\right) = -4$

53. Drawing the diagram:

	x	3
x	x^2	$3x$
3	$3x$	9

55. Drawing the diagram:

	x	1
x	x^2	x
1	x	1

57. Evaluating when $a = 2$: $2a = 2(2) = 4$

59. Evaluating when $a = 2$ and $c = -3$: $4ac = 4(2)(-3) = -24$

61. Evaluating when $a = 2$, $b = 4$, and $c = -3$: $\sqrt{b^2 - 4ac} = \sqrt{(4)^2 - 4(2)(-3)} = \sqrt{16+24} = \sqrt{40} = 2\sqrt{10}$

63. Simplifying the radical: $\sqrt{12} = \sqrt{4 \cdot 3} = \sqrt{4}\sqrt{3} = 2\sqrt{3}$

65. Simplifying the radical: $\sqrt{20x^2y^3} = \sqrt{4x^2y^2 \cdot 5y} = \sqrt{4x^2y^2}\sqrt{5y} = 2xy\sqrt{5y}$

67. Simplifying the radical: $\sqrt{\frac{81}{25}} = \frac{\sqrt{81}}{\sqrt{25}} = \frac{9}{5}$

9.3 The Quadratic Formula

1. Using the quadratic formula with $a = 1$, $b = 3$, and $c = 2$:

$$x = \frac{-b \pm \sqrt{b^2 - 4ac}}{2a} = \frac{-3 \pm \sqrt{(3)^2 - 4(1)(2)}}{2(1)} = \frac{-3 \pm \sqrt{9-8}}{2} = \frac{-3 \pm 1}{2} = -2, -1$$

3. Using the quadratic formula with $a = 1$, $b = 5$, and $c = 6$:

$$x = \frac{-b \pm \sqrt{b^2 - 4ac}}{2a} = \frac{-5 \pm \sqrt{(5)^2 - 4(1)(6)}}{2(1)} = \frac{-5 \pm \sqrt{25-24}}{2} = \frac{-5 \pm 1}{2} = -3, -2$$

5. Using the quadratic formula with $a = 1$, $b = 6$, and $c = 9$:

$$x = \frac{-b \pm \sqrt{b^2 - 4ac}}{2a} = \frac{-6 \pm \sqrt{(6)^2 - 4(1)(9)}}{2(1)} = \frac{-6 \pm \sqrt{36-36}}{2} = \frac{-6 \pm 0}{2} = -3$$

7. Using the quadratic formula with $a = 1$, $b = 6$, and $c = 7$:

$$x = \frac{-b \pm \sqrt{b^2 - 4ac}}{2a} = \frac{-6 \pm \sqrt{(6)^2 - 4(1)(7)}}{2(1)} = \frac{-6 \pm \sqrt{36-28}}{2} = \frac{-6 \pm \sqrt{8}}{2} = \frac{-6 \pm 2\sqrt{2}}{2} = -3 \pm \sqrt{2}$$

9. Using the quadratic formula with $a = 2$, $b = 5$, and $c = 3$:

$$x = \frac{-b \pm \sqrt{b^2 - 4ac}}{2a} = \frac{-5 \pm \sqrt{(5)^2 - 4(2)(3)}}{2(2)} = \frac{-5 \pm \sqrt{25-24}}{4} = \frac{-5 \pm 1}{4} = -\frac{3}{2}, -1$$

11. Using the quadratic formula with $a = 4$, $b = 8$, and $c = 1$:

$$x = \frac{-b \pm \sqrt{b^2 - 4ac}}{2a} = \frac{-8 \pm \sqrt{(8)^2 - 4(4)(1)}}{2(4)} = \frac{-8 \pm \sqrt{64-16}}{8} = \frac{-8 \pm \sqrt{48}}{8} = \frac{-8 \pm 4\sqrt{3}}{8} = \frac{-2 \pm \sqrt{3}}{2}$$

13. Using the quadratic formula with $a = 1$, $b = -2$, and $c = 1$:

$$x = \frac{-b \pm \sqrt{b^2 - 4ac}}{2a} = \frac{-(-2) \pm \sqrt{(-2)^2 - 4(1)(1)}}{2(1)} = \frac{2 \pm \sqrt{4-4}}{2} = \frac{2 \pm 0}{2} = 1$$

15. First write the equation as $x^2 - 5x - 7 = 0$. Using $a = 1$, $b = -5$, and $c = -7$ in the quadratic formula:

$$x = \frac{-b \pm \sqrt{b^2 - 4ac}}{2a} = \frac{-(-5) \pm \sqrt{(-5)^2 - 4(1)(-7)}}{2(1)} = \frac{5 \pm \sqrt{25+28}}{2} = \frac{5 \pm \sqrt{53}}{2}$$

17. Using $a = 6$, $b = -1$, and $c = -2$ in the quadratic formula:

$$x = \frac{-b \pm \sqrt{b^2 - 4ac}}{2a} = \frac{-(-1) \pm \sqrt{(-1)^2 - 4(6)(-2)}}{2(6)} = \frac{1 \pm \sqrt{1+48}}{12} = \frac{1 \pm \sqrt{49}}{12} = \frac{1 \pm 7}{12} = -\frac{1}{2}, \frac{2}{3}$$

19. First simplify the equation:

$$\begin{aligned} (x-2)(x+1) &= 3 \\ x^2 - 2x + x - 2 &= 3 \\ x^2 - x - 5 &= 0 \end{aligned}$$

Using $a = 1$, $b = -1$, and $c = -5$ in the quadratic formula:

$$x = \frac{-b \pm \sqrt{b^2 - 4ac}}{2a} = \frac{-(-1) \pm \sqrt{(-1)^2 - 4(1)(-5)}}{2(1)} = \frac{1 \pm \sqrt{1+20}}{2} = \frac{1 \pm \sqrt{21}}{2}$$

21. First simplify the equation:

$$(2x-3)(x+2)=1$$
$$2x^2-3x+4x-6=1$$
$$2x^2+x-7=0$$

Using $a = 2$, $b = 1$, and $c = -7$ in the quadratic formula:

$$x=\frac{-b\pm\sqrt{b^2-4ac}}{2a}=\frac{-1\pm\sqrt{(1)^2-4(2)(-7)}}{2(2)}=\frac{-1\pm\sqrt{1+56}}{4}=\frac{-1\pm\sqrt{57}}{4}$$

23. First write the equation as $2x^2-3x-5=0$. Using $a = 2$, $b = -3$, and $c = -5$ in the quadratic formula:

$$x=\frac{-b\pm\sqrt{b^2-4ac}}{2a}=\frac{-(-3)\pm\sqrt{(-3)^2-4(2)(-5)}}{2(2)}=\frac{3\pm\sqrt{9+40}}{4}=\frac{3\pm\sqrt{49}}{4}=\frac{3\pm7}{4}=-1,\frac{5}{2}$$

25. First write the equation as $2x^2+6x-7=0$. Using $a = 2$, $b = 6$, and $c = -7$ in the quadratic formula:

$$x=\frac{-b\pm\sqrt{b^2-4ac}}{2a}=\frac{-6\pm\sqrt{(6)^2-4(2)(-7)}}{2(2)}=\frac{-6\pm\sqrt{36+56}}{4}=\frac{-6\pm\sqrt{92}}{4}=\frac{-6\pm2\sqrt{23}}{4}=\frac{-3\pm\sqrt{23}}{2}$$

27. First write the equation as $3x^2+4x-2=0$. Using $a = 3$, $b = 4$, and $c = -2$ in the quadratic formula:

$$x=\frac{-b\pm\sqrt{b^2-4ac}}{2a}=\frac{-4\pm\sqrt{(4)^2-4(3)(-2)}}{2(3)}=\frac{-4\pm\sqrt{16+24}}{6}=\frac{-4\pm\sqrt{40}}{6}=\frac{-4\pm2\sqrt{10}}{6}=\frac{-2\pm\sqrt{10}}{3}$$

29. First write the equation as $2x^2-2x-5=0$. Using $a = 2$, $b = -2$, and $c = -5$ in the quadratic formula:

$$x=\frac{-b\pm\sqrt{b^2-4ac}}{2a}=\frac{-(-2)\pm\sqrt{(-2)^2-4(2)(-5)}}{2(2)}=\frac{2\pm\sqrt{4+40}}{4}=\frac{2\pm\sqrt{44}}{4}=\frac{2\pm2\sqrt{11}}{4}=\frac{1\pm\sqrt{11}}{2}$$

31. Factoring out x results in the equation $x\left(2x^2+3x-4\right)=0$, so $x = 0$ is one solution. The other two solutions are found by using $a = 2$, $b = 3$, and $c = -4$ in the quadratic formula:

$$x=\frac{-b\pm\sqrt{b^2-4ac}}{2a}=\frac{-3\pm\sqrt{(3)^2-4(2)(-4)}}{2(2)}=\frac{-3\pm\sqrt{9+32}}{4}=\frac{-3\pm\sqrt{41}}{4}$$

33. Using $a = 3$, $b = -4$, and $c = 0$ in the quadratic formula:

$$x=\frac{-b\pm\sqrt{b^2-4ac}}{2a}=\frac{-(-4)\pm\sqrt{(-4)^2-4(3)(0)}}{2(3)}=\frac{4\pm\sqrt{16-0}}{6}=\frac{4\pm4}{6}=0,\frac{4}{3}$$

35. Multiplying each side of the equation by 6 results in the equation $3x^2-3x-1=0$. Using $a = 3$, $b = -3$, and $c = -1$ in the quadratic formula: $x=\frac{-b\pm\sqrt{b^2-4ac}}{2a}=\frac{-(-3)\pm\sqrt{(-3)^2-4(3)(-1)}}{2(3)}=\frac{3\pm\sqrt{9+12}}{6}=\frac{3\pm\sqrt{21}}{6}$

37. Multiplying the radicals: $\left(2\sqrt{3}\right)\left(3\sqrt{5}\right)=6\sqrt{15}$

39. Multiplying the radicals: $\left(\sqrt{6}+2\right)\left(\sqrt{6}-5\right)=6+2\sqrt{6}-5\sqrt{6}-10=-4-3\sqrt{6}$

41. Multiplying the radicals: $\left(\sqrt{7}-\sqrt{2}\right)\left(\sqrt{7}+\sqrt{2}\right)=\left(\sqrt{7}\right)^2-\left(\sqrt{2}\right)^2=7-2=5$

43. Rationalizing the denominator: $\frac{2}{3+\sqrt{5}}\bullet\frac{3-\sqrt{5}}{3-\sqrt{5}}=\frac{2\left(3-\sqrt{5}\right)}{9-5}=\frac{2\left(3-\sqrt{5}\right)}{4}=\frac{3-\sqrt{5}}{2}$

9.4 Complex Numbers

1. Combining the complex numbers: $(3-2i)+3i=3+(-2i+3i)=3+i$

3. Combining the complex numbers: $(6+2i)-10i=6+(2i-10i)=6-8i$

5. Combining the complex numbers: $(11+9i)-9i=11+(9i-9i)=11$

7. Combining the complex numbers: $(3+2i)+(6-i)=(3+6)+(2i-i)=9+i$

9. Combining the complex numbers: $(5+7i)-(6+8i)=5+7i-6-8i=-1-i$

11. Combining the complex numbers: $(9-i)+(2-i)=9-i+2-i=11-2i$

13. Combining the complex numbers: $(6+i)-4i-(2-i)=6+i-4i-2+i=4-2i$

15. Combining the complex numbers: $(6-11i)+3i+(2+i)=6-11i+3i+2+i=8-7i$

17. Combining the complex numbers: $(2+3i)-(6-2i)+(3-i)=2+3i-6+2i+3-i=-1+4i$

19. Multiplying the complex numbers: $3(2-i)=6-3i$

21. Multiplying the complex numbers: $2i(8-7i)=16i-14i^2=16i-14(-1)=14+16i$

23. Multiplying the complex numbers: $(2+i)(4-i)=8+4i-2i-i^2=8+2i-(-1)=8+2i+1=9+2i$

25. Multiplying the complex numbers: $(2+i)(3-5i)=6+3i-10i-5i^2=6-7i-5(-1)=6-7i+5=11-7i$

27. Multiplying the complex numbers: $(3+5i)(3-5i)=(3)^2-(5i)^2=9-25i^2=9-25(-1)=9+25=34$

29. Multiplying the complex numbers: $(2+i)(2-i)=(2)^2-(i)^2=4-i^2=4-(-1)=4+1=5$

31. Dividing the complex numbers: $\frac{2}{3-2i}\bullet\frac{3+2i}{3+2i}=\frac{2(3+2i)}{9-4i^2}=\frac{2(3+2i)}{9+4}=\frac{6+4i}{13}$

33. Dividing the complex numbers: $\frac{-3i}{2+3i}\bullet\frac{2-3i}{2-3i}=\frac{-6i+9i^2}{4-9i^2}=\frac{-6i-9}{4+9}=\frac{-9-6i}{13}$

35. Dividing the complex numbers: $\frac{6i}{3-i}\bullet\frac{3+i}{3+i}=\frac{18i+6i^2}{9-i^2}=\frac{18i-6}{9+1}=\frac{6(-1+3i)}{10}=\frac{3(-1+3i)}{5}=\frac{-3+9i}{5}$

37. Dividing the complex numbers: $\frac{2+i}{2-i}\bullet\frac{2+i}{2+i}=\frac{4+2i+2i+i^2}{4-i^2}=\frac{4+4i-1}{4+1}=\frac{3+4i}{5}$

39. Dividing the complex numbers:
$$\frac{4+5i}{3-6i}\bullet\frac{3+6i}{3+6i}=\frac{12+15i+24i+30i^2}{9-36i^2}=\frac{12+39i-30}{9+36}=\frac{-18+39i}{45}=\frac{3(-6+13i)}{45}=\frac{-6+13i}{15}$$

41. Multiplying the expressions: $(x+3i)(x-3i)=x^2-(3i)^2=x^2-9i^2=x^2-9(-1)=x^2+9$

43. Simplifying: $\frac{1}{i}\bullet\frac{i}{i}=\frac{i}{i^2}=\frac{i}{-1}=-i$

45. Solving the equation:
$$\begin{aligned}(x-3)^2&=25\\ x-3&=\pm\sqrt{25}\\ x-3&=\pm 5\\ x&=3\pm 5\\ x&=-2,8\end{aligned}$$

47. Solving the equation:
$$\begin{aligned}(2x-6)^2&=16\\ 2x-6&=\pm\sqrt{16}\\ 2x-6&=\pm 4\\ 2x&=6\pm 4\\ 2x&=2,10\\ x&=1,5\end{aligned}$$

49. Solving the equation:
$$\begin{aligned}(x+3)^2&=12\\ x+3&=\pm\sqrt{12}\\ x+3&=\pm 2\sqrt{3}\\ x&=-3\pm 2\sqrt{3}\end{aligned}$$

51. Simplifying the radical expression: $\sqrt{\frac{1}{2}}=\frac{\sqrt{1}}{\sqrt{2}}\bullet\frac{\sqrt{2}}{\sqrt{2}}=\frac{\sqrt{2}}{2}$

53. Simplifying the radical expression: $\sqrt{\frac{8x^2y^3}{3}}=\frac{\sqrt{8x^2y^3}}{\sqrt{3}}\bullet\frac{\sqrt{3}}{\sqrt{3}}=\frac{\sqrt{24x^2y^3}}{3}=\frac{\sqrt{4x^2y^2\bullet 6y}}{3}=\frac{2xy\sqrt{6y}}{3}$

55. Simplifying the radical expression: $\sqrt[3]{\frac{1}{4}}=\frac{\sqrt[3]{1}}{\sqrt[3]{4}}\bullet\frac{\sqrt[3]{2}}{\sqrt[3]{2}}=\frac{\sqrt[3]{2}}{\sqrt[3]{8}}=\frac{\sqrt[3]{2}}{2}$

9.5 Complex Solutions to Quadratic Equations

1. Writing as a complex number: $\sqrt{-16}=\sqrt{16(-1)}=\sqrt{16}\sqrt{-1}=4i$
3. Writing as a complex number: $\sqrt{-49}=\sqrt{49(-1)}=\sqrt{49}\sqrt{-1}=7i$
5. Writing as a complex number: $\sqrt{-6}=\sqrt{6(-1)}=\sqrt{6}\sqrt{-1}=i\sqrt{6}$
7. Writing as a complex number: $\sqrt{-11}=\sqrt{11(-1)}=\sqrt{11}\sqrt{-1}=i\sqrt{11}$
9. Writing as a complex number: $\sqrt{-32}=\sqrt{-16\bullet 2}=\sqrt{-16}\sqrt{2}=4i\sqrt{2}$
11. Writing as a complex number: $\sqrt{-50}=\sqrt{-25\bullet 2}=\sqrt{-25}\sqrt{2}=5i\sqrt{2}$
13. Writing as a complex number: $\sqrt{-8}=\sqrt{-4\bullet 2}=\sqrt{-4}\sqrt{2}=2i\sqrt{2}$
15. Writing as a complex number: $\sqrt{-48}=\sqrt{-16\bullet 3}=\sqrt{-16}\sqrt{3}=4i\sqrt{3}$
17. First write the equation as $x^2-2x+2=0$. Using $a=1$, $b=-2$, and $c=2$ in the quadratic formula:

$$x=\frac{-b\pm\sqrt{b^2-4ac}}{2a}=\frac{-(-2)\pm\sqrt{(-2)^2-4(1)(2)}}{2(1)}=\frac{2\pm\sqrt{4-8}}{2}=\frac{2\pm\sqrt{-4}}{2}=\frac{2\pm 2i}{2}=1\pm i$$

19. Solving the equation:

$$\begin{aligned}x^2-4x&=-4\\x^2-4x+4&=0\\(x-2)^2&=0\\x-2&=0\\x&=2\end{aligned}$$

21. Solving the equation:

$$\begin{aligned}2x^2+5x&=12\\2x^2+5x-12&=0\\(2x-3)(x+4)&=0\\x&=\frac{3}{2},-4\end{aligned}$$

23. Solving the equation:

$$\begin{aligned}(x-2)^2&=-4\\x-2&=\pm\sqrt{-4}\\x-2&=\pm 2i\\x&=2\pm 2i\end{aligned}$$

25. Solving the equation:

$$\begin{aligned}\left(x+\frac{1}{2}\right)^2&=-\frac{9}{4}\\x+\frac{1}{2}&=\pm\sqrt{-\frac{9}{4}}\\x+\frac{1}{2}&=\pm\frac{3}{2}i\\x&=\frac{-1\pm 3i}{2}\end{aligned}$$

27. Solving the equation:

$$\begin{aligned}\left(x-\frac{1}{2}\right)^2&=-\frac{27}{36}\\ x-\frac{1}{2}&=\pm\sqrt{-\frac{27}{36}}\\ x-\frac{1}{2}&=\pm\frac{3i\sqrt{3}}{6}\\ x-\frac{1}{2}&=\pm\frac{i\sqrt{3}}{2}\\ x&=\frac{1\pm i\sqrt{3}}{2}\end{aligned}$$

29. Using $a=1$, $b=1$, and $c=1$ in the quadratic formula:

$$x=\frac{-b\pm\sqrt{b^2-4ac}}{2a}=\frac{-1\pm\sqrt{(1)^2-4(1)(1)}}{2(1)}=\frac{-1\pm\sqrt{1-4}}{2}=\frac{-1\pm\sqrt{-3}}{2}=\frac{-1\pm i\sqrt{3}}{2}$$

31. Solving the equation:

$$\begin{aligned}x^2-5x+6&=0\\ (x-2)(x-3)&=0\\ x&=2,3\end{aligned}$$

33. First multiply by 6 to clear the equation of fractions:

$$\begin{aligned}6\left(\frac{1}{2}x^2+\frac{1}{3}x+\frac{1}{6}\right)&=6(0)\\ 3x^2+2x+1&=0\end{aligned}$$

Using $a=3$, $b=2$, and $c=1$ in the quadratic formula:

$$x=\frac{-b\pm\sqrt{b^2-4ac}}{2a}=\frac{-2\pm\sqrt{(2)^2-4(3)(1)}}{2(3)}=\frac{-2\pm\sqrt{4-12}}{6}=\frac{-2\pm\sqrt{-8}}{6}=\frac{-2\pm 2i\sqrt{2}}{6}=\frac{-1\pm i\sqrt{2}}{3}$$

35. First multiply by 6 to clear the equation of fractions:

$$\begin{aligned}6\left(\frac{1}{3}x^2\right)&=6\left(-\frac{1}{2}x+\frac{1}{3}\right)\\ 2x^2&=-3x+2\\ 2x^2+3x-2&=0\\ (2x-1)(x+2)&=0\\ x&=\frac{1}{2},-2\end{aligned}$$

37. Solving the equation:

$$\begin{aligned}(x+2)(x-3)&=5\\ x^2-x-6&=5\\ x^2-x-11&=0\end{aligned}$$

Using $a=1$, $b=-1$, and $c=-11$ in the quadratic formula:

$$x=\frac{-b\pm\sqrt{b^2-4ac}}{2a}=\frac{-(-1)\pm\sqrt{(-1)^2-4(1)(-11)}}{2(1)}=\frac{1\pm\sqrt{1+44}}{2}=\frac{1\pm\sqrt{45}}{2}=\frac{1\pm 3\sqrt{5}}{2}$$

39. Solving the equation:

$$\begin{aligned}(x-5)(x-3)&=-10\\ x^2-8x+15&=-10\\ x^2-8x+25&=0\end{aligned}$$

Using $a=1$, $b=-8$, and $c=25$ in the quadratic formula:

$$x=\frac{-b\pm\sqrt{b^2-4ac}}{2a}=\frac{-(-8)\pm\sqrt{(-8)^2-4(1)(25)}}{2(1)}=\frac{8\pm\sqrt{64-100}}{2}=\frac{8\pm\sqrt{-36}}{2}=\frac{8\pm 6i}{2}=4\pm 3i$$

41. Solving the equation:

$$(2x-2)(x-3)=9$$
$$2x^2-8x+6=9$$
$$2x^2-8x-3=0$$

Using $a=2$, $b=-8$, and $c=-3$ in the quadratic formula:

$$x=\frac{-b\pm\sqrt{b^2-4ac}}{2a}=\frac{-(-8)\pm\sqrt{(-8)^2-4(2)(-3)}}{2(2)}=\frac{8\pm\sqrt{64+24}}{4}=\frac{8\pm\sqrt{88}}{4}=\frac{8\pm2\sqrt{22}}{4}=\frac{4\pm\sqrt{22}}{2}$$

43. Substituting $x=2+2i$ into the equation:

$$x^2-4x+8=(2+2i)^2-4(2+2i)+8=4+8i+4i^2-8-8i+8=4+8i-4-8-8i+8=0$$

Yes, $x=2+2i$ is a solution to the equation.

45. The other solution is $3-7i$.

47. Graphing the line:

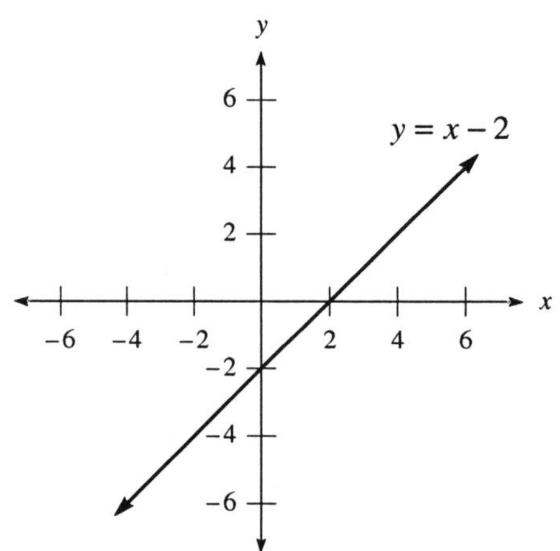

49. Graphing the line:

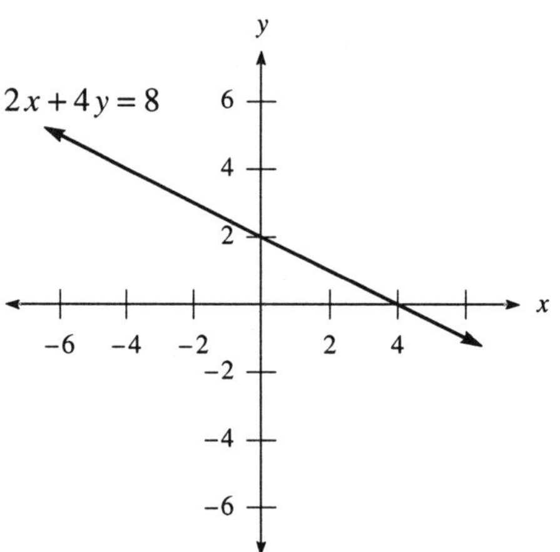

51. Simplifying the radical expression: $3\sqrt{50}+2\sqrt{32}=3\sqrt{25\cdot2}+2\sqrt{16\cdot2}=3\cdot5\sqrt{2}+2\cdot4\sqrt{2}=15\sqrt{2}+8\sqrt{2}=23\sqrt{2}$

53. Simplifying the radical expression: $\sqrt{24}-\sqrt{54}-\sqrt{150}=\sqrt{4\cdot6}-\sqrt{9\cdot6}-\sqrt{25\cdot6}=2\sqrt{6}-3\sqrt{6}-5\sqrt{6}=-6\sqrt{6}$

55. Simplifying the radical expression:

$$2\sqrt{27x^2}-x\sqrt{48}=2\sqrt{9x^2\cdot3}-x\sqrt{16\cdot3}=2\cdot3x\sqrt{3}-x\cdot4\sqrt{3}=6x\sqrt{3}-4x\sqrt{3}=2x\sqrt{3}$$

9.6 Graphing Parabolas

1. Graphing the parabola:

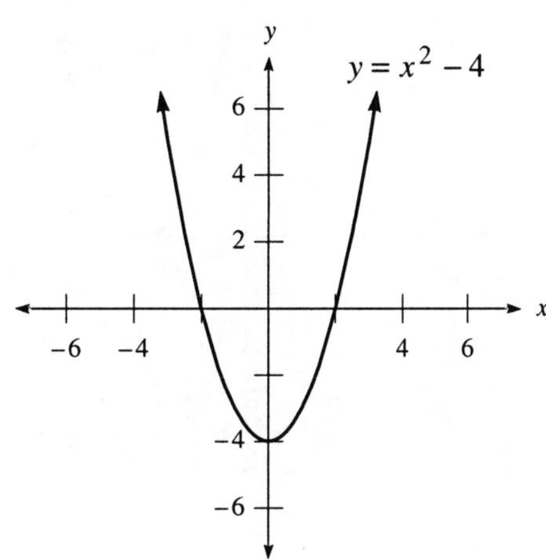

3. Graphing the parabola:

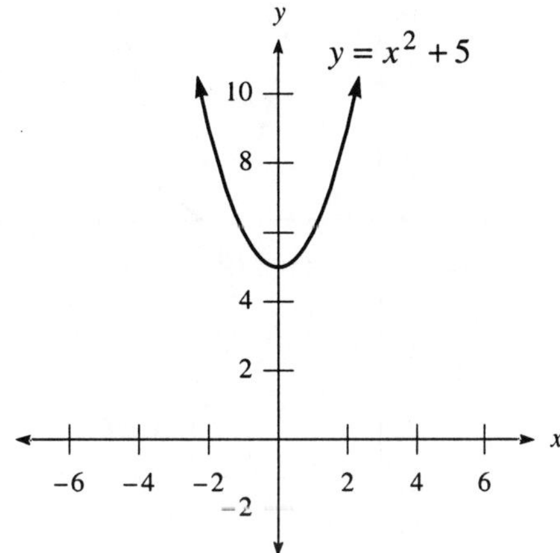

5. Graphing the parabola:

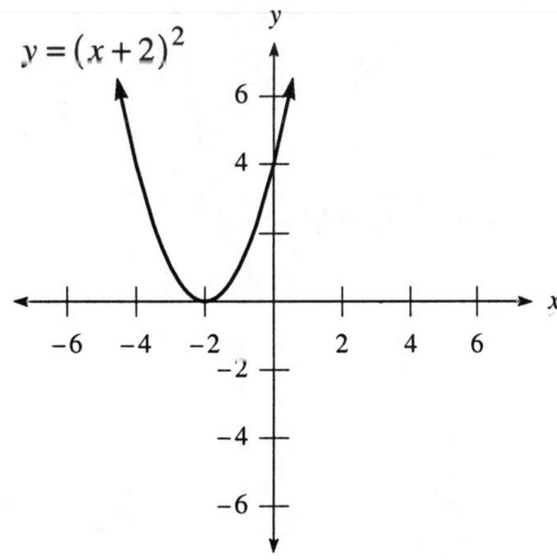

7. Graphing the parabola:

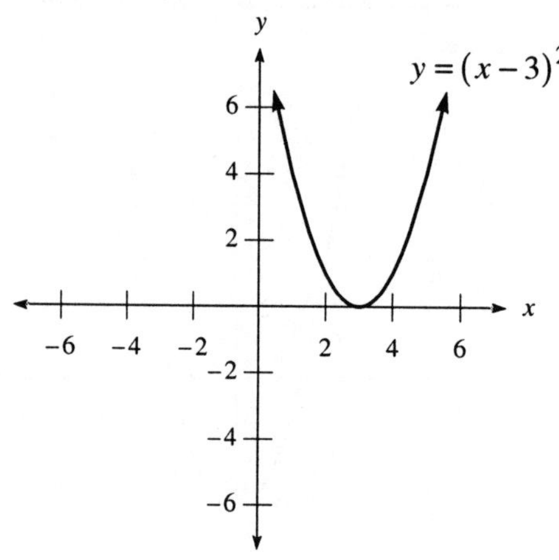

9. Graphing the parabola:

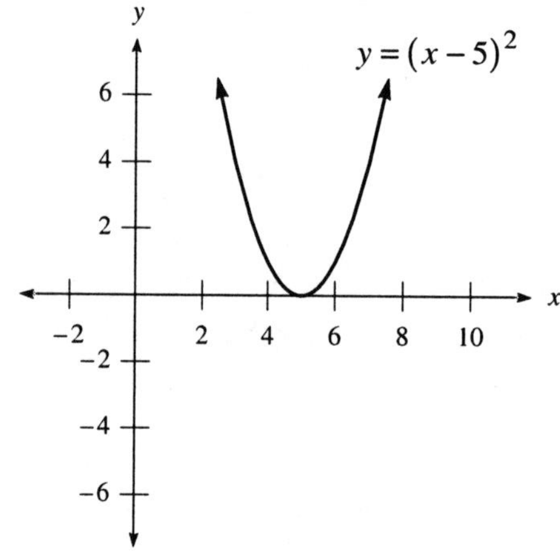

11. Graphing the parabola:

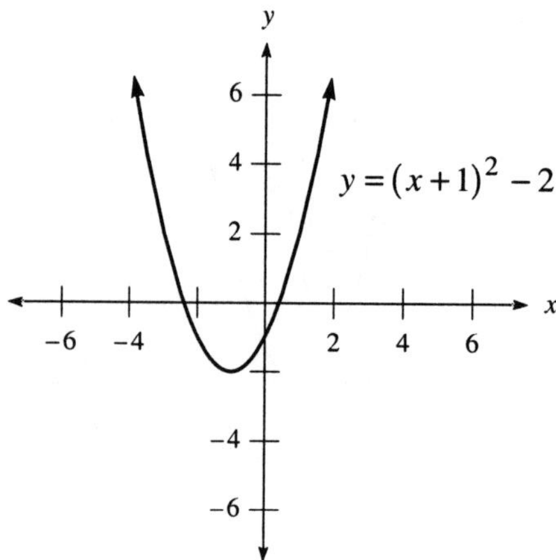

13. Graphing the parabola:

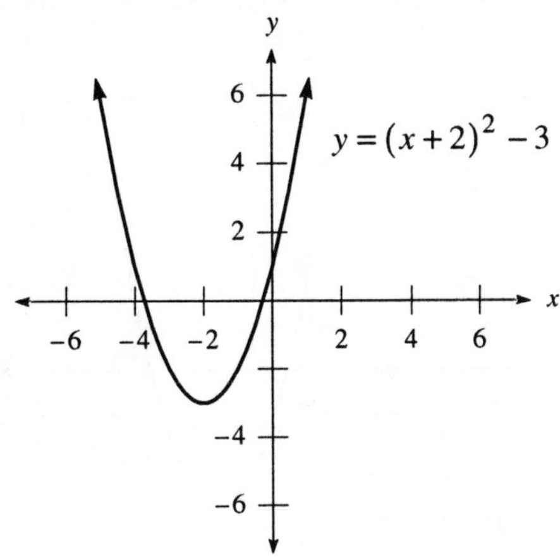

15. Graphing the parabola:

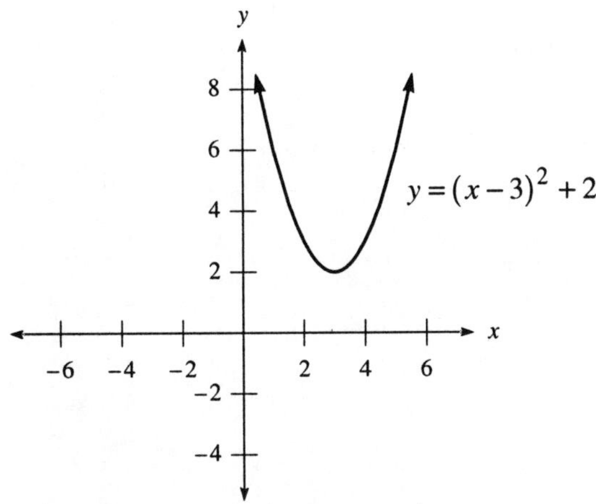

17. Begin by completing the square: $y = x^2 + 6x + 5 = \left(x^2 + 6x + 9\right) + 5 - 9 = (x+3)^2 - 4$

Graphing the parabola:

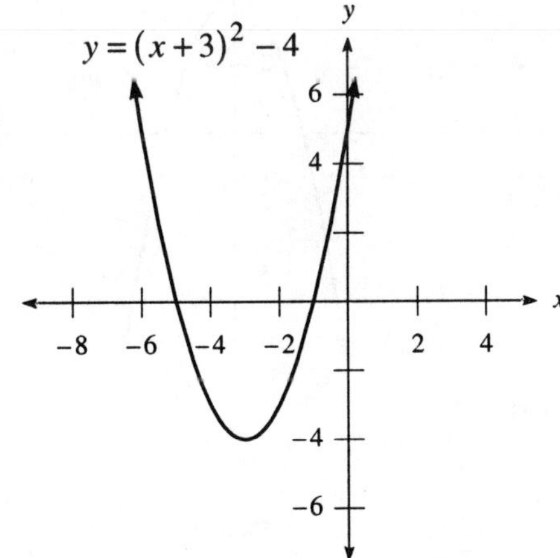

19. Begin by completing the square: $y = x^2 - 2x - 3 = \left(x^2 - 2x + 1\right) - 3 - 1 = (x-1)^2 - 4$

Graphing the parabola:

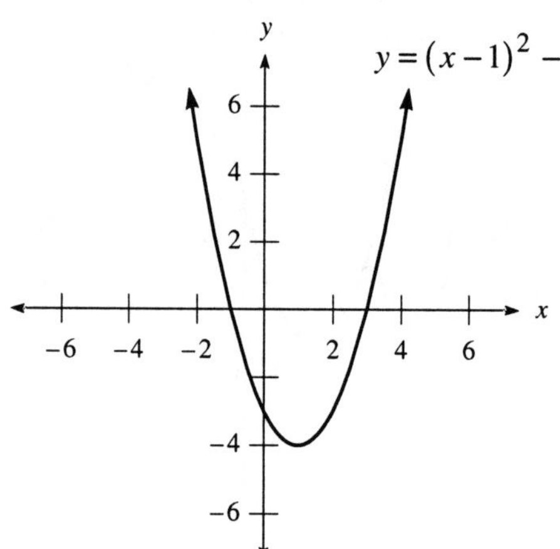

21. Begin by making a table of values:

x	y
–3	–5
–2	0
–1	3
0	4
1	3
2	0
3	–5

Graphing the parabola:

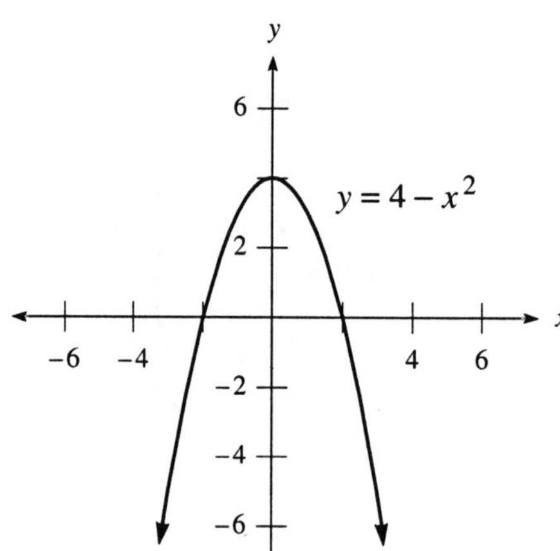

23. Begin by making a table of values:

x	y
–2	–5
–1	–2
0	–1
1	–2
2	–5

Graphing the parabola:

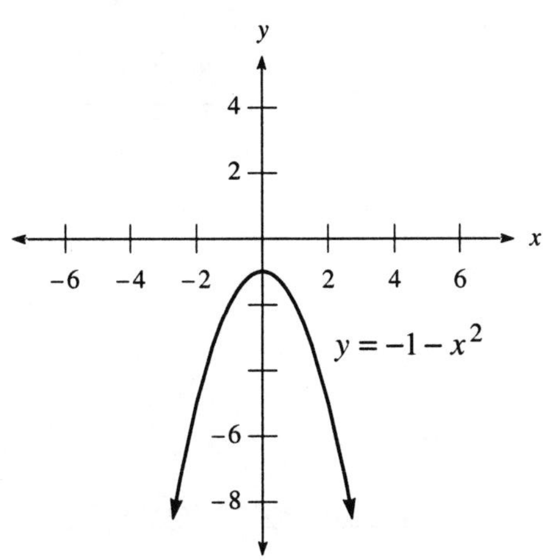

25. Graphing the line and the parabola:

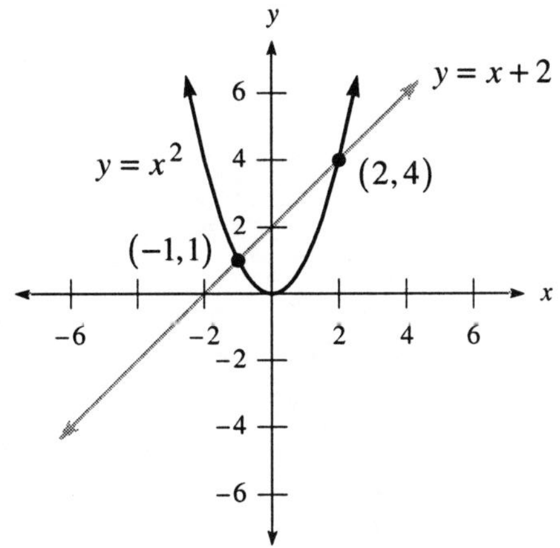

The intersection points are $(-1,1)$ and $(2,4)$.

27. Graphing the two parabolas:

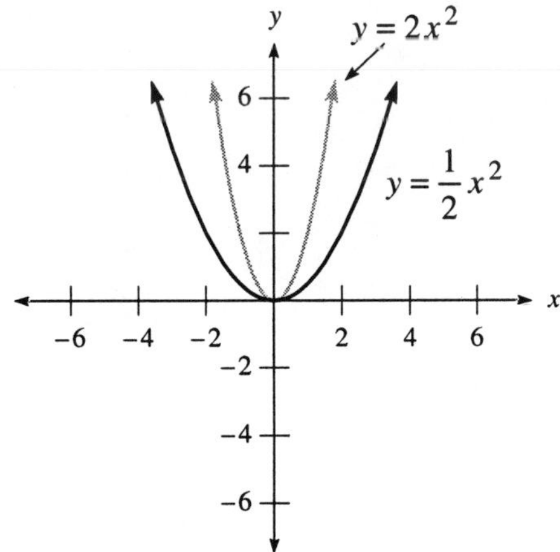

29. Finding the root: $\sqrt{49} = 7$

31. Simplifying the radical: $\sqrt{50} = \sqrt{25 \cdot 2} = 5\sqrt{2}$

33. Simplifying the radical: $\sqrt{\frac{2}{5}} = \frac{\sqrt{2}}{\sqrt{5}} \cdot \frac{\sqrt{5}}{\sqrt{5}} = \frac{\sqrt{10}}{5}$

35. Performing the operations: $3\sqrt{12} + 5\sqrt{27} = 3\sqrt{4 \cdot 3} + 5\sqrt{9 \cdot 3} = 3 \cdot 2\sqrt{3} + 5 \cdot 3\sqrt{3} = 6\sqrt{3} + 15\sqrt{3} = 21\sqrt{3}$

37. Performing the operations: $\left(\sqrt{6}+2\right)\left(\sqrt{6}-5\right)=\sqrt{36}+2\sqrt{6}-5\sqrt{6}-10=6-3\sqrt{6}-10=-4-3\sqrt{6}$

39. Rationalizing the denominator: $\dfrac{8}{\sqrt{5}-\sqrt{3}}\bullet\dfrac{\sqrt{5}+\sqrt{3}}{\sqrt{5}+\sqrt{3}}=\dfrac{8\left(\sqrt{5}+\sqrt{3}\right)}{5-3}=\dfrac{8\left(\sqrt{5}+\sqrt{3}\right)}{2}=4\left(\sqrt{5}+\sqrt{3}\right)=4\sqrt{5}+4\sqrt{3}$

41. Solving the equation:
$$\begin{aligned}\sqrt{2x-5}&=3\\ \left(\sqrt{2x-5}\right)^2&=3^2\\ 2x-5&=9\\ 2x&=14\\ x&=7\end{aligned}$$
This solution checks in the original equation.

Chapter 9 Review

1. Solving the quadratic equation:
$$\begin{aligned}a^2&=32\\ a&=\pm\sqrt{32}\\ a&=\pm4\sqrt{2}\end{aligned}$$

3. Solving the quadratic equation:
$$\begin{aligned}2x^2&=32\\ x^2&=16\\ x&=\pm\sqrt{16}\\ x&=\pm4\end{aligned}$$

5. Solving the quadratic equation:
$$\begin{aligned}(x-2)^2&=81\\ x-2&=\pm\sqrt{81}\\ x-2&=\pm9\\ x&=2\pm9\\ x&=-7,11\end{aligned}$$

7. Solving the quadratic equation:
$$\begin{aligned}(2x+5)^2&=32\\ 2x+5&=\pm\sqrt{32}\\ 2x+5&=\pm4\sqrt{2}\\ 2x&=-5\pm4\sqrt{2}\\ x&=\frac{-5\pm4\sqrt{2}}{2}\end{aligned}$$

9. Solving the quadratic equation:
$$\begin{aligned}\left(x-\frac{2}{3}\right)^2&=-\frac{25}{9}\\ x-\frac{2}{3}&=\pm\sqrt{-\frac{25}{9}}\\ x-\frac{2}{3}&=\pm\frac{5i}{3}\\ x&=\frac{2}{3}\pm\frac{5}{3}i\end{aligned}$$

11. Solving by completing the square:

$$\begin{aligned} x^2-8x&=4\\ x^2-8x+16&=4+16\\ (x-4)^2&=20\\ x-4&=\pm\sqrt{20}\\ x-4&=\pm2\sqrt{5}\\ x&=4\pm2\sqrt{5} \end{aligned}$$

13. Solving by completing the square:

$$\begin{aligned} x^2+4x+3&=0\\ x^2+4x&=-3\\ x^2+4x+4&=-3+4\\ (x+2)^2&=1\\ x+2&=\pm\sqrt{1}\\ x+2&=\pm1\\ x&=-2\pm1\\ x&=-3,-1 \end{aligned}$$

15. Solving by completing the square:

$$\begin{aligned} a^2&=5a+6\\ a^2-5a&=6\\ a^2-5a+\frac{25}{4}&=6+\frac{25}{4}\\ \left(a-\frac{5}{2}\right)^2&=\frac{49}{4}\\ a-\frac{5}{2}&=\pm\sqrt{\frac{49}{4}}\\ a-\frac{5}{2}&=\pm\frac{7}{2}\\ a&=\frac{5}{2}\pm\frac{7}{2}\\ a&=-1,6 \end{aligned}$$

17. Solving by completing the square:

$$\begin{aligned} 3x^2-6x-2&=0\\ x^2-2x-\frac{2}{3}&=0\\ x^2-2x&=\frac{2}{3}\\ x^2-2x+1&=\frac{2}{3}+1\\ (x-1)^2&=\frac{5}{3}\\ x-1&=\pm\sqrt{\frac{5}{3}}\\ x-1&=\pm\frac{\sqrt{5}}{\sqrt{3}}\bullet\frac{\sqrt{3}}{\sqrt{3}}\\ x-1&=\pm\frac{\sqrt{15}}{3}\\ x&=1\pm\frac{\sqrt{15}}{3}\\ x&=\frac{3\pm\sqrt{15}}{3} \end{aligned}$$

19. Using $a=1$, $b=-8$, and $c=16$ in the quadratic formula:

$$x=\frac{-b\pm\sqrt{b^2-4ac}}{2a}=\frac{-(-8)\pm\sqrt{(-8)^2-4(1)(16)}}{2(1)}=\frac{8\pm\sqrt{64-64}}{2}=\frac{8\pm 0}{2}=4$$

21. First write the equation as $2x^2+8x-5=0$. Using $a=2$, $b=8$, and $c=-5$ in the quadratic formula:

$$x=\frac{-b\pm\sqrt{b^2-4ac}}{2a}=\frac{-8\pm\sqrt{(8)^2-4(2)(-5)}}{2(2)}=\frac{-8\pm\sqrt{64+40}}{4}=\frac{-8\pm\sqrt{104}}{4}=\frac{-8\pm2\sqrt{26}}{4}=\frac{-4\pm\sqrt{26}}{2}$$

23. First multiply by 10 to clear the equation of fractions:

$$\begin{aligned} 10\left(\frac{1}{5}x^2-\frac{1}{2}x\right)&=10\left(\frac{3}{10}\right)\\ 2x^2-5x&=3\\ 2x^2-5x-3&=0 \end{aligned}$$

Using $a=2$, $b=-5$, and $c=-3$ in the quadratic formula:

$$x=\frac{-b\pm\sqrt{b^2-4ac}}{2a}=\frac{-(-5)\pm\sqrt{(-5)^2-4(2)(-3)}}{2(2)}=\frac{5\pm\sqrt{25+24}}{4}=\frac{5\pm\sqrt{49}}{4}=\frac{5\pm7}{4}=-\frac{1}{2},3$$

25. Combining the complex numbers: $(4-3i)+5i=4+2i$

27. Combining the complex numbers: $(5+6i)+(5-i)=10+5i$

29. Combining the complex numbers: $(3-2i)-(3-i)=3-2i-3+i=-i$

31. Combining the complex numbers: $(3+i)-5i-(4-i)=3+i-5i-4+i=-1-3i$

33. Multiplying the complex numbers: $2(3-i)=6-2i$

35. Multiplying the complex numbers: $4i(6-5i)=24i-20i^2=24i-20(-1)=20+24i$

37. Multiplying the complex numbers: $(3-4i)(5+i)=15-20i+3i-4i^2=15-17i-4(-1)=15-17i+4=19-17i$

39. Multiplying the complex numbers: $(4+i)(4-i)=16-i^2=16-(-1)=16+1=17$

41. Dividing the complex numbers: $\frac{i}{3+i}\bullet\frac{3-i}{3-i}=\frac{3i-i^2}{9-i^2}=\frac{3i+1}{9+1}=\frac{1+3i}{10}$

43. Dividing the complex numbers: $\frac{5}{2+5i}\bullet\frac{2-5i}{2-5i}=\frac{10-25i}{4-25i^2}=\frac{10-25i}{4+25}=\frac{10-25i}{29}$

45. Dividing the complex numbers: $\frac{-3i}{3-2i} \bullet \frac{3+2i}{3+2i} = \frac{-9i-6i^2}{9-4i^2} = \frac{-9i+6}{9+4} = \frac{6-9i}{13}$

47. Dividing the complex numbers: $\frac{4-5i}{4+5i} \bullet \frac{4-5i}{4-5i} = \frac{16-20i-20i+25i^2}{16-25i^2} = \frac{16-40i-25}{16+25} = \frac{-9-40i}{41}$

49. Writing as a complex number: $\sqrt{-36} = 6i$

51. Writing as a complex number: $\sqrt{-17} = i\sqrt{17}$

53. Writing as a complex number: $\sqrt{-40} = \sqrt{-4 \bullet 10} = 2i\sqrt{10}$

55. Writing as a complex number: $\sqrt{-200} = \sqrt{-100 \bullet 2} = 10i\sqrt{2}$

57. Graphing the parabola:

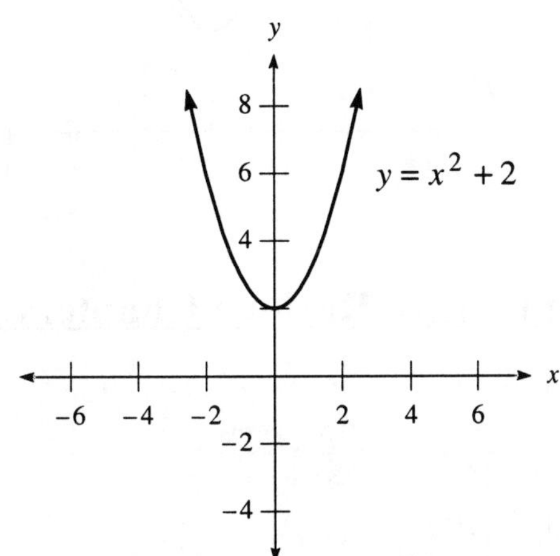

59. Graphing the parabola:

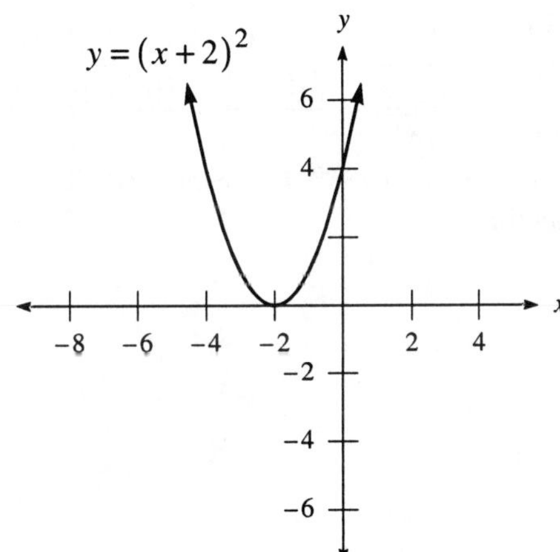

61. Begin by completing the square: $y=x^2+4x+7=\left(x^2+4x+4\right)+7-4=(x+2)^2+3$
Graphing the parabola:

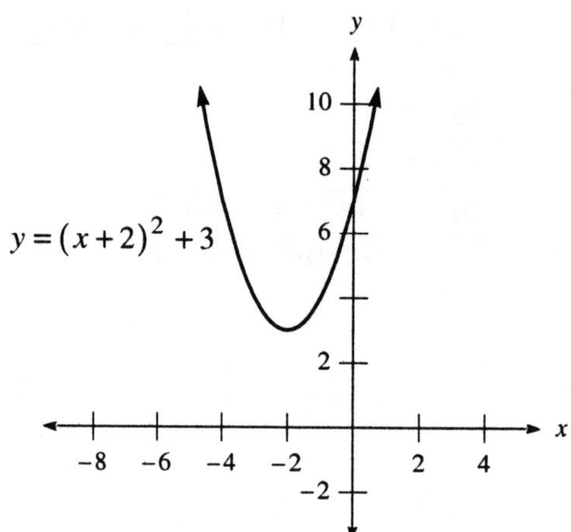

Cumulative Review: Chapters 1-9

1. Simplifying the expression: $10-8-11=10+(-8)+(-11)=-9$
3. Simplifying the expression: $-\frac{4}{5}\div\frac{8}{15}=-\frac{4}{5}\bullet\frac{15}{8}=-\frac{60}{40}=-\frac{3}{2}$
5. Simplifying the expression: $\frac{\left(3x^5\right)\left(20x^3\right)}{15x^{10}}=\frac{60x^8}{15x^{10}}=4x^{-2}=\frac{4}{x^2}$
7. Simplifying the expression: $\sqrt{81}=9$
9. Simplifying the expression: $\sqrt{\frac{200}{81}}=\frac{\sqrt{200}}{\sqrt{81}}=\frac{\sqrt{100\bullet 2}}{9}=\frac{10\sqrt{2}}{9}$
11. Simplifying the expression: $(3+3i)-7i-(2+2i)=3+3i-7i-2-2i=1-6i$
13. Multiplying using the column method:

$$\begin{array}{r} a^2-6a+7 \\ a-2 \\ \hline a^3-6a^2+7a \\ -2a^2+12a-14 \\ \hline a^3-8a^2+19a-14 \end{array}$$

15. Solving the equation:

$$\begin{aligned} 7-4(3x+4)&=-9x \\ 7-12x-16&=-9x \\ -12x-9&=-9x \\ -9&=3x \\ x&=-3 \end{aligned}$$

17. Solving the equation:

$$\begin{aligned} 5x^2&=-15x \\ 5x^2+15x&=0 \\ 5x(x+3)&=0 \\ x&=0,-3 \end{aligned}$$

19. Solving the equation:

$$\begin{aligned}\sqrt{6x-2}&=3x-5\\ \left(\sqrt{6x-2}\right)^2&=(3x-5)^2\\ 6x-2&=9x^2-30x+25\\ 0&=9x^2-36x+27\\ 0&=x^2-4x+3\\ 0&=(x-3)(x-1)\\ x&=1,3\end{aligned}$$

Note that $\sqrt{6\cdot 1-2}=\sqrt{6-2}=\sqrt{4}=2$, while $3\cdot 1-5=3-5=-2$, so $x=1$ does not check. The solution is $x=3$, which checks.

21. First multiply by 6 to clear the equation of fractions:

$$\begin{aligned}6\left(\frac{1}{2}x^2-\frac{1}{3}x\right)&=6\left(-\frac{1}{6}\right)\\ 3x^2-2x&=-1\\ 3x^2-2x+1&=0\end{aligned}$$

Using $a=3$, $b=-2$, and $c=1$ in the quadratic formula:

$$x=\frac{-b\pm\sqrt{b^2-4ac}}{2a}=\frac{-(-2)\pm\sqrt{(-2)^2-4(3)(1)}}{2(3)}=\frac{2\pm\sqrt{4-12}}{6}=\frac{2\pm\sqrt{-8}}{6}=\frac{2\pm 2i\sqrt{2}}{6}=\frac{1\pm i\sqrt{2}}{3}$$

23. Solving the inequality:

$$\begin{aligned}-5&\le 2x-1\le 7\\ -4&\le 2x\le 8\\ -2&\le x\le 4\end{aligned}$$

Graphing the solution set:

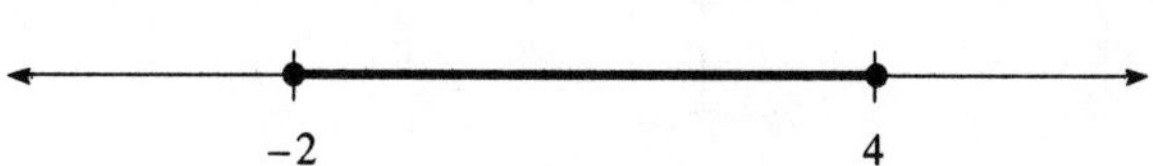

25. Graphing the line:

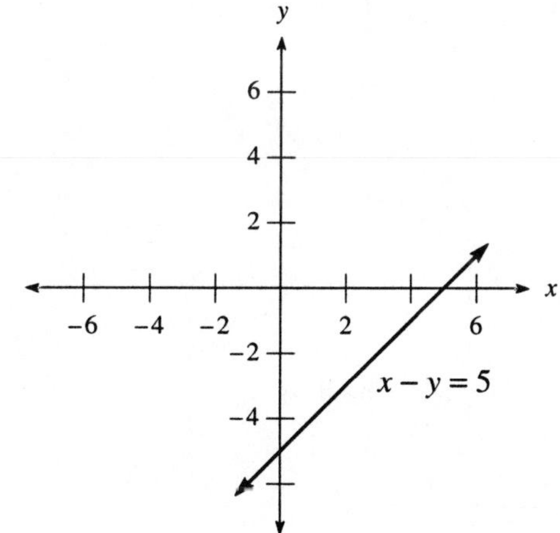

27. Graphing the line:

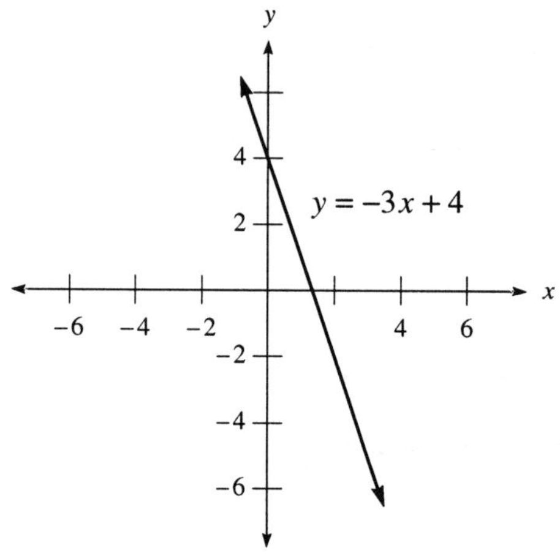

29. Graphing the parabola:

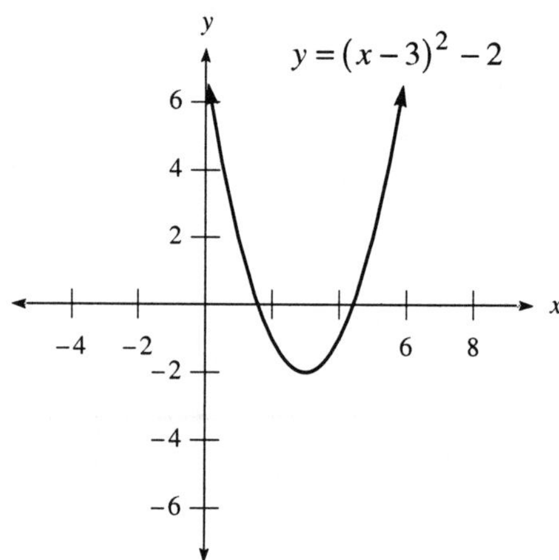

31. Substituting into the second equation:

$$\begin{aligned} 2(y-3)+3y &= 4 \\ 2y-6+3y &= 4 \\ 5y-6 &= 4 \\ 5y &= 10 \\ y &= 2 \\ x &= 2-3=-1 \end{aligned}$$

The solution is $(-1,2)$.

33. Factoring the polynomial: $x^2-5x-24=(x-8)(x+3)$

35. Factoring the polynomial: $25-y^2=(5+y)(5-y)$

37. Finding the slope: $m=\frac{8-(-2)}{1-3}=\frac{8+2}{-2}=\frac{10}{-2}=-5$

39. Subtracting: $6-(-2)=6+2=8$

41. Dividing by the monomial: $\frac{15a^3b-10a^2b^2-20ab^3}{5ab}=\frac{15a^3b}{5ab}-\frac{10a^2b^2}{5ab}-\frac{20ab^3}{5ab}=3a^2-2ab-4b^2$

43. Simplifying the rational expression: $\frac{x^2+5x-24}{x+8}=\frac{(x+8)(x-3)}{x+8}=x-3$

45. Simplifying the rational expression:

$$\begin{aligned}\frac{-1}{x^2-4}-\frac{-2}{x^2-4x-12}&=\frac{-1}{(x+2)(x-2)}+\frac{2}{(x-6)(x+2)}\\&=\frac{-1(x-6)}{(x+2)(x-2)(x-6)}+\frac{2(x-2)}{(x+2)(x-2)(x-6)}\\&=\frac{-x+6+2x-4}{(x+2)(x-2)(x-6)}\\&=\frac{x+2}{(x+2)(x-2)(x-6)}\\&=\frac{1}{(x-2)(x-6)}\end{aligned}$$

47. Combining the radicals: $5\sqrt{200}+9\sqrt{50}=5\sqrt{100\bullet 2}+9\sqrt{25\bullet 2}=5\bullet 10\sqrt{2}+9\bullet 5\sqrt{2}=50\sqrt{2}+45\sqrt{2}=95\sqrt{2}$

49. Solving for y:

$$\begin{aligned}3x-8y&=24\\-8y&=-3x+24\\y&=\frac{3}{8}x-3\end{aligned}$$

Chapter 9 Test

1. Solving the equation:

$$\begin{aligned}x^2-7x-8&=0\\(x-8)(x+1)&=0\\x&=-1,8\end{aligned}$$

2. Solving the equation:

$$\begin{aligned}(x-3)^2&=12\\x-3&=\pm\sqrt{12}\\x-3&=\pm 2\sqrt{3}\\x&=3\pm 2\sqrt{3}\end{aligned}$$

3. Solving the equation:

$$\begin{aligned}\left(x-\frac{5}{2}\right)^2&=-\frac{75}{4}\\x-\frac{5}{2}&=\pm\sqrt{-\frac{75}{4}}\\x-\frac{5}{2}&=\pm\frac{5i\sqrt{3}}{2}\\x&=\frac{5\pm 5i\sqrt{3}}{2}\end{aligned}$$

4. First multiply by 6 to clear the equation of fractions:

$$\begin{aligned}6\left(\frac{1}{3}x^2\right)&=6\left(\frac{1}{2}x-\frac{5}{6}\right)\\2x^2&=3x-5\\2x^2-3x+5&=0\end{aligned}$$

Using $a=2$, $b=-3$, and $c=5$ in the quadratic formula:

$$x=\frac{-b\pm\sqrt{b^2-4ac}}{2a}=\frac{-(-3)\pm\sqrt{(-3)^2-4(2)(5)}}{2(2)}=\frac{3\pm\sqrt{9-40}}{4}=\frac{3\pm\sqrt{-31}}{4}=\frac{3\pm i\sqrt{31}}{4}$$

5. Solving the equation:

$$\begin{aligned} 3x^2 &= -2x+1 \\ 3x^2+2x-1 &= 0 \\ (3x-1)(x+1) &= 0 \\ x &= \frac{1}{3},-1 \end{aligned}$$

6. Simplifying the equation:

$$\begin{aligned} (x+2)(x-1) &= 6 \\ x^2+x-2 &= 6 \\ x^2+x-8 &= 0 \end{aligned}$$

Using $a=1$, $b=1$, and $c=-8$ in the quadratic formula:

$$x=\frac{-b\pm\sqrt{b^2-4ac}}{2a}=\frac{-1\pm\sqrt{(1)^2-4(1)(-8)}}{2(1)}=\frac{-1\pm\sqrt{1+32}}{2}=\frac{-1\pm\sqrt{33}}{2}$$

7. Solving the equation:

$$\begin{aligned} 9x^2+12x+4 &= 0 \\ (3x+2)^2 &= 0 \\ 3x+2 &= 0 \\ 3x &= -2 \\ x &= -\frac{2}{3} \end{aligned}$$

8. Solving by completing the square:

$$\begin{aligned} x^2-6x-6 &= 0 \\ x^2-6x &= 6 \\ x^2-6x+9 &= 6+9 \\ (x-3)^2 &= 15 \\ x-3 &= \pm\sqrt{15} \\ x &= 3\pm\sqrt{15} \end{aligned}$$

9. Writing as a complex number: $\sqrt{-9}=3i$

10. Writing as a complex number: $\sqrt{-121}=11i$

11. Writing as a complex number: $\sqrt{-72}=\sqrt{-36\bullet 2}=6i\sqrt{2}$

12. Writing as a complex number: $\sqrt{-18}=\sqrt{-9\bullet 2}=3i\sqrt{2}$

13. Combining the complex numbers: $(3i+1)+(2+5i)=3+8i$

14. Combining the complex numbers: $(6-2i)-(7-4i)=6-2i-7+4i=-1+2i$

15. Combining the complex numbers: $(2+i)(2-i)=4-i^2=4-(-1)=4+1=5$

16. Combining the complex numbers: $(3+2i)(1+i)=3+2i+3i+2i^2=3+5i-2=1+5i$

17. Combining the complex numbers: $\frac{i}{3-i}\bullet\frac{3+i}{3+i}=\frac{3i+i^2}{9-i^2}=\frac{3i-1}{9+1}=\frac{-1+3i}{10}$

18. Combining the complex numbers: $\frac{2+i}{2-i}\bullet\frac{2+i}{2+i}=\frac{4+2i+2i+i^2}{4-i^2}=\frac{4+4i-1}{4+1}=\frac{3+4i}{5}$

19. Graphing the parabola:

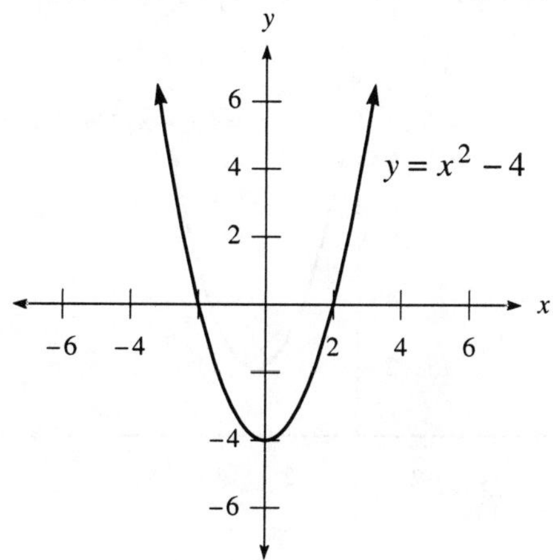

20. Graphing the parabola:

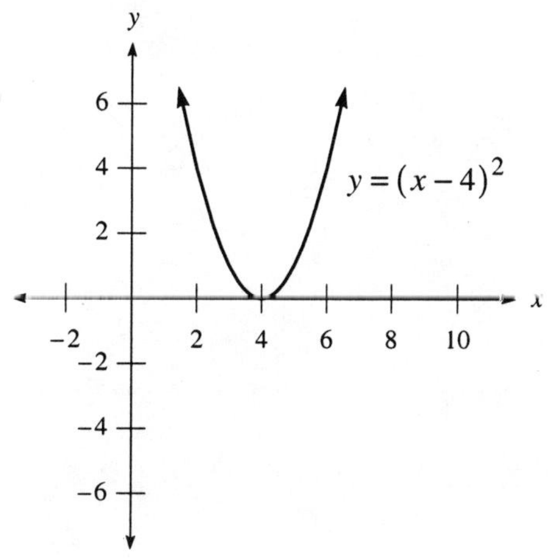

21. Graphing the parabola:

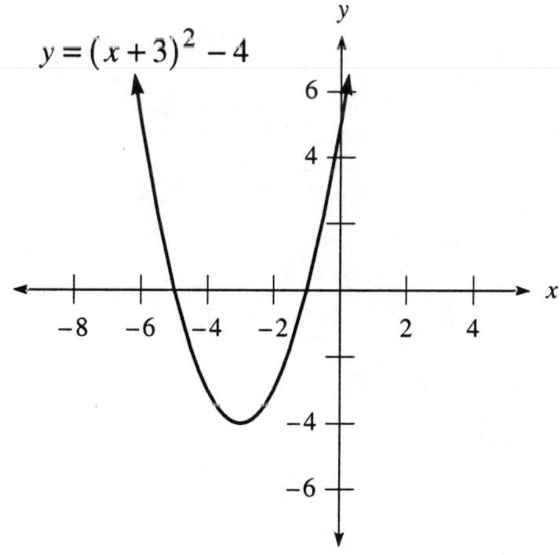

22. First complete the square: $y = x^2 - 6x + 11 = \left(x^2 - 6x + 9\right) + 11 - 9 = (x-3)^2 + 2$

Graphing the parabola:

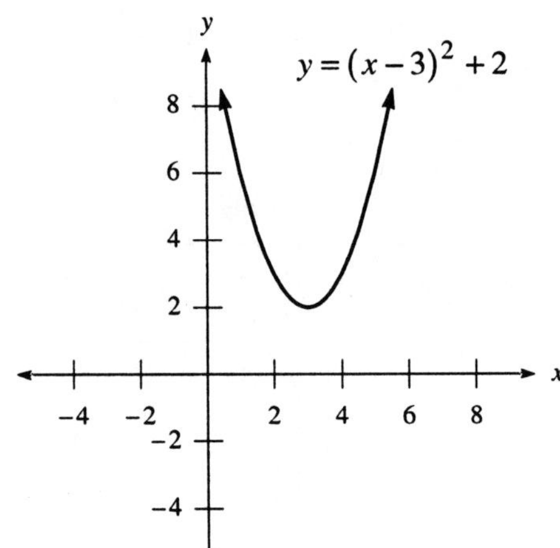